THE CHORDATES

THE CHORDATES

R. McNEILL ALEXANDER

Department of Pure and Applied Zoology
University of Leeds

CAMBRIDGE UNIVERSITY PRESS

Published by the Syndics of the Cambridge University Press
Bentley House, 200 Euston Road, London NW1 2DB
American Branch: 32 East 57th Street, New York, NY 10022

© Cambridge University Press 1975

Library of Congress Catalogue Card Number: 74–76580

ISBNs: 0 521 20472 0   hard covers
0 521 09857 2   paperback

First published 1975

Printed in Great Britain by
William Clowes & Sons, Limited, London, Beccles and Colchester

Contents

Preface

This book is designed for undergraduates, but I hope other people will read it as well. It is about the vertebrates and the few other animals which are included with them in the phylum Chordata. It will be obvious that this is too small a book to contain a complete account of modern knowledge of the chordates. Prospective readers may like to be told in general terms what sorts of topics have been included, and what has been left out.

Most of the book is about animal structures, how they seem to have evolved and how they work. A great many experiments are described because it is often as interesting and worthwhile to know how an item of knowledge was obtained as to know the item itself. Since the book is about structures, many of the explanations involve physics or engineering. There are many simple calculations designed either to throw light on the dimensions of a structure, or to test the plausibility of an explanation.

Few of the structures which are described are very small: most are visible with the naked eye or at least with the low power of a microscope. Readers who want to know about the structure of the cells of chordates, and about the molecules they contain, will need other books.

There is information in this book about locomotion, respiration, blood circulation, feeding, osmotic and ionic regulation and sense organs. These topics relate well to gross structure. There is relatively little about digestion, endocrine organs and the central nervous system because most of the really interesting things which could be written on these topics concern cells and molecules rather than larger structures.

Many authors and publishers have given permission for reproduction of illustrations and tables. The sources are indicated in the captions. Several of my colleagues in the University of Leeds have helped me with information. Dr John Clevedon Brown read the typescript, rescued me from a number of errors and made many very helpful suggestions.

I hope that readers who detect errors in this book or find passages obscure will let me know, so that I can improve the book if demand for it justifies a second edition.

University of Leeds   R. McNeill Alexander
October 1973

1

Introduction

This is the only chapter in the book which is not about a group of chordates. It contains rather curiously varied information which should be helpful to readers when they come to read the chapters which follow. A large part of the book is about experiments which help us to understand the chordates, and the first section of this chapter is about instruments and techniques which have been particularly useful in recent research. Most of them are referred to again in several of the later chapters. Parts of the book are anatomical and a section of this chapter explains some of the fundamental concepts used by anatomists. The fossil record of the chordates is much better than that of any other phylum, and tells us a lot about the course of evolution. Many fossils are described in the book, and a section of this chapter gives a rudimentary account of how fossils are formed and studied. The final section of the chapter states some of the principles of animal classification, and presents the classification of the chordates which is used in this book.

TECHNIQUES

I have aimed in this book not only to summarize current knowledge of the chordates but also to explain in outline how the knowledge has been obtained. Many experiments are therefore described. Some of them involve special equipment designed for the purpose or at least not used in other experiments described in the book. Such equipment is described at the same time as the experiments. There are, however, a small number of techniques which are referred to frequently because they have been used in many of the experiments. They are described in this section. Many of them have come into general use in zoology only since 1960 or thereabouts.

Cinematography plays an important part in the study of animal movement. Films are normally taken and shown at 16 or 24 frames (pictures) per second: this is just fast enough to avoid a flickering effect. Very fast movements can only be studied on film taken at higher rates and either shown at normal rates to slow down the motion or examined frame-by-frame. Many cameras can take film at up to 64 frames s^{-1} and special cameras are capable of very much higher rates. Rates up to at least 900 frames s^{-1} have been used in research on vertebrates, for instance in an investigation of the hovering flight of hummingbirds. Considerably higher rates have been used in studies of insects.

X-ray cinematography is sometimes much more useful than light cinematography. It has been used to show how the limb bones of mammals move when they walk (Figs. 12-3 to 12-6) and how teeth move over each other in chewing (Fig. 14-5). A visible image can be obtained by projecting a beam of X-rays through the specimen on to a fluorescent screen. A film of the screen can be taken with an ordinary cine camera, but this is not generally a satisfactory procedure because the image is rather dim. It is usual to obtain a very much brighter image by means of an electronic device known as an image intensifier, and to take a film of that. Framing rates up to about 60 frames s^{-1} are practicable.

The opacity of materials to X-rays depends on their density and on the atomic number of the atoms they contain. Dense materials, and elements of high atomic number, are relatively opaque. Bone contains a high proportion of calcium phosphate, which makes it denser than other tissues. Calcium and phosphorus have higher atomic numbers than the elements which are most plentiful in other tissues. Hence bone is relatively opaque and can be distinguished in X-ray images. Air, on the other hand, has a low density and is relatively transparent. X-ray films which have been taken of lungfish breathing show both the X-ray-opaque bones, and the X-ray-transparent air in the lung (p. 221). Structures which would not normally be distinguishable in X-ray images can be made visible by injecting fluids containing elements of high atomic number. This method has been used, for instance, in investigations of the flow of blood through the hearts of frogs.

Zoologists often want to know which muscles are responsible for particular movements, and the sequence in which they act. This information can be obtained by the technique of electromyography, which takes advantage of the electrical changes which occur within muscles when they are active. Vertebrate striated muscle fibres contract in response to action potentials in their motor nerves. Most have only one motor nerve ending, from which the stimulus to contract spreads as an action potential along the muscle fibre. These muscle action potentials can be detected by electrodes in the muscles. To be sure that any potentials that are recorded actually arise in the muscle being investigated, two electrodes should be placed in the muscle and the potential difference between them recorded. Concentric electrodes are often used, made from hypodermic needles. The needle is filled with an insulating material, with a wire running down its centre. The bare tip of the wire is exposed at the point of the needle, and the potential difference between wire and needle is recorded. Alternatively, two fine wire electrodes, each insulated except at the tip, are placed in the muscle. Very fine, flexible electrodes can be inserted with the help of a hypodermic needle. The needle is slipped over the electrode, pushed into the muscle and then withdrawn leaving the electrode in place. The potentials detected by the electrodes must be amplified and can then be displayed on a cathode-ray oscilloscope screen (which can be photographed) or can be recorded directly on sensitive paper by an ultraviolet recorder. Electromyography shows when a muscle is active and gives some indication of how active it is, but does not give any quantitative indication of the force that is being exerted. It has been used in many investigations of vertebrates, for instance in investigations of the mechanisms of breathing used by turtles (Fig. 10-3) and of chewing by bats (p. 390).

Transducers are devices which translate one type of signal into another. For instance, a microphone is a transducer which translates sound into electrical signals and a loudspeaker is a transducer which has the reverse effect. Experimental zoologists have made a great deal of use, in recent years, of transducers which produce an electrical signal that can be displayed on an oscilloscope screen or recorded by a pen recorder or ultraviolet recorder.

Fig. 1-1a shows a transducer which can be used (in different circumstances) to sense forces or movements. It is simply a flexible beam fixed at one end, with devices called strain gauges glued to its upper and lower surfaces. When its free end is pulled down the upper strain gauge is stretched a little and the lower one is compressed, and this alters their electrical resistances. There are two types of strain gauge. One type consists of a thin wire or a strip of metal foil. Stretching makes it longer and thinner so that its electrical resistance increases. The other type is a slice of semiconductor material, which undergoes a *decrease* in resistance when it is stretched, and is much more sensitive than the wire or foil type. By incorporating strain gauges in a Wheatstone bridge circuit an electrical output proportional to their extension or compression can be obtained. Thus a record of any bending of the beam shown in Fig 1-1a can be obtained. This transducer can be used to sense forces if its beam is relatively stiff, so that the forces which act on it bend it only a little. A transducer of this type was used as a force transducer to measure the stiffness of the notochord of amphioxus (Fig. 2-9b). Alternatively, if the beam is so flexible that it does not limit the movements of the organ to which it is attached, it can be used as a displacement transducer, to sense movement. This type of transducer was used as a displacement transducer, to sense movements of the heart of a shark, in the experiment illustrated in Fig. 4-17.

There are also transducers designed to sense pressure. Steady pressures, or pressures which change only slowly, can be measured with U-tube manometers, but such manometers do not respond very quickly to a

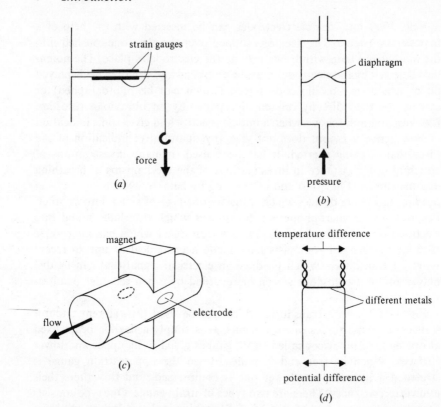

Fig. 1-1. Diagrams illustrating (*a*) a force/displacement transducer, (*b*) a pressure transducer, (*c*) an electromagnetic flowmeter and (*d*) a thermocouple.

change of pressure. They take an appreciable time to settle at the new equilibrium position, and cannot follow rapid fluctuations of pressure. Often, the pressures in which zoologists are interested fluctuate rather rapidly: the pressures in arteries, which fluctuate as the heart beats, are one example. Such pressures can generally be recorded satisfactorily by means of a pressure transducer incorporating a stiff metal diaphragm (Fig. 1-1*b*). The diaphragm is forced into a domed shape by the pressure which is to be measured. The higher the pressure, the more the diaphragm is distorted. The diaphragm has to be stiff to give a really fast response, so some sensitive device is needed to register its distortion. Various electrical devices have been used, including strain gauges attached to the diaphragm. Pressure transducers have been used in many of the experiments described in this book, for instance to obtain records of the pressures which drive water over the gills of fishes (Fig. 4-13).

Velocity of air flow has sometimes to be measured, for instance in an investigation of flow in the lungs and air passages of birds (p. 354). The hot-wire anemometer is a convenient instrument for the purpose. The electrical resistance of a wire increases as its temperature rises. The anemometer has an electrically heated filament which is held in the air current. The moving air tends to cool it, so the faster the flow the cooler the wire and the lower its resistance.

It is often useful to be able to record the velocity of flow of blood in a blood vessel. It is undesirable to introduce any instrument into the vessel where it might interfere with the flow. There is no need to do this because flowmeters are available which fit like a cuff around the vessel, without any need to cut or pierce it. One type is the electromagnetic flowmeter. It depends on the principle of electromagnetic induction: blood is a conductor of electricity so blood moving across the lines of force of a magnetic field will set up a potential difference. The flowmeter contains an electromagnet which is fitted around the blood vessel, and potential differences across the vessel are recorded (Fig. 1-1*c*). The ultrasonic flowmeter is another instrument which can be used in the same way but depends on the Doppler effect. An electromagnetic flowmeter has been used to record flow in the ventral aorta of a shark (Figs. 4-16, 4-17), and ultrasonic ones to record flow in various blood vessels of seals during diving (Fig. 13-10).

It is often useful to have small devices to measure temperature changes, because small devices are easiest to fit into animals and because they heat and cool quickly. One such device is the thermocouple. If a conductor runs through a gradient of temperature, a potential difference is set up between its ends. This cannot be measured in a circuit made entirely of one metal, because the effect on one wire running down the gradient is cancelled out by the effect on the other wire running up the gradient to complete the circuit. This difficulty can be overcome by using two metals which are affected to different extents, as shown in Fig. 1-1*d*. A potential difference is developed proportional to the difference in temperature between the two junctions where one metal joins the other. Thermocouples are incorporated in the device for studying air flow in bird lungs which is illustrated in Fig. 11-15.

Another small device for measuring temperature is the thermistor, which is simply a small bead of a metal oxide or oxide mixture with electrical leads attached. The electrical resistance of a thermistor decreases as its temperature rises. (Metals also change their resistance as temperature changes but not nearly as much and, as it happens, in the opposite direction.) Thermistors have been used to measure the body temperatures of lizards in their natural habitat (p. 284).

Data on oxygen concentrations are often wanted, particularly in studies of respiration. The concentration of oxygen in a gas mixture, or dissolved in water, can be measured by chemical methods, but there are other methods which are often more convenient. One uses the Clark oxygen electrode (Fig. 1-2*a*). This can be used to measure the partial pressure of oxygen either in a gas mixture or in solution. It actually contains two electrodes,

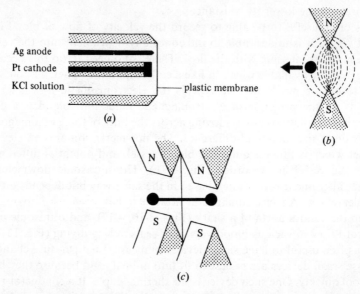

Fig. 1-2. Diagrams of equipment for measuring partial pressures of oxygen. (*a*) represents a Clark oxygen electrode; (*b*) shows the force which acts on a diamagnetic body in a non-uniform magnetic field in the presence of oxygen; and (*c*) shows a paramagnetic oxygen analyser which utilizes this effect.

a platinum cathode and a chloride-coated silver anode. The cathode is kept around 0.5 V negative to the anode. Cathode and anode are in a potassium chloride solution, behind a plastic membrane. The device senses the partial pressure of oxygen in the gas or solution immediately *outside* the membrane. Oxygen diffuses through the membrane to the cathode where it is immediately reduced

$$O_2 + 2H_2O + 4e = 4OH^-$$

or $$O_2 + 2H_2O + 2e = 2OH^- + H_2O_2.$$

The current depends on the rate at which the oxygen diffuses in, which in turn depends on its partial pressure outside the membrane. Thus the current indicates the partial pressure. Oxygen electrodes have been used to record the rate at which swimming fish remove oxygen from the water

(Fig. 4-5) and also (with a broadly similar device sensitive to carbon dioxide) in an investigation of the diffusion of gases through eggshells (Fig. 9-3).

Paramagnetic oxygen analysers are devices which measure the proportion of oxygen in gas mixtures by quite different means. We are all familiar with the forces which act on ferromagnetic materials (such as iron) in a magnetic field. Most materials are not ferromagnetic but diamagnetic, and some are paramagnetic. A diamagnetic body in a non-uniform magnetic field experiences a force which is relatively small if it is surrounded by a diamagnetic gas but much stronger if it is surrounded by oxygen, which is strongly paramagnetic. The oxygen tends to displace the body from the strongest part of the field (Fig. 1-2b). One type of paramagnetic oxygen analyser contains a diamagnetic dumb-bell suspended between the poles of a powerful magnet in such a way that its ends are in non-uniform parts of the field (Fig. 1-2c). When oxygen is present forces act on the dumbbell tending to make it turn on its suspension. Electrostatic forces are applied automatically to balance these forces, and the potential required is proportional to the oxygen concentration. Oxides of nitrogen are paramagnetic and would affect readings if they were present, but all other common gases are diamagnetic and have very little effect on the instrument. Paramagnetic oxygen analysers have been used in many experiments with vertebrates, for instance to measure the oxygen consumption of running lizards (p. 280).

Research on the physiology of kidneys and of other organs involved in controlling the salt and water content of the body requires means of measuring osmotic and ionic concentrations. It is often necessary to make these measurements on very small quantities of fluid, for instance on samples taken from individual kidney tubules. Osmotic concentration has usually been calculated from the freezing point of the fluid, which can be determined for very small samples. The freezing point of an aqueous solution of osmotic concentration x osmol l^{-1} is $-1.86x$ °C. Obviously, very sensitive thermometers are needed to measure the osmotic concentration of dilute solutions. The concentration of sodium and potassium ions in very small samples of fluid can be measured by flame photometry. Vaporized atoms in a flame emit light at particular wavelengths (the yellow light emitted by sodium will be familiar to most readers). The concentration of sodium or potassium in a solution can be measured by spraying the solution into a flame under controlled conditions and measuring the intensity of the light of appropriate wavelength which is emitted. The method can also be used for many other elements, but it is particularly sensitive for sodium and potassium. Atomic absorption spectrophotometry is a related technique which is much more sensitive for some elements. Chloride concentrations in small samples are often measured by a titration method, with the end-point detected by electrical means.

Many of the experiments described in this book were performed on living animals. The information they provide could not have been obtained in any other way. It is important that zoologists should take all reasonable precautions to avoid causing pain or distress to experimental animals. Some information cannot be obtained without risk of causing pain or discomfort, and in such cases zoologists should consider carefully whether an experiment is justified before performing it. In Britain there are legal restrictions as well as moral ones. Potentially painful experiments on living vertebrates may only be performed by holders of a licence from the Home Office. This applies even if the entire experiment is performed under anaesthetic and the animal is killed before it comes round: indeed experiments without anaesthetic, or in which the animal is allowed to recover from the anaesthetic, require certificates in addition to the basic licence.

CONCEPTS IN MORPHOLOGY

There are some concepts considered basic by most zoologists concerned with the structure of animals, which are used in this book and had better be explained. One is the concept of homology. It is seldom mentioned explicitly in this book, but it is often implied. For instance, Figs. 10-7 and 12-10 show the skulls of a python and an opossum, with many of the bones labelled. In many cases, the same name is given to a bone in each skull. As a general rule, bones in comparable positions are given the same name: for instance, the bones labelled maxilla both run along the side of the mouth and bear the main teeth of the upper jaw. There are, however, exceptions. The bone next posterior to the maxilla is called the ectopterygoid in the python, but the jugal in the opossum. This implies that the maxilla of the python and the maxilla of the opossum are considered homologous, but the ectopterygoid of the python and the jugal of the opossum are not.

What does this mean? It means that the python and the opossum are believed to have had a common ancestor that had a structure identifiable as a maxilla, and that their maxillae have been derived from this ancestral maxilla by the process of evolution. Further, there was no single structure in any ancestor that gave rise to both the ectopterygoid of the python and the jugal of the opossum. The most recent common ancestor of the python and the opossum was probably an early reptile with a skull much like that of *Palaeothyris*, which is illustrated in Fig. 9-10e. This particular illustration is not labelled but more distant ancestors are believed to have included fish with skulls like the one shown in Fig. 7-15. Notice that bones labelled maxilla, ectopterygoid and jugal appear in this illustration.

Identical names do not always imply homology. Insects and cows both have structures called legs, but no modern zoologist would suggest that they had a common ancestor with legs. On the other hand, structures given different names are sometimes claimed to be homologous. The hyomandibular cartilage of sharks, the hyomandibular bone of teleosts and the tiny bone called the stapes in the ears of mammals are considered homologous. The reasons will become apparent later in this book.

There is a severe difficulty implicit in this evolutionary concept of homology. There is no complete and unambiguous record of the course of evolution. The fossil record is fragmentary. For instance, to refer back to the example which has just been used, there is no known fossil which seems at all likely to resemble a common ancestor of sharks and teleosts, which has any structure identifiable as a hyomandibular cartilage or bone. The conclusion that the hyomandibular cartilage of sharks and the hyomandibular bone of teleosts are homologous has not been reached by tracing their ancestry, but by comparing the structure of both adult and embryo sharks and teleosts. Indeed the concept of homology was introduced before it was generally accepted that evolution had occurred, and it did not at first have evolutionary overtones. It implied correspondence of a rather abstract type.

Much thinking about homology is based on the assumption that evolution proceeds by distortion, not by processes which involve breaking and re-making of connections. Thus, if, for instance, an artery initially crosses a cartilage lateral to it, it is assumed that it cannot in the course of evolution come to lie median to it (unless perhaps by a complex process of distortion which takes it round the end of the cartilage). In mathematical terms, it is assumed that ancestor and descendants are topologically isomorphic. This assumption is probably not always justified, but it is often useful. It is often invoked by stating that two structures must be homologous because they have 'the same relations'. An example will illustrate what this means. Fig. 1-3 shows the cartilages which lie immediately anterior to the first gill slit of (a) a dogfish and (b) a ray. The ray has a large cartilage X which was considered by one zoologist to be a ceratohyal which had extended dorsally in the course of evolution, and by another to have formed by coalescence of hyal rays. The question was resolved by Sir Gavin de Beer, who found that the ray has a very small cartilage Y. Both this cartilage and the ceratohyal of the dogfish are attached ventrally to the hypohyal and dorsally to the hyomandibular. In each case the latter attachment (which is made by a long strand of connective tissue in the ray) is median to the afferent hyoidean artery. Thus Y has the same relations

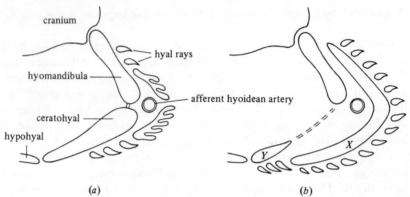

cranium

hyal rays

hyomandibula

afferent hyoidean artery

ceratohyal

hypohyal

(a)   *(b)*

Fig. 1-3. Diagrammatic transverse sections of (*a*) a shark and (*b*) a ray, showing the cartilages of the hyoid arch. Further explanation is given in the text. Redrawn from G. R. de Beer (1932). *Q. Jl microsc. Sci.* **75** NS, 307–20.

as the ceratohyal of the ray, and is judged to be homologous with it. *X*, however, is lateral to the artery, does not have the same relations as the ceratohyal and is not homologous with it.

Problems of homology are particularly difficult in some skulls where bones are apt to coalesce or disappear in the course of evolution. A computer program has been used to compare fossil fish skulls and identify homologous bones, but it does not give unambiguous results. The results depend on initial assumptions about the types of change which are most likely to have occurred.

Structures are often described as adaptations. For instance, the very strong forelimbs of moles may be described as adaptations for burrowing. This means that they are believed to have evolved because ancestral moles with stronger forelimbs were better able to burrow and were consequently favoured by natural selection. They were more likely to leave offspring than their fellows with weaker forelimbs, and so forelimbs in the population as a whole became progressively stronger.

Zoologists discussing adaptations are apt to make statements such as 'Moles have strong forelimbs for burrowing'. Statements like this which indicate that something has a purpose or function are called teleological. Many teachers of biology condemn teleological statements, perhaps because they feel they imply *conscious* purpose; that they imply for instance, that God provided moles with strong forelimbs so that they could burrow. Others contend that teleological statements are proper and useful in biology, and have no metaphysical implications. If strong forelimbs evolved because they gave a selective advantage attributable to their use in burrowing, surely they are *for* burrowing. I support this view, and have made no conscious effort to avoid teleological language in this book.

Animals, or features of animals, are often described as primitive or advanced. The opossum is a primitive mammal, and its molar teeth are among its primitive features. This means that it has many features in which it resembles the ancestral mammals rather than many present-day mammals, and that its molar teeth are among these features. Though it is a primitive mammal it is not a primitive vertebrate, for the mammals are more different from the ancestral vertebrates than any other vertebrates except perhaps the birds: mammals and birds are advanced vertebrates.

A number of quantitative comparisons are made in this book, which are complicated by the need to compare animals of different size. For instance, measurements on dinosaur skulls show that the weights of dinosaur brains must have been tiny fractions of total body weight. Does this indicate that dinosaurs were less intelligent than modern reptiles, or would crocodiles and lizards of the same size (if they existed) have brains just as small? What is the usual relationship between brain weight and body weight, for similar animals? Fig. 10-16 is a graph which provides an answer. It is plotted on logarithmic co-ordinates: in effect the logarithm of brain weight has been plotted against the logarithm of body weight. This is more convenient than a straightforward graph on linear co-ordinates, for two reasons. First, it has made it feasible to display data on lizards, mice, dinosaurs and whales on the same graph. If linear co-ordinates had been used either the whales would have had to be omitted or the points for mice and lizards would be squeezed so close into the bottom left hand corner that they could not be distinguished from each other. Secondly, this method of plotting has arranged the points for mammals and the points for reptiles in bands which approximate to straight lines. It seems clear that dinosaurs had brains of about the size one would expect by extrapolation of the trend shown by modern reptiles. It is much easier to extrapolate from the straight bands than from the curved ones which would be obtained by plotting on linear co-ordinates.

The straight bands indicate a mathematical relationship between brain weight and body weight. If a graph of $\log y$ against $\log x$ is a straight line of gradient m and intercept c

$$\log y = m \log x + c.$$

Taking antilogarithms

$$y = kx^m$$

where $\log k = c$. In other words a straight line on logarithmic co-ordinates indicates a power law relationship. Fig. 10-16 shows a tendency for the weights of the brains of similar animals to be proportional to (body weight)$^{0.65}$. The width of each band represents deviation from a precise power law relationship.

Physiological quantities can be studied in the same way. Fig. 11-4 shows graphs on logarithmic co-ordinates of oxygen consumption against body weight. These, too, approximate to straight lines, but plotting on logarithmic co-ordinates is not a magical procedure guaranteed to produce straight lines.

FOSSILS

Fossils provide the most direct evidence of the course of evolution, and many fossils are described in this book. They are remains of animals preserved in sediments which have generally, in the course of time, become rock.

The surface of the earth is continually being crinkled by the processes which produce mountains, and levelled again by processes of erosion and sedimentation. A variety of processes break down rocks, particularly on land. Heating of their surfaces by day and cooling by night sets up stresses due to thermal expansion and breaks fragments off. Water expands when it freezes; so water freezing in cracks in rocks is apt to split them. Streams carrying abrasive particles such as sand scour and erode the rocks over which they run. Water containing dissolved carbon dioxide removes calcium and other elements from rocks, carrying them away as a solution of carbonates and bicarbonates.

Materials removed from rocks in these ways are deposited as sediments in other places. Particles which are carried along by fastflowing water settle out where flow is slower, for instance on the flood plains of rivers and around their mouths. The smaller the particles, the slower flow must be before they will settle, so relatively large particles settle as gravel or sand in different places from small particles which form mud. Dry sand may be blown by the wind, and accumulate as dunes. Dissolved calcium carbonate is apt to be precipitated out of water in areas where algae are removing carbon dioxide from the water by photosynthesis. There are also sediments which are not formed from products of erosion, but from animal or plant remains. Shell gravel and the ooze formed on the ocean floor by accumulation of the shells of planktonic Foraminifera are two examples of calcium carbonate sediments of animal origin. Peat is a deposit of incompletely decomposed plant material (in modern times mainly mosses).

Sediments are generally soft when they are formed but if they are not disturbed they tend in time to become rocks. Mud becomes shale, sand becomes sandstone and deposits of calcium carbonate become limestone. The change is partly due to the particles becoming more tightly packed and partly to processes which cement them together. Mud is a mixture of fine mineral particles and water. Initially up to 90% of its volume may be water, the particles are not in contact and it is sloppy. As more mud accumulates on top of it, it is subjected to pressure and water is squeezed out. Adjacent particles make contact when the water content is about 45% by volume and further compaction involves rearrangement and crushing of particles. Note that complete compaction of a layer of mud which initially contained 90% water involves reduction to one-tenth of its initial thickness. Settled sand contains only about 37% water by volume, so compaction cannot reduce its thickness much. The grains in sandstones do not generally seem to have been crushed, but to have dissolved at the points where they touch other grains, so that the grains fit more closely together. Generally silica (perhaps from the dissolved corners of the grains) or a deposit of calcium carbonate cements sandstone together.

An animal which dies where a sediment is being deposited, or is carried there by currents after death, is liable to become embedded in the sediment. If it does not decay completely, it becomes a fossil. Most vertebrate fossils consist only of bones, teeth and scales. The soft parts of the body are much more susceptible to decay, but traces of them occasionally survive. For instance, black marks in fossils of the fish *Jamoytius* (Fig. 3-18c) are believed to be carbon from the organic matter of cartilage. The feathers of the bird *Archaeopteryx* have decayed completely, but the sediment was moulded to their shape and impressions of them survive (Fig. 11-19).

It is unlikely enough that an individual animal will be preserved as a fossil, but it is far more unlikely that the fossil will be found. Fossils are formed where sediment is being deposited, generally under water. They are nearly always found on land, generally where they have been exposed by erosion. Fossils generally only become accessible if the sediments which contain them are raised by earth movements and subsequently eroded.

Small fossils in reasonably coherent rocks are generally fairly easy to quarry out, in a slab of rock, and carry home to the laboratory. It is often advisable to harden and protect bones by coating them with plastic cement. Fossils in soft or crumbly rock must be encased in a jacket of burlap bandages and plaster of Paris before they can be moved safely. If the fossil is large this can be a formidable task. A trench must be excavated all round the fossil. The top and sides of the fossil are encased in plaster, and then the fossil is freed by undercutting. It is turned over, and the underside is jacketed with burlap and plaster.

Sometimes a rock splits so that a fossil is beautifully exposed. Often it is necessary to clear away rock that is hiding parts of the specimen. The simplest way of doing this is by chipping the unwanted rock away with a needle. There is a useful tool which makes a needle vibrate up and down like a miniature pneumatic drill. Miniature sandblasting equipment is also made which produces a fine jet of compressed carbon dioxide and abrasive powder. This can be directed at rock which has to be removed. Rock containing calcium carbonate can be softened or broken up by dilute acetic or formic acid. These dissolve calcium carbonate but not calcium phosphate, which is the main mineral constituent of bone. It is even possible to clean one side of a delicate fossil with acid, embed that side in clear plastic and dissolve away the rock from the other side. This process leaves a skeleton firmly attached to the plastic and visible from both sides.

Most illustrations of fossil vertebrates in this and other books, show complete skeletons in life-like positions. Fossils are seldom found like this. They are often incomplete, and even if they are reasonably complete their positions are those of dead animals rather than lives ones (see, for instance, the fossil of the early bird, *Archaeopteryx*, shown in Fig. 11-19). Fossils are often distorted or crushed: fossils in shale are particularly apt to be flattened since compaction of mud to shale reduces its thickness so much. The original shape of a crushed skull can often be discovered by making paper cut-outs of the shapes of the individual bones and fitting them together to make a solid model. A great deal of inference is needed to produce reconstructions of skeletons such as one shown, for instance, in Fig. 10-14a. Even more inference is needed when the intact animal is drawn (Fig. 10-14b), unless scales are preserved which show the shape of the body surface. Sometimes bones and sutures which have not been found are drawn in broken lines (Fig. 8-1, 8-2a).

The posture of an extinct tetrapod can only be guessed, or inferred from the structure of the joints, unless footprints have been preserved and can be identified as having been made by the same animal. Footprints are most likely to be preserved if they are made in damp mud and quickly covered by a layer of contrasting sediment. A skeleton of a dinosaur with corresponding footprints is shown in Fig. 10-13.

Sections of fossils can be made for microscopic examination. A fragment of fossil is embedded in Canada balsam, or in a plastic embedding medium. One side is ground away to the level of the section required. The ground surface is then cemented to a glass slide and the other surface of the fossil is ground till all that remains is a section of the required thickness, stuck to the slide. Sections of fossil bones and scales are shown in Fig. 3-17.

When evolutionary history is being traced it is important to know the relative ages of rocks in which fossils have been found. Sediments are formed in layers. Layers of the same material may be distinct because sedimentation was interrupted. Layers of different sediments may be formed on top of one another because local conditions changed. The layers mark time intervals. Successive layers sometimes followed one another immediately, but sometimes there were extremely long gaps while no sediment was formed, and erosion may even have occurred, at the locality in question. The order in which the sediments were formed at any particular locality is generally obvious since later sediments are on top of earlier ones, though later earth movements may fold sediments so that part of the sequence is upside-down. Sediments formed simultaneously at different places can often be matched, particularly if they contain similar fossils. Thus the relative ages of fossils which are being studied can generally be established.

Fossils are very rare in the oldest sedimentary rocks. The time spanned by rocks in which fossils are reasonably plentiful is divided into three eras and eleven periods, as shown in Table 1-1. It is usually possible to determine the period in which a particular sedimentary rock was formed, and the approximate position within the period. It is very much more difficult to decide the age of the rock in years. Different methods are apt to give

TABLE 1-1 *The main divisions of time since the beginning of the Palaeozoic era*

The Present is at the top of the table. Age is the approximate time since the *beginning* of the period, estimated mainly from the decay of radioactive elements

Era	Period	Age million years
CENOZOIC	Quaternary	2
	Tertiary	70
MESOZOIC	Cretaceous	140
	Jurassic	190
	Triassic	230
PALAEOZOIC	Permian	280
	Carboniferous	350
	Devonian	400
	Silurian	440
	Ordovician	500
	Cambrian	570

very different dates for the same rock, and the dates given in the table are not necessarily reliable. This is why the ages of fossils are generally not given in years, but by naming the period.

Such estimates of ages in years as we have are based on the decay of radioactive elements in the rocks. For instance, many rock-forming minerals contain potassium which occurs as three isotopes, K^{39}, K^{40} and K^{41}. Potassium-40 is radioactive and decays very slowly over periods of hundreds of millions of years. Two alternative reactions are involved, one producing calcium-40 and the other argon-40. Calcium is a normal constituent of many minerals and calcium-40 is the common isotope. There is no obvious way of distinguishing calcium produced by decay of potassium-40 from calcium which has been in the rock ever since it was formed. Any argon in the rock must presumably have been produced since the rock was formed. It diffuses out of some rocks but is well retained by others such as mica. The age of such rocks can be calculated from the relative proportions of potassium-40 and argon-40 which they contain.

It must not be assumed that because a fossil is found, for instance, in Britain that the animal experienced what we think of as a British climate. There is often evidence that it did not. Lower (i.e. early) Carboniferous limestones containing fossil corals are plentiful in Britain and suggest a tropical or semitropical climate. British coal is an upper (i.e. late) Carboniferous sediment, believed to have been formed in tropical forest. Indian rocks of similar date have marks which indicate the presence of glaciers at the time when they were formed.

Geographical changes which explain the climatic ones are indicated by study of the magnetic properties of rocks. Many rocks which contain magnetic compounds of iron and nickel are permanently magnetized. The direction of their magnetic field seldom corresponds with the present direction of the earth's magnetic field, but it must have done so when the rock was formed: magnetized materials in a new, loose sediment (or in a molten lava) would tend to be aligned with the earth's magnetic field. The earth's magnetic poles do not coincide with the geographic poles but there are theoretical reasons for believing that their average positions, over any period of several thousand years, do. Hence the latitude of magnetized rocks at the time they were formed can be determined from the direction of their magnetism. Allowance has to be made for tilting by subsequent earth movements. Such evidence indicates that England was in the southern hemisphere in the early Palaeozoic and crossed the equator on its journey north during the Carboniferous. Europe and Asia (excluding India) apparently moved as a unit, and the other continents also moved extensively.

Even an exceptionally good sequence of fossils does not provide an unambiguous record of the course of evolution because fossils do not have birth certificates naming their ancestors. All tracing of the course of evolution involves assumptions. One cannot even be sure that the earliest fossil is the most primitive. It may happen that early, primitive members of the group have not been found and that one of the later members is more primitive than the earliest known members.

One criterion which seems implicit in the thinking of many zoologists is that if two evolutionary trees linking the same animals are equally acceptable on other grounds, the one involving the smaller *total* amount of

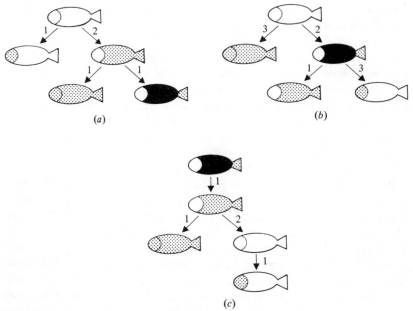

Fig. 1-4. This diagram is explained in the text.

change is the more probable. The meaning of this is illustrated by Fig. 1-4 which shows three possible evolutionary trees for a group of five fishes. These fish differ only in colouring. The numbers beside the arrows indicate the amount of change involved in each step: a change in colour of one section of the body counts as one unit of evolutionary change. Tree (*a*) involves a total of 5 units of change but tree (*b*) involves 9, so tree (*a*) is judged more probable. Tree (*c*) involves as little change as (*a*) and indeed shows the fish connected together in the same way, with some of the arrows reversed. Choice between (*a*) and (*c*) would have to depend on other criteria: if the all-white fish was an early fossil (*a*) would presumably be preferred.

The criterion of minimum total change has been used as the basis of computer programs, which have been used to construct evolutionary trees from data on living animals but not, so far as I am aware, on fossils.

CLASSIFICATION

This book is full of generalizations, some about large groups of animals (such as the vertebrates) and some about small ones (such as the members of a single species). Names are needed for all these groups. A system of classification, if it is well designed, helps zoologists to marshal their knowledge of animals and provides them with names for most of the groups about which they wish to make generalizations.

The smallest unit of classification with which we are concerned is the species. This unit is notoriously difficult to define, but most readers will have at least an intuitive understanding of its scope. All lions are placed in a single species, and all tigers in another. Every species is assigned to a genus, and is referred to by a formal name which includes the name of the genus. Thus the lion is the species *Panthera leo* within the genus *Panthera*. The tiger is *Panthera tigris* and the leopard *Panthera pardus* but the cat is considered sufficiently different from them to be placed (with the other small cats) in a separate genus, *Felis*.

Precise rules are needed to ensure (as far as possible) that all zoologists call the lion *Panthera leo* and that none use this name for other animals. These rules are incorporated in the *International Code of Zoological Nomenclature* and are administered by an international commission. Their general effect is to prevent chaos, but their application leads occasionally to abandonment of familiar names. For instance, the genus *Branchiostoma* (discussed in Chapter 2) used to be better known under the name *Amphioxus*, which was given to it in 1836. However, it had been described by another zoologist in 1834 as *Branchiostoma*, so the latter name takes priority. The lion was originally *Felis leo* but became *Panthera leo* when it was decided that the large cats are too different from the small ones to be kept in the same genus.

Genera are grouped in families, families in orders, orders in classes and classes in phyla. Thus the lion belongs (with the other big cats) to the genus *Panthera*. All the cats (including *Panthera*) are included in the family Felidae. Cats, dogs, bears, weasels and the many other mammals commonly referred to as the carnivores are grouped together in the order Fissipedia. All mammals belong to the class Mammalia. They and all the other animals with which this book is concerned belong to the phylum Chordata.

Every species of animal belongs to a genus, family, order, class and phylum, but these are not the only groupings used in classification. Others are introduced into the system as and when convenient. For instance, the Chordata are divided into three subphyla which are in turn divided into classes. Much the largest of these subphyla is the subphylum Vertebrata, the vertebrates.

The names used in classification are Latin in form. Many are actually Latin words, or have been coined from Latin words which describe the group. Others have been derived from the names of places or people and often look rather strange in their latinized form. For instance, the fossil genera *Jamoytius* (Fig. 3-18c) and *Moythomasia* (Fig. 5-14a) were both named in compliment to the distinguished palaeontologist J. A. Moy-Thomas.

The classification used in this book is set out below. Appropriate extracts from it are repeated under the headings of sections of chapters throughout the book. Diligent students will quickly discover discrepancies between this classification and those adopted in other textbooks. They are at liberty to decide they prefer one classification to the others, but not that one is right and the others wrong. Classification is a matter of opinion, not of fact. A classification is not right or wrong, but it may be good or bad. The best classification is generally the most useful one. It helps zoologists to marshal their thoughts in the most profitable way, it provides the names they need when they wish to make generalizations, and it does not unnecessarily change old names which they have grown accustomed to use.

Some differences between classifications are mere differences of name. Such inconsistencies are more or less inevitable because names of groups above the level of family are not subject to the *International Code*. Others reflect differences or changes of opinion. For instance, the two groups of extinct fishes which are referred to in this book as the classes Acanthodii and Placodermi were for a long time generally put together in a single class Aphetohyoidei. It was thought that both had jaws constructed in a particular way, and that the jaws of all later fish had evolved from jaws of this type. It was thought that this and other less striking similarities showed that they were fairly closely related. Recent detailed investigations of high quality fossils have shown that the jaws of the two groups are in fact strikingly different, so that there is no longer any substantial reason for putting the groups together. There are however similarities between placoderms and sharks and between acanthodians and bony fishes. Some zoologists consider that these are sufficient evidence of relationship to justify putting the placoderms and sharks together in a class Elasmobranchiomorphi, and the acanthodians and bony fishes in a class Teleostomi.

It is generally agreed that a classification should not be inconsistent with existing knowledge of the course of evolution. In practice this means that if group *a* is believed to have evolved directly from class (or order or family) *A* and group *b* from class (or order or family) *B*, *a* and *b* will not be put together in the same class (or order or family) *C*. However, if *a* is believed to have evolved from order *x* of class *A* and *b* from order *y* of the same class, it would be permissible to include *a* and *b* in a single class *C*. For instance, the teleosts are all included in a single infraclass Teleostei although it is thought likely that three or more groups of teleosts have evolved independently from holostean ancestors. The immediate ancestors of all the groups are supposed to have belonged to the same order within the infraclass Holostei.

PHYLUM CHORDATA

SUBPHYLUM UROCHORDATA

Class Ascidiacea	sea squirts
Class Larvacea	*Oikopleura*, etc.
Class Thaliacea	salps

SUBPHYLUM CEPHALOCHORDATA

Amphioxus (*Branchiostoma* and *Asymmetron* only).

SUBPHYLUM VERTEBRATA

Class Agnatha
  Order Petromyzoniformes lampreys
  Order Myxiniformes hagfishes
  Order Heterostraci ⎤
  Order Osteostraci ⎟
  Order Anaspida   ⎬ ostracoderms (extinct)
  Order Thelodonti   ⎦

Class Acanthodii (extinct)

Class Placodermi (extinct)

Class Holocephali (rabbit fishes)

Class Selachii
  Order Heterodontiformes the Port Jackson shark (*Heterodontus*)
  Order Hexanchiformes
  Order Lamniformes most dogfishes and sharks
  Order Squaliformes the Piked dogfish (*Squalus*), etc.
  Order Raiiformes skates and rays
  Order Torpediniformes electric rays

Class Osteichthyes
Subclass Actinopterygii
Infraclass Palaeoniscoidei extinct except *Polypterus* and *Calamoichthys*
Infraclass Chondrostei sturgeons
Infraclass Holostei extinct except *Amia* and *Lepisosteus*
Infraclass Teleostei
Superorder Elopomorpha eels, tarpons, etc.
Superorder Clupeomorpha herrings, etc.
Superorder Osteoglossomorpha

Superorder Protacanthopterygii   salmon, lantern fishes, etc.
Superorder Ostariophysi
   Order Gonorhynchiformes
   Order Cypriniformes
      Suborder Characinoidei   characins
      Suborder Gymnotoidei   American knife-fishes
      Suborder Cyprinoidei   carps and minnows
   Order Siluriformes   catfishes
Superorder Paracanthopterygii   cod, etc.
Superorder Atherinomorpha   toothcarps, etc.
Superorder Acanthopterygii
   Order Perciformes   perches, etc.
   Order Pleuronectiformes   flatfishes
   and other orders

Subclass Dipnoi   lungfishes

Subclass Crossopterygii
   Order Rhipidistia   extinct
   Order Coelacanthini   extinct except *Latimeria*

Class Amphibia
Subclass Labyrinthodontia ⎱
Subclass Lepospondyli ⎰   extinct
Subclass Lissamphibia
   Order Salientia   frogs and toads
   Order Urodela   newts and salamanders
   Order Apoda

Class Reptilia
Subclass Anapsida
   Order Cotylosauria   extinct
   Order Chelonia   tortoises and turtles
Subclass Lepidosauria
   Order Eosuchia   extinct
   Order Rhynchocephalia   extinct except the tuatara (*Sphenodon*)
   Order Squamata
      Suborder Lacertilia   lizards
      Suborder Amphisbaenia
      Suborder Ophidia   snakes
Subclass Archosauria
   Order Thecodontia   extinct
   Order Saurischia ⎱
   Order Ornithischia ⎰   dinosaurs (extinct)

Order Pterosauria   extinct
Order Crocodilia   Crocodiles, etc.
Subclass Euryapsida   plesiosaurs, etc. (extinct)
Subclass Ichthyopterygia   ichthyosaurs (extinct)
Subclass Synapsida
   Order Pelycosauria ⎫
   Order Therapsida   ⎭   extinct

Class Aves
Subclass Archaeornithes   *Archaeopteryx* (extinct) only
Subclass Neornithes   all other birds. Many orders including
   Order Passeriformes   crows, warblers, finches, etc.

Class Mammalia
Subclass uncertain
   Order Monotremata   monotremes
   several extinct orders
Subclass Theria
Infraclass Metatheria
   Order Marsupialia
      Suborder Polyprotodonta   opossums, etc.
      Suborder Diprotodonta   kangaroos, etc.
Infraclass Eutheria
   Order Insectivora   shrews, etc.
   Order Chiroptera   bats
      Suborder Megachiroptera
      Suborder Microchiroptera
   Order Fissipedia   dogs, cats, etc.
   Order Pinnipedia   seals
   Order Cetacea
      Suborder Odontoceti   toothed whales
      Suborder Mysticeti   whalebone whales
   Order Rodentia   rodents
   Order Lagomorpha   rabbits, etc.
   Order Artiodactyla
      Suborder Palaeodonta   extinct
      Suborder Suina   pigs, hippopotamus, etc.
      Suborder Tylopoda   camels, etc.
      Suborder Ruminantia   deer, antelopes, etc.
   Order Perissodactyla
      Suborder Hippomorpha   horses

Suborder Ancylopoda   extinct
Suborder Ceratomorpha   rhinoceroses and tapirs
Order Proboscidea   elephants
Order Primates
Suborder Prosimii   tree shrews, lemurs, etc.
Suborder Anthropoidea   monkeys, apes and man
and other orders

FURTHER READING

General books on the chordates

Goodrich, E. S. (1930). *Studies on the structure and development of vertebrates.* Macmillan, London.
Jarvik, E. (1960). *Théories de l'évolution des vertébrés reconsidérées à la lumière des récentes découvertes sur les vertébrés inférieurs.* Masson, Paris.
Romer, A. S. (1966). *Vertebrate palaeontology,* 3rd edit. University of Chicago Press.
Schmidt-Nielsen, K. (1972). *How animals work.* Cambridge University Press, London.
Young, J. Z. (1963). *The life of vertebrates,* 2nd edit. Oxford University Press, London.

General books, largely on chordates

Alexander, R. McN. (1968). *Animal mechanics.* Sidgwick & Jackson, London.
Gray, J. (1968). *Animal locomotion.* Weidenfeld & Nicolson, London.
Hoar, W. S. (1966). *General and comparative physiology.* Prentice-Hall, Englewood Cliffs, New Jersey.

Techniques

Giles, A. F. (1966). *Electronic sensing devices.* Newnes, London.
Kay, R. H. (1964). *Experimental biology. Measurement and analysis.* Chapman & Hall, London.
Neubert, H. K. P. (1963). *Instrument transducers. An introduction to their performance and design.* Clarendon Press, Oxford.

Concepts in morphology

Ayala, F. J. (1970). Teleological explanations in evolutionary biology. *Phil. Sci.* **37**, 1–15.
de Beer, G. R. (1971). *Homology, an unsolved problem.* Oxford University Press, London.
Gould, S. J. (1971). Geometric similarity in allometric growth: a contribution to the problem of scaling in the origin of size. *Am. Nat.* **105**, 113–36.
Jardine, N. (1969). The observational and theoretical components of homology: a study on the morphology of the dermal skull-roofs of rhipidistian fishes. *Biol. J. Linn. Soc.* **1**, 327–61.
Munson, R. (1961). Biological adaptation. *Phil. Sci.* **38**, 200–15.

Fossils

Kummel, B. (1970). *History of the earth*, 2nd edit. Freeman, San Francisco.
Kummel, B. & D. Raup, (eds.) (1965). *Handbook of palaeontological techniques.*
  Freeman, San Francisco.
Weller, J. M. (1960). *Stratigraphic principles and practice.* Harper, New York.

Classification

Blackwelder, R. E. (1967). *Taxonomy. A text and reference book.* Wiley, New
  York.
Savory, T. (1970). *Animal taxonomy.* Heinemann, London.

2

Sea squirts and amphioxus

This chapter is about the animals which are included among the chordates, but not among the vertebrates. The phylum Chordata is divided into three subphyla, the Urochordata and Cephalochordata which are described in this chapter and the Vertebrata which occupy the rest of the book.

The fish, amphibians, reptiles, birds and mammals resemble each other in many striking ways, and form an obvious group. They are the vertebrates. The urochordates (the tunicates or sea squirts) and the cephalochordates (amphioxus) are clearly distinct from the vertebrates and from each other, but they have some features which are generally accepted as strong evidence of relationship to the vertebrates. The most obvious of these are the notochord (which will be described), a tubular dorsal nerve cord and perforations in the pharynx comparable to the gill slits of fishes. Many adult vertebrates have no notochord or gill slits, but these structures can nevertheless be found in their embryos.

SEA SQUIRTS

Phylum Chordata, subphylum Urochordata, class Ascidiacea

The urochordates all live in the sea. Their larvae have features which seem to show that they are related to the vertebrates, but typical adult urochordates have very little in common with vertebrates. Nevertheless, these adults will be described first. They are sack-shaped animals, commonly a few centimetres long. They do not move about but remain firmly anchored to a rock or, by root-like structures, to mud. The body is enclosed in a protective tunic which consists of protein and polysaccharide (including cellulose) with only a few cells embedded in it. The only openings in the tunic are the two siphons at the upper end (Fig. 2-1). These openings are inlet and outlet for water which the animal pumps through its body and filters to obtain its food. Examination of gut contents shows that unicellular algae (including diatoms) are the main food.

A large proportion of the space within the tunic is occupied by the pharynx, which has so many small perforations that its wall is in effect a fine network. Water entering the inhalent siphon passes through this network into a surrounding cavity known as the atrium, which leads to the exhalent siphon. It is propelled by cilia. Fig. 2-1c shows that the bars of

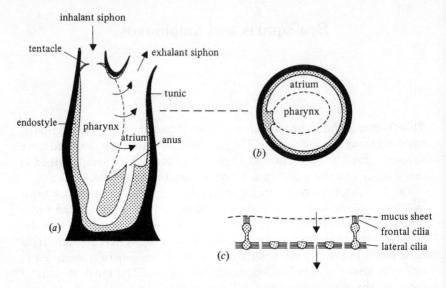

Fig. 2-1. Diagrammatic sections of a typical sea squirt such as *Ciona*. (*a*) Longitudinal section, (*b*) transverse section and (*c*) a greatly magnified section through part of the wall of the pharynx. Arrows indicate flow of water.

the pharyngeal network bear two sets of cilia: the frontal cilia which project into the pharyngeal cavity and the lateral cilia which project across the perforations. The lateral cilia propel the water through the animal. They beat in such a way as to drive water from the pharynx into the atrium.

A ring of tentacles arranged radially in the inhalent siphon may prevent excessively large particles from entering. Particles that do enter, down to very small dimensions, are retained in mucus on the pharynx wall. The perforations are typically about 50 μm across but colloidal graphite particles only 1 μm in diameter are retained. Plainly the pharynx wall is not a simple sieve. By careful observation of sea squirts with transparent tunics it can be seen that the perforations are spanned by a continuous sheet of mucus which covers the inner face of the wall. It is in this sheet that the particles are trapped. At least most of the mucus is secreted by gland cells in the endostyle, a structure which lies ventrally in the pharynx. (It is by no means obvious from Fig. 2-1 that this position is ventral, but comparison of the adult with the larva makes it quite clear that it is.) The pressure difference between the pharynx and atrium which is produced by the lateral cilia not only drives water through the mucus filter, but also keeps the mucus firmly pressed against the frontal cilia. These move it dorsally until it reaches the dorsal mid-line, where there are other cilia which transport it posteriorly to the digestive parts of the gut. Particles trapped in the mucus are carried with it, and both the mucus and the food particles are digested.

A sea squirt placed in a dish containing a suspension of food or other particles in seawater, filters the water and so clears it. The rate at which the water is filtered can be calculated from the rate of clearing. For instance, sea squirts have been put in a suspension of colloidal graphite and the concentrations of the suspension measured from time to time by means of a colorimeter. Let the concentration at time t be C. Then if the animal is pumping water through itself at a rate W (volume per unit of time) and removing all the particles from it, it is removing the suspended matter at a rate CW. If the volume of the suspension is V the rate of change of concentration, dC/dt, is $-CW/V$ and

$$dt = -(V/CW)\,dC.$$

Let the concentration be C_1 at time t_1 and C_2 at time t_2. Then

$$t_2 - t_1 = -(V/W) \int_{C_1}^{C_2} (dC/C)$$
$$= (V/W) \ln (C_1/C_2). \tag{2.1}$$

Since C_1, C_2, t_1, t_2 and V are all measured in the course of the experiment, W can be calculated. It has been shown in this way that the sea squirt *Ciona* can pump at least $3\,l$ water h^{-1}. (The size of the specimens was not recorded, but they are unlikely to have been more than 15 cm long.) It is possible that some of the particles may escape the filter, in which case the actual rate of flow would be higher than this. However, attempts to measure the rate of flow more directly, by means of tubes attached to the siphons, have given lower, not higher, values. The direct method probably gave abnormally low values because the cilia had to drive water through the apparatus as well as through the animal.

The suspension clearance method has been modified to find out whether many fine particles do escape the filter. *Ciona* was placed in a mixed suspension, containing diatom colonies of diameter 200 μm and colloidal graphite of diameter 1–2 μm. The concentrations of both were measured periodically and equation 2.1 was used to calculate the flow rate (W). Identical values were obtained whether the calculation was based on diatom concentration or graphite concentration. The diatom colonies were not stopped by the tentacles in the inhalent siphon, but were too large to pass through the perforations in the pharynx. It could be assumed that all the diatoms in the water passing through the animal were filtered out. Since flow rates calculated from graphite concentrations were as high, graphite was filtered out just as effectively. It seems that no appreciable quantity of graphite passed through the filter without being caught. Further experiments with suspensions containing haemoglobin showed that haemoglobin molecules (diameter about 3 nm) pass fairly freely through the filter. Less than 10% are captured as the suspension passes through.

It might be expected that sea squirts would filter graphite suspensions less rapidly than suspensions of food material from which they can get nourishment. However, *Ciona* pumps suspensions of graphite (alone) about as fast as it pumps suspensions of diatoms (alone).

The pressure in the pharynx of *Ascidia* (another sea squirt) has been recorded. One end of a fine cannula was attached to a pressure transducer and the other was put into the pharynx of the animal, through the inhalent siphon. The pressure difference between the pharynx and the same level in the water outside proved to be only about 0.3 mm water (3 N m⁻²) for most of the time during normal feeding. However, the animal periodically contracted its body by muscular action and squirted out water. When it did this the pressure inside rose briefly by about 20 mm water.

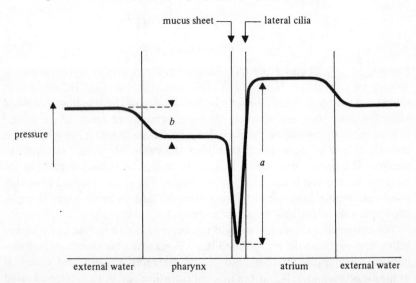

Fig. 2-2. A diagram showing the pressure changes which water is presumed to experience as it passes through a sea squirt. The pressure difference a which is produced by beating of the lateral cilia must be sufficient to drive the water through the inhalent siphon, the mucus sheet and the exhalent siphon. It is probably substantially greater than the pressure difference b, which has been measured.

The pressure of 0.3 mm water is only the pressure drop required to drive the water through the inhalant opening (Fig. 2-2). There must be a similar pressure drop at the exhalant opening. There must also be a pressure drop across the sheet of mucus in the pharynx, forcing the water through the filter. This pressure drop has not been measured, and it is hard to see how it could be. It would be necessary to measure the pressure in the narrow and inaccessible gap between the mucus sheet and the lateral cilia. The driving pressure produced by the cilia must equal the sum of the three pressure drops and it, too, would be exceedingly difficult to measure during normal, unhindered feeding. However, it is known from experiments in which tubes were attached to their exhalant siphons that *Phallusia* (yet another sea squirt) are able to drive water slowly through their bodies against a pressure of 2 mm water. Since they can only pump slowly against this pressure it seems likely that the driving pressure is only a little more than 2 mm water.

A filter-feeding animal requires energy for pumping water through its filter, for other metabolic processes and for growth. It cannot grow unless the energy obtained from the food exceeds the requirements for pumping and metabolism. If digestible suspended matter is sparse, filter feeding may be unprofitable: it might yield less energy than it consumed.

The rate at which *Ciona* use oxygen while feeding has been measured. Animals were put in jars of seawater containing suspended diatoms or graphite. Samples were removed from time to time and analysed chemically for dissolved oxygen content. The rate at which the suspension cleared was also measured, so that the filtration rate could be calculated. In this way it was shown that about 0.08 cm³ of oxygen was used for every litre of water filtered. How much food must the water contain, to support metabolism at this rate? Consider the oxidation of polysaccharide,

$$(C_6H_{10}O_5)_n + 6n\,O_2 = 6n\,CO_2 + 5nH_2O$$

(n is the number of monosaccharide units in the molecule). The equation shows that 1 gramme molecule of polysaccharide ($162n$ g) is oxidized by $6n$ gramme molecules of oxygen ($6n \times 22.4 = 134n$ l). Hence 1 cm³ oxygen combines with 1.2 mg polysaccharide. Alternatively, it could combine with about the same quantity of protein or with about 0.5 mg fat. Hence if 0.08 cm³ oxygen is used for every litre of water pumped, that water must yield more than about 40–100 µg dry organic matter (depending on the proportion of fat) if the animal is to grow. The concentration of planktonic algae seems generally to be lower than this in oceanic water, but higher in coastal waters. For instance, in the English Channel it fluctuates with the seasons between about 100 and 1000 µg dry organic matter l^{-1}. Sea squirts are found mainly in coastal waters.

How much of the energy used by a sea squirt is needed simply to maintain the feeding current? The energy needed to drive a pump is the volume of fluid pumped multiplied by the pressure difference across the pump. Unfortunately we do not know the pressure difference, but we have seen that it is probably around 2 mm water (20 N m^{-2}). If it has this value, 20 J is needed to pump each cubic metre of water, and 0.02 J to pump each litre. This is the minimum requirement, if the pump and the metabolic processes were 100% efficient. It is more likely that they are around 10% efficient, making the requirement $0.2 \, \mathrm{J} \, \mathrm{l}^{-1}$. Metabolism involving $1 \, \mathrm{cm}^3$ oxygen yields about 20 J (whatever food is being used) so the part of the oxygen consumption attributable to the needs of pumping is of the order of $0.01 \, \mathrm{cm}^3$ oxygen for every litre of water pumped. This is about 13% of the oxygen which is actually used. It must be emphasized that this is a very rough estimate, based on unreliable data.

The energy used for pumping is not the whole of the energy required for the feeding process. Energy is also needed for secretion of enzymes and mucus, and for other purposes. The mucus which is secreted in the pharynx and forms the filter is digested again in the gut, but presumably some energy is lost in the process.

Apart from the structures used in feeding, the anatomy of sea squirts is remarkably simple. There is a single nerve ganglion in the body wall between the siphons. There are no special respiratory organs: the large surface area of the pharynx must be ample to enable the animal to take up the oxygen it needs from the feeding current. There seem to be no special organs of excretion or of osmotic and ionic regulation. Ammonia is formed as a waste product of protein metabolism and simply diffuses out of the body. The blood of *Phallusia* has been analysed and found to have the same osmotic concentration as the seawater the animal lives in, but not quite the same ionic composition: the main difference is that there is only about half as much sulphate in the blood as in seawater.

There is a system of blood vessels, and a heart which is simply a U-shaped tube. Muscular constrictions travel along the heart, driving the blood through. For a few minutes, all the constrictions travel in the same direction and the blood flows one way round the circulation. Then constrictions start travelling in the opposite direction and flow is reversed for a few minutes. The rate at which blood is pumped by *Ciona* has been calculated from the dimensions of the heart and the rate at which it beats. It seems that the blood makes about twenty circuits of the body in one direction before reversing.

SWIMMING LARVAE

Though the adults of typical sea squirts are stationary, anchored to rocks, they develop from swimming larvae. The larvae of most species are about a millimetre long. They do not feed, but generally settle and start metamorphosis to the adult form after only a few hours.

The larvae look like tiny tadpoles (Fig. 2-3). They swim in the manner of tadpoles, at speeds up to about $3 \, \mathrm{cm} \, \mathrm{s}^{-1}$, by tail movements. Running the length of the tail is the notochord, which is a row of 40–42 cylindrical cells with large vacuoles enclosed in a sheath of connective tissue fibres. On either side of it are a few (usually three) rows of muscle cells. These are large cells with myofilaments in their outer layers, and a central core of vacuolated cytoplasm. Swimming must involve alternate contraction of the cells on the left and on the right, bending the tail from side to side. The notochord presumably functions in the same way as the vertebral column vertebrates, as a structure which is flexible but of fixed length. It cannot shorten because the sheath prevents it from swelling. If it were not there

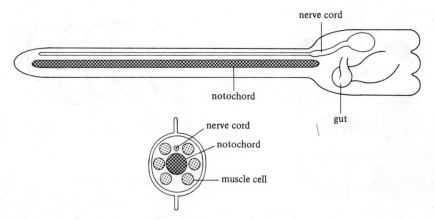

Fig. 2-3. Diagrammatic lateral view of the larva of a typical sea squirt, and a transverse section through its tail.

the muscles on both sides of the tail could contract simultaneously, shortening the tail. Because it is there shortening of the muscles on one side lengthens those on the other. The notochord makes the muscles on either side of the tail antagonistic to one another.

Dorsal to the notochord is a tubular nerve cord which is swollen at the anterior end, where it contains two simple organs. These seem from their structure to be a light detector and a sense organ sensitive to tilting. The part of the nerve cord which lies within the trunk of the tadpole is relatively stout and seems to be nervous tissue. The part in the tail is slim, and its cells do not look like nerve cells. It has no obvious function.

The larvae of different ascidians behave in ways which seem likely to take them to suitable sites for settlement, and metamorphosis to the stationary adult. Most swim upwards at first, and later downwards and away from light. This tends to take them to sites such as overhanging rock faces where they can attach themselves firmly and where they are unlikely to get covered by sediment.

VARIETY IN STRUCTURE AND HABITS

Subphylum Urochordata, classes Ascidiacea, Larvacea and Thaliacea

The typical sea squirts are included in the class Ascidiacea. Besides solitary sea squirts such as *Ciona* the class includes colonial ones. A colony is founded by a single larva which metamorphoses and then produces a number of individuals asexually, by budding. These remain connected together. Some sea squirts form loose, irregular colonies but others form

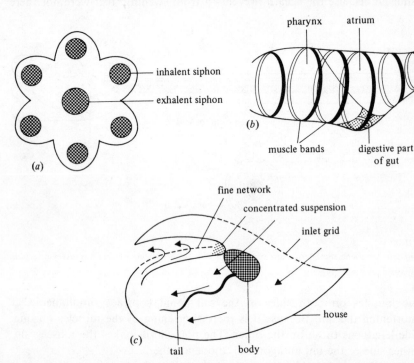

Fig. 2-4. (*a*) A sketch of *Botryllus*, a colonial ascidiacean. (*b*) A sketch of a salp. (*c*) A simplified section of the 'house' of *Oikopleura*, showing the path of water through it.

more regular ones. For instance, *Botryllus* forms a rosette of individuals contained in a single tunic and sharing a single central exhalant siphon (Fig. 2-4*a*). It is common on stones and seaweeds on British shores. *Octacnemus* is a peculiar ascidiacean found in samples from the bottom of the oceans at depths of about 2000 to 4000 m. It has no openings in the wall of its pharynx so cannot filter seawater like *Ciona*. It is apparently a predator, for nematodes and small crustaceans have been found in its gut. It is not known how it catches its prey: perhaps it can crawl.

The salps (Salpidae) belong to the class Thaliacea. They are planktonic animals with an inhalant siphon at one end of the body and an exhalant one at the other (Fig. 2-4*b*). Most are 25–150 mm long. The wall of the pharynx is not a closely spaced mesh with numerous small openings, as in typical sea squirts, but has only two very large openings. It does not support the mucus sheet in the way the close mesh of sea squirts does. Instead the sheet forms a funnel: mucus is added to the sheet anteriorly, around the rim of the funnel, and is drawn into the digestive part of the gut from the point of the funnel. The mucus is not kept moving by frontal cilia, as in sea squirts, but by cilia in the digestive part of the gut which pull the sheet in. Water is filtered through the mucus sheet as in sea squirts, but it is not propelled by cilia. Rings of muscle encircle the body, and their rhythmic contractions pump the water through. There are also muscles around the siphons. They close the siphons alternately, so that the water is always drawn in through the inhalant siphon and driven out through the exhalant one. Salps have no tadpole-like larva, but develop directly to the adult form. The salps are not the only Thaliacea. There are others which resemble them in many ways but use cilia, not muscles, to drive the feeding current. All of them are planktonic and some are colonial.

The Larvacea are also planktonic. They develop from tadpole-like larvae like sea squirts, but do not lose their tails: the tail is retained as a permanent organ. They are very small, seldom more than 5 mm long including the tail. There is only one pair of openings in the wall of the pharynx and since there is no atrium they open directly to the exterior. The mucus forms a funnel inside the pharynx and water is driven through it by cilia. There is however an extraordinary refinement of the filter-feeding mechanism. The whole animal is enclosed in a gelatinous 'house', so delicate and transparent as to be extremely difficult to study. The house is believed to be homologous with the tunic of sea squirts. It has openings at either end and water is kept flowing through by undulation of the tail. This drives the house slowly through the water.

The best known of the larvaceans is *Oikopleura*. The structure of its house is shown diagrammatically in Fig. 2-4*c*. The water enters through a relatively coarse grid. Within the house it encounters a large sheet of very much finer mesh, fine enough to stop particles less than 1 μm in diameter. The water itself flows through but the suspended matter accumulates in the position shown in the diagram, close to the animal's mouth. The animal feeds on concentrated suspension, taken from here, and it is this concentrated suspension which is filtered through the mucus sheet in the pharynx.

It has been suggested that the vertebrates may have evolved from urochordates by a process of neoteny: that is, by retention of larval features in adult animals. The notochord and tail muscles are found in the larvae of typical sea squirts, but not in the adults. They may have been evolved initially in urochordates, as larval features which disappeared immediately after settlement. Neoteny may have occurred subsequently, producing adult filter-feeding animals which retained the notochord and the tail, with its muscles. Such animals may have been the ancestors of amphioxus (which will be described next) and the vertebrates. The Larvacea seem much too peculiar to be likely ancestors of the vertebrates, but they show that neotenous urochordates are feasible.

AMPHIOXUS

Phylum Chordata, subphylum Cephalochordata

The urochordates are quite a large and diverse group but the cephalochordates include only the twenty-odd species of *Branchiostoma* and *Asymmetron*. The familiar name *Amphioxus* has had to be abandoned because the name *Branchiostoma* was used in the original description, but amphioxus (without an initial capital, and not italicized) is often used as the common English name.

Amphioxus are small cigar-shaped animals, up to about 7 cm long. They spend their adult life in the sea bottom, buried in sand or shell gravel. They are very common in some places: a handful of sand from the Lagos area may contain 30–40 *Branchiostoma nigeriense*, and *B. belcheri* is common enough on the Chinese coast to be collected as food.

Careful studies of the distribution of *B. nigeriense* around Lagos showed that it lives only in areas where there is coarse sand with little silt in it. Laboratory experiments were made as well as field observations. Specimens were put in dishes of seawater, with various grades of sand and mud on the bottom. Fig. 2-5 shows how they behaved. When given mud, they generally rested on the surface. When given fine sand, they buried only the hind part of the body. In either case they were restless, apt to swim if disturbed. This behaviour must tend to make them move away from fine sediments in nature. When given coarse sand they burrowed, penetrating the sand head first and moving through it by eel-like undulations. They either remained under the sand or adopted a position with the mouth, and very little else, above the surface. Burrowing must make them inconspicuous and inaccessible to predators. They cannot easily penetrate excessively coarse sand.

Amphioxus are filter feeders, using a mechanism very like that of sea squirts. Water enters by the mouth and leaves by the atrial opening, which is just clear of the surface of the fine sand in Fig. 2-5. Feeding is possible in coarse sand even when the animal is completely buried, because water can permeate easily between the grains.

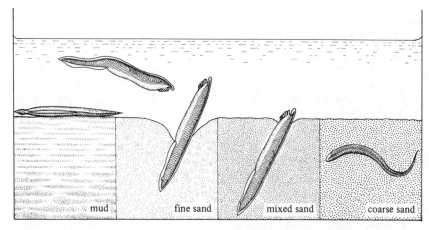

Fig. 2-5. Amphioxus on mud, and in several grades of sand. Further explanation is given in the text. From J. E. Webb & M. B. Hill (1958). *Phil. Trans. Roy. Soc. Ser.* B, **241**, 355–91.

The main structures used in feeding are shown in Fig. 2-6. Two sets of tentacles keep excessively large particles out of the mouth. The wall of the pharynx is not a network as in sea squirts, but is perforated by a series of narrow, sloping slits. The bars between the slits have lateral and frontal

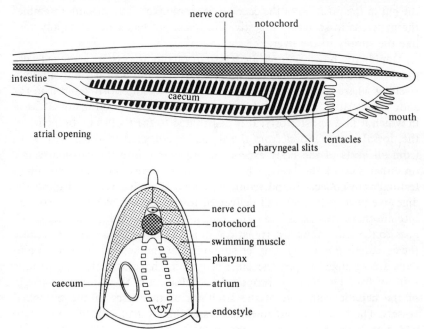

Fig. 2-6. A diagrammatic lateral view of amphioxus, and a transverse section.

cilia, as in sea squirts. The lateral cilia drive water through from the mouth to the atrium. The frontal ones move a sheet of mucus which covers the slits on the inner side.

Small amphioxus are remarkably transparent, and if they are fed a suspension of carmine particles, for instance, the particles trapped in the mucus can be seen through the body wall. The movements of the mucus, both in the pharynx and in the gut, have been observed in this way. The mucus is secreted by a ventral endostyle. It is moved dorsally by the frontal cilia to a median groove in the roof of the pharynx, crossing the pharyngeal slits obliquely and trapping food particles from the feeding current which is filtered through it. Cilia in the median groove move the mucus and trapped particles posteriorly to the digestive parts of the gut where a ring of cilia makes the string of mucus rotate.

A deep narrow pouch, the caecum, branches off the gut immediately posterior to the pharynx and runs forward on the right side of the pharynx. Cilia beat anteriorly on its roof, and posteriorly on its floor. Small particles become detached from the string of mucus in the gut and move into the caecum. The cells of the caecum secrete digestive enzymes and also ingest fragments of food material which are digested further in vacuoles. Both the liver and pancreas of vertebrates develop as pouches which grow out of the gut in the position of the caecum of amphioxus. The caecum resembles the pancreas in secreting digestive enzymes, but has a blood supply very like the supply to the liver.

Amphioxus has a complicated system of blood vessels, arranged very much as in fish (Fig. 2-7). The sinus venosus is not a heart, but simply a swelling where several vessels meet. Blood flows anteriorly from it in the endostylar artery, then dorsally in the branchial vessels to the dorsal aorta in which it flows posteriorly. It is distributed to the tissues by branches of the dorsal aorta. Blood from the tissues is collected in veins. The veins from all parts of the body except the gut empty into the cardinal veins, on either side of the body, which lead to the sinus venosus. The subintestinal vein collects blood from the network of fine vessels around the digestive part of the gut and delivers it to the caecum where it breaks up into another network before re-forming as the so-called hepatic vein which goes to the sinus venosus. Hepatic vein is an inappropriate name because the caecum is not a liver, but the subintestinal and hepatic veins of amphioxus are arranged in just the same way as the hepatic portal and hepatic veins of fish. The blood is propelled by pulsation of the endostylar artery, of the hepatic vein and of the swellings at the bases of the branchial vessels. The blood has no corpuscles, and the layer of cells (endothelium) which encloses the vessels is incomplete.

Amphioxus has nothing comparable to the kidneys of vertebrates, but

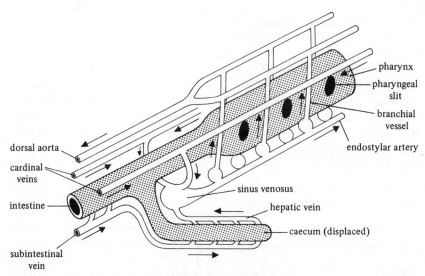

Fig. 2-7. A diagram showing the main features of the blood circulation of amphioxus.

has numerous bunches of solenocytes very like those found in the nephridia of some polychaete worms (Fig. 2-8). Each solenocyte is a cell formed as a blind-ended tube with a flagellum beating within it. The wall of the tube is not continuous, but consists of ten rods with slits between. The solenocyte is surrounded by fluid in a small coelomic cavity, and leads to a

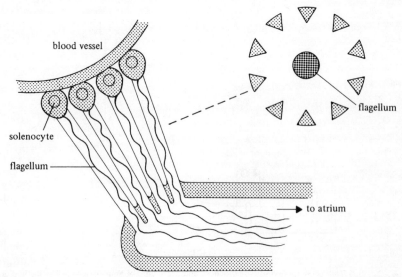

Fig. 2-8. A diagram of a group of solenocytes of amphioxus, with a section (at a higher magnification) through the tube of one of them.

sac which opens into the atrium. There are about 100 of these sacs on each side of the body, each with a bunch of solenocytes. It seems likely that they excrete a urine, but this has not been demonstrated experimentally.

The notochord is enclosed in a sheath of collagen fibres. It consists of a series of discs of muscle, with fluid-filled spaces between them. The muscle fibres run across the notochord, and have extensions which run dorsally to end near nerve endings (Fig. 2-9a). The collagen partition is not actually perforated, but it is very thin where it separates the neurones from the

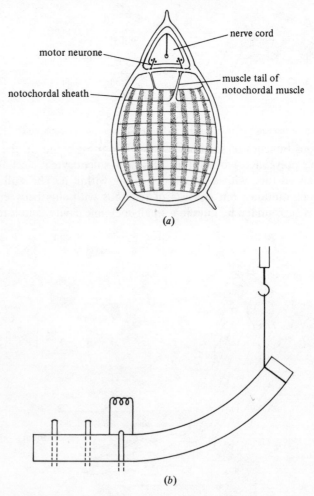

(a)

(b)

Fig. 2-9. (a) A transverse section of the notochord and nerve cord of amphioxus. (b) The method of measuring the stiffness of the notochord, which is described in the text. From D. M. Guthrie & J. R. Banks (1970). *J. exp. Biol.* **52**, 125–38.

muscle tails. The muscle fibres are presumably stimulated to contract by activity in the neurones. They resemble some invertebrate muscles and differ from vertebrate ones in containing the protein paramyosin.

When the muscles contract they must tend to make the notochord narrower from side to side, and they must tend to increase the pressure in it. Narrowing would tend to make the notochord more flexible. An increase in pressure would tend to make it less flexible. Which effect predominates? Fig. 2-9*b* shows an experiment designed to find out. A notochord was dissected out and one end was fixed. The other was attached to a force transducer and the force required to bend it to a given extent was measured. The muscle was made to contract by electrical stimulation, and the change in force was noted. It was found that contraction of the muscles increased the stiffness of the notochord substantially. Stiffening could possibly aid fast swimming by increasing the natural frequency of side-to-side vibration of the animal. Stiffening may also aid burrowing. The notochord extends further forward than in fishes, almost to the tip of the snout. It therefore stiffens the snout, which may aid penetration of sand.

The muscles which are used for swimming and burrowing are divided into myomeres (muscle segments) by partitions of collagenous connective tissue. The myomeres resemble those of fishes but are simpler in shape.

There is a hollow dorsal nerve cord contained in a tube of collagen fibres which runs along the top of the notochord sheath (Fig. 2-9*a*). Though there are no vertebrae, these collagenous structures enclose the nerve cord and notochord in essentially the same way as do the vertebrae of many fish (see Fig. 8-16*a*). There is no swelling of the anterior end of the nerve cord which might be compared with the brains of vertebrates, but the central canal is enlarged at the anterior end. The nerve cord sends a motor nerve to each myomere, and gives rise to an equal number of sensory nerves. The motor nerves leave the cord ventrally and the sensory ones dorsally, as in vertebrates.

POSSIBLE CHORDATES

There are two groups of animals which some zoologists would include among the chordates but which I prefer to exclude. They are the living Hemichordata and the extinct Calcichordata. Both have features which are strikingly reminiscent of echinoderms and provide strong evidence of relationship to echinoderms. They also have features which suggest relationship to the chordates, though not particularly strongly. Most modern zoologists would probably omit both from the phylum Chordata. It is in any case convenient to avoid describing them in this book, since it would be impossible to discuss their relationships at all fully without giving a lot of information about echinoderms.

FURTHER READING

General

Barrington, E. J. W. (1965). *The biology of Hemichordata and Protochordata.* Oliver & Boyd, Edinburgh.
Berrill, N. J. (1955). *The origin of vertebrates.* Clarendon Press, Oxford.
Grassé, P.-P. (1948). *Traité de zoologie,* vol. **11,** *Echinodermes, stomocordes, procordes.* Masson, Paris.

Sea squirts

Goodbody, I. & Trueman, E. R. (1969). Observations on the hydraulics of *Ascidia. Nature, Lond.* **224,** 85–6.
Jørgensen, C. B. (1966). *Biology of suspension feeding.* Pergamon, Oxford.
Kriebel, M. E. (1968). Studies on cardiovascular physiology of tunicates. *Biol. Bull. mar. biol. Lab.,* Woods Hole, **134,** 434–55.
Millar, R. H. (1953), *Ciona.* Liverpool University Press, Liverpool.
Millar, R. H. (1971). The biology of ascidians. *Adv. mar. Biol.* **9,** 1–100.

Amphioxus

Brandenberg, J. & Kummel, G. (1961). Die Feinstruktur der Solenocyten. *J. Ultrastruct. Res.* **5,** 437–52.
Guthrie, D. M. & Banks, J. R. (1970). Observations on the function and physiological properties of a fast paramyosin muscle – the notochord of amphioxus (*Branchiostoma lanceolatum*). *J. exp. Biol.* **52,** 125–38.
Webb, J. E. (1958). The ecology of Lagos Lagoon. III. The life history of *Branchiostoma nigeriense* Webb. *Phil. Trans. Roy. Soc. Ser. B,* **241,** 335–53.
Webb, J. E. & Hill, M. B. (1958). The ecology of Lagos Lagoon. IV. On the reactions of *Branchiostoma nigeriense* to its environment. *Phil. Trans. Roy. Soc. Ser. B,* **241,** 355–91.

Possible chordates

Jefferies, R. P. S. (1968). The subphylum Calcichordata (Jefferies 1967). Primitive fossil chordates with echinoderm affinities. *Bull. Br. Mus. nat. Hist. Geol.* **16,** 243–339.

3

Fish without jaws

This chapter is about the fish of the class Agnatha, which have no jaws. The modern members of the class are the lampreys and hagfishes, which are highly specialized for peculiar ways of life. The extinct members include the earliest known vertebrates.

The vertebral column is not the most characteristic feature of the vertebrates, in spite of their name. Indeed, hagfishes have no vertebral column and lampreys have only nodules of cartilage alongside the notochord. Most vertebrates have a well formed vertebral column which has taken over the function of the notochord, but these ones do not. More characteristic of the vertebrates are the brain, the skull, the ears, the kidneys and various other organs which are basically similar in all vertebrates and quite different from any organs found in invertebrates. Some zoologists have recognized this by using the term Craniata instead of Vertebrata.

LAMPREYS

Subphylum Vertebrata, class Agnatha, order Petromyzoniformes
The lampreys are considered first in this chapter, and at more length than any of the other orders of Agnatha. This is because more is known about them than about the others and because they have a larval stage (the ammocoete) which is believed to be more like the ancestral vertebrates than any other living fish.

Lampreys lay their eggs in fresh water. The ammocoete hatches out and grows there, but many species of lamprey spend their adult life in the sea. Both the ammocoete and the adult have slender, eel-like bodies but the adult has a characteristic sucker around its mouth and the larva does not. Ammocoetes are small, up to about 20 cm long, but adults of some species grow to lengths approaching 1 m and weights over 2.5 kg. The larval stage lasts several years, often about five. The duration of the adult stage varies greatly between species but death generally occurs soon after breeding.

FEEDING AND RESPIRATION

Ammocoetes live in much the same way as *Branchiostoma*, but in fresh water (Fig. 3-1). They bury themselves in sediment on the bottom of slow-flowing streams and seldom leave their burrows. They filter small organisms from water which is drawn in at the mouth and passed out through

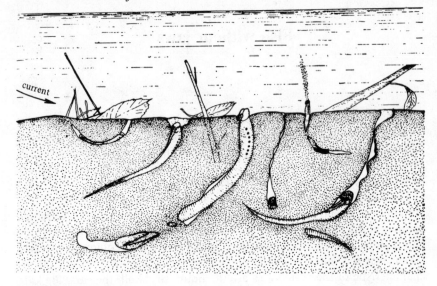

Fig. 3-1. Ammocoetes in their natural habitat. From V. C. Applegate (1950). *Spec. scient. Rep. U.S. Fish Wildl. Serv.* **55**, 1–237.

the gills. Since diatoms and detritus (dead organic matter) tend to settle on the surface of the sediment, and the water is drawn in from just above the sediment, the water that is filtered probably contains far more diatoms and detritus than the bulk of the stream water. Examination of the contents of ammocoete guts shows that diatoms make up a large proportion of their food but that protozoa and occasionally other small animals, and detritus, are also eaten. It is not known to what extent the detritus is digested. The incoming water is strained through finger-like cirri, as in *Brachiostoma*. The large particles that are prevented in this way from entering the mouth are blown clear from time to time by a sort of coughing action (one of the ammocoetes in Fig. 3-1 is doing this).

The gill openings are not long clefts like those of *Branchiostoma* but are small and round. There are only seven on each side of the body. Because the openings are few and relatively small, the gill apparatus is far less delicate than in *Branchiostoma* and is not protected in an atrium. There are flaps of skin over the openings which act as valves (Fig. 3-2a). Water can leave through them but is prevented from entering: any inward movement of water through the openings makes the flaps close over them. Similarly, water can enter through the mouth but is prevented from leaving that way (except in coughing) by the valve-like action of the velum, just inside the mouth. Thus flow is normally kept going in one direction, in at the mouth and out through the gills.

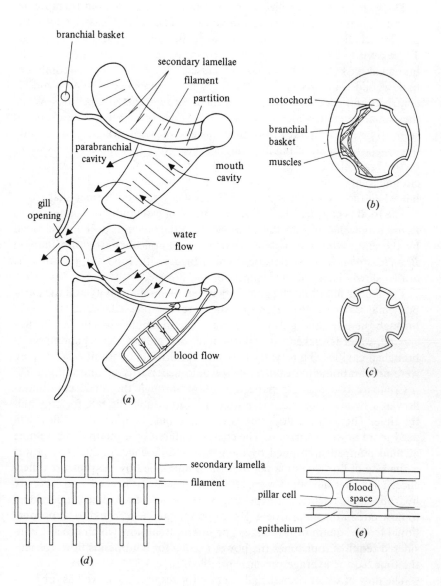

Fig. 3-2. Diagrams of the gills of ammocoetes. (*a*) Two gill arches in horizontal section. Arrows indicate the path of water between the secondary lamellae. The path of the blood is indicated on the filament at the bottom. (*b*) A transverse section of the gill region, with branchial basket expanded. (*c*) The branchial basket compressed. (*d*) A section through two gill filaments, at right angles to the secondary lamellae. (*e*) A section through a secondary lamella.

The water is not propelled by cilia as in *Branchiostoma* but by muscular action. It is pumped by changes in volume of the mouth cavity and of the parabranchial cavities which lie between the gills and the gill openings. These cavities are enclosed in a framework of cartilage bars, known as the branchial basket, and by a sheet of muscle fibres which run circumferentially round the body (Fig. 3-2*b*). They are compressed when the muscle contracts and water is driven out of them through the gill openings. When they enlarge again water is drawn in through the mouth. There do not seem to be any muscles to produce this enlargement, which is apparently due to elastic recoil of the branchial basket. Note that the cartilage hoops of the branchial basket have inwardly directed kinks in them. When the circumferential muscles contract these kinks become more bent (Fig. 3-2*c*), but when they relax the kinks recoil elastically to their former shape.

The food is captured in a sheet of mucus, in essentially the same way as in *Branchiostoma* and the tunicates. Some of the mucus is probably secreted by the endostyle, which is a much more complicated structure than in *Branchiostoma*, and some elsewhere. Mucus with food trapped in it is drawn slowly back into the intestine, and digested.

The mucus filter and the gills lie between the mouth cavity and the parabranchial cavities. When the cavities are compressed, water is driven through the filter and gills from the contracting mouth cavity. When they expand, water is sucked through the filter and gills by the expanding parabranchial cavities. Thus water may be kept flowing continuously. If there were no parabranchial cavities flow would necessarily be intermittent.

Compare two possible patterns of flow through the filter: continuous flow at a steady rate, and intermittent flow at double the rate for only half the time. The average flow rate is the same but the intermittent flow will need more power to drive it. The energy required by a pump is the volume of fluid pumped multiplied by the pressure difference (p. 29). The power is the rate of flow (volume per unit time) multiplied by the pressure difference. The faster the pump works the more pressure is needed, and there are hydrodynamic reasons for expecting pressure difference to be proportional to flow rate in the ammocoete. Hence power can be expected to be proportional to the square of flow rate. The intermittent flow which is being considered requires four times the power needed for continuous flow, for half the time, so the average power is doubled.

The flow of water over the gills serves for respiration as well as for feeding. In other fish it serves only for respiration but the advantage of continuous flow remains: less energy is needed to pump water over the gills at a given rate, if it is kept flowing continuously, than if it flows intermittently. Continuous flow is also more favourable for exchange of oxygen and carbon dioxide between the water and the blood.

The gills have a complicated structure which gives them a large surface area. A transverse partition separates each gill slit from the next. From it project horizontal gill filaments, like so many shelves (Fig. 3-2*d*). The filaments in turn bear secondary lamellae which project up from their upper surfaces and down from the lower ones. The secondary lamellae are closely spaced so their total surface area is large, and it is through their surfaces that most of the respiratory exchange of gases probably occurs. It has been estimated that their total area in a 1 g ammocoete of *Lampetra planeri* is about 0.7 cm², or about one tenth of the whole external area of the body. This may seem large, but it is much less than one would expect to find in a teleost. For instance, measurements on small specimens of the teleost *Micropterus* indicate that a 1 g specimen would have a gill area of about 7 cm².

The structure of a secondary lamella is shown in Fig. 3-2*e*. Its two faces are formed by thin epithelia. If these were not held together in some way they would be pushed apart by the pressure of the blood in the spaces between them, so that the lamellae were not thin and flat-faced but were inflated to a bulbous shape. They are held together by pillar cells which extend (like so many pillars) from one epithelium to the other. Extensions of the pillar cells underlie the whole of the epithelia so that the blood is everywhere separated from the water outside the lamella by two layers of cells. However, the layers are together only about 4 μm thick, so the water and blood are brought very close to each other. The thickness of the lamellae and the distance between pillar cells are just big enough to allow red blood corpuscles (diameter 10 μm) to pass.

The tips of the anterior and posterior filaments of a gill slit almost touch. So do the secondary lamellae of successive filaments. Most of the water passing through the gills must therefore take the route shown in Fig. 3-2*a*, crossing the filaments in the channels between the secondary lamellae and then travelling laterally in the space between the lamellae and the transverse partitions. Thus the water between the lamellae is constantly renewed, and the distances which oxygen must travel by diffusion are small.

Hydrodynamic considerations make it plain that flow in the spaces between the lamellae must be laminar, not turbulent. Since the lamellae are about 30 μm apart, oxygen in the water must diffuse 0–15 μm to reach the nearest lamella. It must then diffuse through the cells of the lamella wall. After this, little diffusion is probably needed: it is probable that the plasma is kept thoroughly stirred as a consequence of the close fit of the red corpuscles in the blood spaces. The average total distance which the oxygen must diffuse, through water and tissue, can be estimated as only about 10 μm.

The blood flows through the secondary lamellae in the direction opposite to the flow of water over them (Fig. 3-2a). The consequences of this can be inferred from the properties of industrial heat exchangers. Two examples of heat exchangers are illustrated in Fig. 3-3. Each consists of two pipes, running alongside and in contact with each other. In both, a fluid entering one pipe at 100 °C is cooled by a fluid entering the other pipe at 0 °C. The dimensions of the exchangers and the rates of flow through them are supposed to be identical. In the parallel flow exchanger (a) the fluids run in the same direction. Heat passes from the hot fluid to the cold

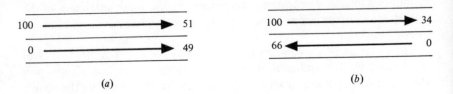

Fig. 3-3. Diagrams of (a) a parallel-flow heat exchanger, and (b) a counter-flow exchanger.

one so that their temperatures come closer and closer together. In the particular example illustrated, they leave the exchanger at 51 °C and 49 °C. In the counter-flow exchanger (b) the fluids flow in opposite directions. It can be calculated that if this is the only difference between our two exchangers, the hot fluid will be cooled, in our particular example, to 34 °C and the cold one will be warmed to 66 °C. The hot fluid could be cooled to 34 °C in the parallel-flow exchanger if the rate of flow of the cold fluid were doubled, but the counter-flow arrangement increases the rate of heat transfer at given rates of flow. Gills are not concerned with conduction of heat between two fluids but with diffusion of oxygen. However, these processes tend to follow mathematically similar laws. The counter-flow arrangement of the gills can be expected to increase the rate of transfer of oxygen from the water to the blood, for given rates of flow of the two fluids.

The discussion so far indicates that the gills are constructed in such a way as to allow rapid uptake of oxygen from the water. Their area is large, diffusion distances are small, flow is probably more or less continuous and there is a counter-flow arrangement. How well do they meet the needs of the ammocoete?

Fig. 3-4 shows apparatus which was used to measure the oxygen consumption of resting *Ichthyomyzon* ammocoetes. Ammocoetes tend to swim about when removed from their burrows so a layer of fine glass beads was provided for the ammocoete to burrow in. Well-aerated water entered at the top of the apparatus and was drawn down through the beads by the

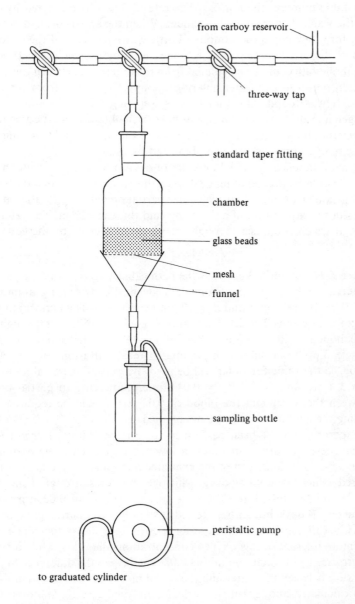

from carboy reservoir

three-way tap

standard taper fitting

chamber

glass beads

mesh

funnel

sampling bottle

peristaltic pump

to graduated cylinder

Fig. 3-4. Apparatus used for measuring the oxygen consumption of ammocoetes. From B. J. Hill & I. C. Potter (1970). *J. exp. Biol.* **53**, 47–58.

peristaltic pump at the bottom. The rate of flow was measured by collecting the water in a graduated cyclinder. When the apparatus had been running for a few hours the sampling bottle was removed and the concentration of dissolved oxygen in it was measured by a chemical method. The metabolic rate could be calculated from the difference between this and the concentration in the water entering the apparatus, and from the rate of flow. It was found that ammocoetes weighing 1 g used about 0.05 cm^3 oxygen h^{-1} at 15 °C. Similar results have been obtained in experiments with *Lampetra* ammocoetes. Typical teleosts of the same weight, resting at the same temperature, use oxygen about ten times as fast.

What difference in partial pressure of oxygen between the water and the blood would be needed to make oxygen diffuse into the secondary lamellae at this rate? Consider a gas which has partial pressure P_1, P_2 atm on opposite sides of a partition of area A cm^2 and thickness d cm. The rate, J cm^3 h^{-1}, at which it diffuses through the partition is given by the equation

$$J = AD(P_1 - P_2)/d, \qquad (3.1)$$

where D is a quantity known as the permeability constant of the gas in the material of the partition. From data given already for 1 g ammocoetes $J = 0.05$, $A = 0.7$ cm^2 and $d = 10$ μm $= 10^{-3}$ cm. The permeability constant for oxygen is 2×10^{-3} cm^2 atm^{-1} h^{-1} if it is diffusing through water and about 8×10^{-4} cm^2 atm^{-1} h^{-1} if it is diffusing through tissues such as muscle. Less than half of the estimated mean diffusion path is through tissue, so the value for water will be used. By putting these values in equation 3.1 we find $(P_1 - P_2) = 0.04$ atm. A difference in partial pressure between the water and the blood of 0.04 atm should be required, if the resting metabolic rate is to be maintained.

Experiments have been performed to test the ability of *Ichthyomyzon* ammocoetes to survive in water of low oxygen content. The partial pressures of dissolved oxygen in the experimental aquaria were kept at the required values by bubbling oxygen/nitrogen mixtures through them. Nearly all the animals tested at 16 °C at a partial pressure of 0.02 atm survived for at least 4 days, but animals tested at slightly lower partial pressures died quickly. (Similar tests on fresh water teleosts have shown that some species are more tolerant of low oxygen concentrations than this, while others are less tolerant.) In water containing 0.02 atm oxygen the difference in partial pressure between the water and the blood must be much less than the 0.04 atm calculated above, but that calculation was based on anatomical data for a different species, and on the assumption that oxygen consumption had to be maintained at the level usual in well-aerated water.

Adult lampreys have suckers around their mouths. They use them to anchor themselves to stones in streams, or to attach themselves to prey.

Some lampreys attach themselves to fish and feed on their blood. They may remain attached (at least in aquaria) for several days, and their victims frequently die. Pieces of fish muscle and other tissues are found in the guts of such lampreys, as well as blood. The other species of lamprey do not feed after they have become adults, but simply breed and die.

The sucker is round when spread open for use, but its lateral margins are generally drawn together when the lamprey swims. The edges of the sucker are studded with tooth-like knobs of keratin which must help to prevent it from slipping when attached to the slippery surface of a fish. In the floor of the mouth cavity is a structure known as the tongue which is stiffened by a long central cartilage. Smaller cartilages at the tip of the tongue bear horny teeth, which are used to rasp wounds in prey. This is done by the muscles of the tongue; alternate contraction of the dorsal and ventral tongue muscles makes the teeth rock up and down (Fig. 3-5).

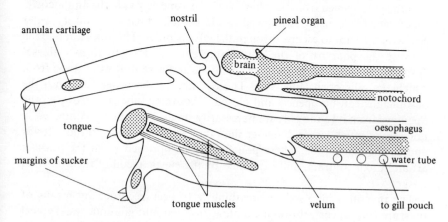

Fig. 3-5. A diagrammatic median section of the head of an adult lamprey.

The sucker is muscular, and also contains a ring of cartilage. It seems to work in essentially the same way as a rubber sucker. It is spread open by its muscles, and placed against the surface to which it is to be attached. Any force which tends to enlarge the space between the sucker and the surface will then fix it. Elastic forces which remain after the spreading muscles have relaxed may suffice for this (rubber suckers depend on elasticity), but muscles may also play an active part in maintaining suction. When the lamprey is feeding, blood is passed from the sucker into the gut without detaching the sucker. The mechanism has not been fully explained.

The blood of the prey is prevented from clotting by the secretion of two large glands which open into the mouth cavity. The secretion also contains enzymes which attack the tissues of the prey.

Lampreys readily attach themselves to the walls of the aquaria in which they are kept. The pressure under the sucker has been measured by inducing lampreys to attach to a sheet of Perspex over a hole leading to a pressure transducer. Pressures of -0.5 to -6 cmH$_2$O (relative to the surrounding water) were usual, with occasional brief stronger sucks. A pressure of -120 cmH$_2$O was recorded when attempts were made to pull a lamprey off the Perspex.

While the sucker is attached the lamprey cannot breathe in through its mouth. Indeed, adult lampreys do not normally breathe in through their mouths even when they are not attached. This can be demonstrated by releasing a suspension of carmine particles or Indian ink from a pipette near the animal. Particles released near the mouth do not generally enter the mouth but particles released near the gill openings enter the openings as the lamprey breathes in and emerge from them again as it breathes out.

If the gills opened from the pharynx in the same way as in the ammocoete, blood which was being swallowed would get mixed with respiratory water and would be apt to escape through the gill openings. This does not happen because the pharynx of the adult lamprey is divided to form a dorsal oesophagus which is the passage for food, and a ventral water tube from which the gills open (Fig. 3-5). The opening of the water tube into the oesophagus can be opened and closed. The water tube is narrow, so the openings from it to the gills are necessarily small. The gills are contained in pouches of much larger diameter than their small openings to the exterior and to the water tube. These pouches are compressed when the muscles around them contract, and expanded by the elastic recoil of the branchial basket.

The pressures involved in breathing have been recorded by means of pressure transducers, through cannulae inserted into gill pouches. Typical records show the pressure fluctuating between about -0.3 cmH$_2$O (as the lamprey breathed in) and $+1$ cmH$_2$O (as it breathed out).

Breathing movements serve also for sniffing. There is a single nostril on top of the head (Fig. 3-5). A tube leads down from it to the olfactory epithelium (where the odour-sensitive cells are) immediately anterior to the brain. It carries on into a sac which lies within the branchial basket and is compressed and expanded by the breathing movements, in the same way as the gill pouches. Water is sucked into the nostril whenever the lamprey breathes in, and expelled when it breathes out. A flap tends to deflect some of this water so that it passes over the olfactory epithelium.

It will be evident from the above that ammocoetes undergo a remarkable metamorphosis when they change to the adult form. A sucker and tongue develop and the water tube becomes separated from the oesophagus. The eyes move from their original position deep under the skin to the surface of the head. These and other changes fit the animal for a drastic change in feeding habits. There is a great advantage in having them occur rapidly, for while they are in progress the lamprey is not well fitted for either mode of feeding. The change in external appearance takes only about a month but internal changes may take longer and feeding is generally interrupted for several months.

SENSE ORGANS

Some information has already been given about the organ of smell in lampreys. This section of the chapter is mainly about the lateral line sense organs and the ear, which have in common a distinctive type of sensory cell and are referred to together as the acoustico-lateralis system. There is also some information about eyes.

The characteristic cells of the acoustico-lateralis system are the hair cells, of which two are shown in Fig. 3-6*b*. These cells are incorporated in an epithelium and each has a tuft of processes projecting above the epithelial surface. Within the tuft, the longest process is a cilium with the same structure as cilia which beat: it contains filaments arranged in the well-known 9 + 2 pattern. The rest have a different structure, quite unlike cilia. Sensory neurones synapse with the hair cells; each generally synapses with several hair cells.

Hair cells are found in groups called neuromasts. Typically the tufts of processes of all the cells in the group are embedded in the base of a jelly-like structure called the cupula (Fig. 3-6*a*). There are neuromasts of this sort on the external surfaces of fish (including lampreys) and aquatic amphibians. Fish generally have a line of them running along either side of the trunk: this is the lateral line, and these neuromasts are called lateral line organs. There is a more complicated pattern of lateral line organs on the head. In each neuromast, half the hair cells have the cilium on one side of the bundle of processes and half on the diametrically opposite side. Wherever they lie on the body, lateral line neuromasts are served by cranial nerves, not by spinal nerves.

The physiology of lateral line organs has been studied in experiments both with fishes and with amphibians. Some particularly thorough experiments have been done on the clawed toad (*Xenopus*). The neuromasts were stimulated in various ways and action potentials were recorded from their nerves. When no stimulus was applied the neurones carried a steady succession of action potentials, following each other at a constant frequency. The frequency could be altered by pushing the cupula with a microneedle or by a jet of water from a pipette. The effect depends on the direction of movement. Suppose the neuromast is placed so that all the cilia lie either

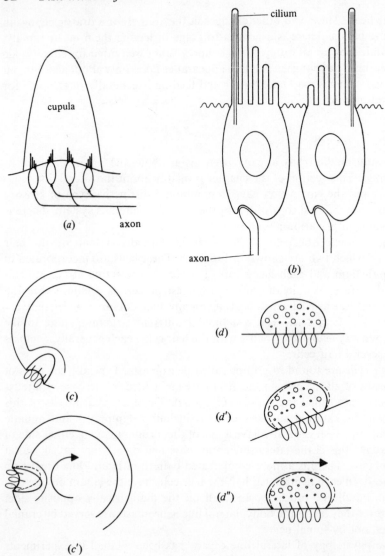

Fig. 3-6. Diagrams of (*a*) a lateral line neuromast; (*b*) two hair cells; (*c*) a semicircular canal, which is rotating in (*c'*); (*d*) an otolith organ, which is tilted in (*d'*) and is accelerating to the right in (*d"*).

on the north side or on the south side of their tufts. North–south movements will affect the frequency of action potentials. There are two sensory neurones to the neuromast. One has the frequency increased by northward and decreased by southward movements. The other is affected in the opposite way. Presumably one is connected only to hair cells with north cilia and the other to hair cells with south cilia. East–west movements do not affect the frequency in either.

The lateral line neuromasts also respond to vibration of the water. They can thus detect sounds, at least sounds of low frequency. This does not, however, seem to be an important function. An object vibrating in a fluid has two effects. It pushes the nearby fluid out of its way as it moves, and it sets up sound waves. The first effect (the near-field effect) would occur even if the fluid were totally incompressible. The second (the far-field effect) depends on the compressibility of the fluid: sound waves are alternate compressions and rarefactions of the fluid. Close to the vibrating object, the fluid movements are mainly due to the near-field effect: far from it, they are mainly due to the far-field effect. The two effects are equal at a distance of about 0.2 wavelengths. Lateral line neuromasts have been shown to respond to vibrations of up to 200 Hz (cycles s^{-1}): that is, to frequencies at which the wavelength of sound in water is 7 m or more. There does not seem to be much evidence for the use of lateral line neuromasts to detect water movements produced by objects more than a few centimetres away. The neuromasts apparently serve as detectors of acoustic near fields, rather than of true sound.

Blinded fish can be trained to locate moving or vibrating glass rods, but if the lateral line nerve to part of the body is cut, that part becomes insensitive. This suggests uses for the lateral line system, helping fish to find moving prey in dark or turbid water and warning them of approaching predators. They probably also help the fish to avoid obstacles: as a fish approaches a rock or the wall of its aquarium water will tend to flow faster over its head because the obstacle obstructs the flow of water displaced by the fish. The blind cave fish *Anoptichthys* moves competently around an aquarium without collisions, and may well depend more on its lateral line neuromasts than do fish which can see.

There are two main types of neuromast in the ears of fishes. The neuromasts of the semicircular canals are very like those of the lateral line system. Those of the otolith organs have a mass of crystals of calcium carbonate, or even a single solid mass, embedded in the cupula. The ear is filled with a fluid called endolymph which resembles blood plasma but contains less protein and more potassium.

Each semicircular canal is a curved tube connected at both ends to the main cavity of the ear (Fig. 3-6c). It has a swelling (ampulla) at one end, which contains the neuromast. The cupula is almost invisible because it has the same refractive index as the endolymph, and it tends to be shrivelled in histological preparations. However, its size has been demonstrated by injecting ink into the endolymph, so that it becomes conspicuous as a clear region in the black fluid. It reaches right across the ampulla, more or less completely blocking it. Any movement of fluid along the canal must move the cupula.

Semicircular canals are sensitive to rotation. If the canal shown in Fig. 3-6c is rotated clockwise the endolymph lags behind, due to its inertia, so that it is flowing anticlockwise relative to the canal wall (Fig. 3-6c'). The cupula is deflected as shown.

A lot of our knowledge of the physiology of semicircular canals (and of otolith organs) comes from experiments on rays (*Raia*) done by Professor O. Lowenstein and his collaborators. They found that they could remove the part of the skull containing an ear from the body of a freshly killed ray, and still record action potentials from the branches of the auditory nerve within it. Isolated ears conveniently remained responsive for several hours. Experiments could be performed with them which would have been awkward with complete fish.

In one set of experiments, isolated ears were fixed on a turntable. Action potentials were recorded from the nerve to a semicircular canal. Their frequency was constant while the turntable was stationary. When it was rotated the frequency increased or decreased, according to the direction of rotation.

The mechanics of semicircular canals is quite complicated. The endolymph has inertia and viscosity and the cupula has stiffness. All these properties have to be taken into account. Consider a fish which rotates through an angle α in time t. Its angular acceleration is $\ddot{\alpha}$. Let this make the endolymph move round the canal through an angle θ. The angular velocity and angular acceleration of the endolymph, *relative to the canal wall*, are $\dot{\theta}$ and $\ddot{\theta}$ respectively. The total angular acceleration of the endolymph is $(\ddot{\alpha} + \ddot{\theta})$ (if $\ddot{\alpha}$ is positive $\ddot{\theta}$ will generally be negative). If the moment of inertia of the endolymph in the canal is I the moment required to give it this angular acceleration is $I(\ddot{\alpha} + \ddot{\theta})$. Similarly a moment $K\dot{\theta}$, where K is a constant, is needed to overcome the viscosity of the endolymph (viscous resistance is proportional to velocity of flow) and a moment $S\theta$ is required to overcome the elastic restoring force exerted by the deflected cupula. These are the only forces on the endolymph so its equation of motion is

$$I(\ddot{\alpha} + \ddot{\theta}) + K\dot{\theta} + S\theta = 0. \qquad (3.2)$$

Suppose the fish is given a constant angular acceleration $\ddot{\alpha}$ which persists for a considerable time. The response of the canal has three phases:

(i) If the endolymph is initially at rest in the canal, θ and $\dot{\theta}$ are initially zero. Consequently the terms $K\dot{\theta}$ and $S\theta$ are small at first, and

$$I(\ddot{\alpha} + \ddot{\theta}) \simeq 0. \qquad (3.2a)$$

(ii) In the next phase of response viscosity becomes dominant. $\dot{\theta}$ has become large enough to play the major role in limiting flow but θ is still too small for the elasticity of the cupula to be important. At this stage the endolymph flows at a fairly steady velocity relative to the canal wall (so $\dot\theta$ is small) and

$$I\ddot{\alpha} + K\dot\theta \simeq 0. \tag{3.2b}$$

(iii) As time goes on the cupula is deflected further and further so that its elasticity comes to have an important effect. Eventually the endolymph becomes stationary relative to the canal ($\dot\theta = 0$ and $\ddot\theta = 0$), having moved through the angle θ given by the equation

$$I\ddot{\alpha} + S\theta = 0. \tag{3.2c}$$

The duration of phases (i) and (ii) depends on the ratios K/I and K/S. These have been determined for ray semicircular canals by mounting the ear on the torsion pendulum shown in Fig. 3-7. This is a platform suspended from two wires so that if it is set rotating it will go into simple harmonic motion, turning alternately clockwise and anticlockwise. The frequency can be altered over a wide range, by changing the spacing of the wires and the mass of the platform. Action potentials were recorded from the nerve of the semicircular canal, as the platform rotated. A slotted disc suspended below the platform lay between a lamp and a photocell, so that the photocell was illuminated only when there was a slot over it. The output of the

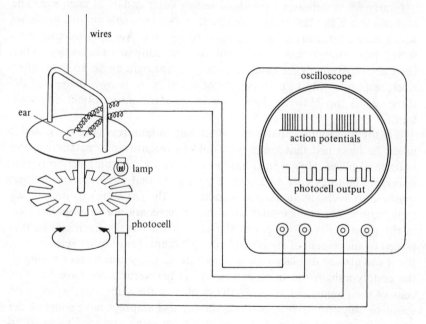

Fig. 3-7. A diagram of apparatus used in the experiments with ray semicircular canals, which are described in the text.

photocell was thus a series of pulses, of which the frequency was proportional to the angular velocity of the platform. It was displayed on an oscilloscope, with the action potentials. As the platform rotated, clockwise and anticlockwise, the frequency of the action potentials fluctuated. The phase difference between these fluctuations and the motion of the platform could be determined from the oscilloscope record. From the results of such measurements at high frequencies and at low ones, K/I and K/S were calculated. (It was assumed that the fluctuations of frequency of action potentials were in phase with the movements of the endolymph in the canal.) The values obtained indicate that phase (i) of the response to a constant acceleration must be completed in about 30 ms, and that phase (iii) is only reached after about 30 s. The phases merge into each other, but it seems that in movements lasting about 30 ms to 30 s the canal must behave essentially according to equation 3.2b. θ must be roughly proportional to $\ddot{\alpha}$ so, by integration, θ must be roughly proportional to $\dot{\alpha}$. The deflection of the cupula must be roughly proportional to the angular velocity of the fish. Most turns made by a ray would last between 30 ms and 30 s, so the semicircular canals are in effect indicators of angular velocity. The same is apparently true of the semicircular canals of vertebrates generally.

Nearly all vertebrates have three semicircular canals in each ear. The exceptions are the lampreys which have only two canals and the hagfishes which have only one (though with two neuromasts). Any rotation can be resolved into components about three axes mutually at right angles. When there are three canals they are set more or less at right angles to each other. Each responds only to components of rotation in its own plane and the three canals can between them provide full information about any rotation.

It is the horizontal canal which is missing in lampreys (Fig. 3-8), and it might be imagined that lampreys would be insensitive to rotation in the horizontal plane. However, it has been shown by recording action potentials from lamprey ears on a torsion pendulum that lampreys are, in fact, sensitive to such rotations. The structure of the ear seems to provide an explanation. The semicircular canal neuromasts are not in ampullae, but in the main cavity of the ear at the ends of the canals. Each can be displaced by movement of endolymph along its canal (as in other vertebrates), but it can also be displaced (in another direction) by horizontal swirling of the endolymph in the main ear cavity. Other vertebrates have the hair cells of their semicircular canal all oriented in the same way, so as to be sensitive only to movements along the canal. Lampreys have some of the hair cells oriented in this way, but also have some arranged so as to be stimulated by cupula movements at right angles to the canal. These are apparently the hair cells which respond to rotation in the horizontal plane.

Otoliths behave quite differently from semicircular canal neuromasts because they are denser than endolymph. They tend to sink in the endolymph under the influence of gravity, so when the head is tilted their hair cells are affected (Fig. 3-6*d'*). They also tend to lag behind when the head accelerates (Fig. 3-6*d"*). It has been shown in experiments with ray ears that the frequency of action potentials in neurones from the otoliths is

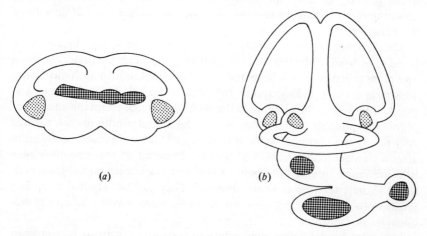

(a)   (b)

Fig. 3-8. Sketches of the ear, removed from the skull, of (*a*) a lamprey and (*b*) a teleost. The positions of the otoliths and cupulae within them are indicated by coarse and fine stipple, respectively.

affected by tilting, and also by vibration, which is of course acceleration alternately in one direction and in the opposite one. Note that it is to linear, not angular, accelerations that the otolith organs respond. Sensitivity to vibration implies that the otoliths may function in hearing.

There is no way in which an otolith organ can distinguish between tilting and acceleration: the distinction can only be made with the help of other sense organs such as the eyes. This ambiguity seems to have been the cause of many aircraft accidents. A pilot, flying near the ground, accelerates (for instance, on deciding not to land). The acceleration gives him a spurious sensation of tilting, as if the aircraft were tilted nose-up. He moves the stick as if to level the aircraft, and crashes it into the ground.

The equation of motion of an otolith is very like the equation for a semicircular canal (equation 3. 2). It includes terms representing the inertia of the otolith, the viscous damping imposed by the surrounding endolymph and the stiffness of the attachment of the otolith to the wall of the ear. However, the viscous damping is much smaller (i.e. K is smaller relative to I and S) so if the ear is subjected to a constant acceleration the otoliths quickly reach equilibrium positions corresponding to phase (iii) of the response of semicircular canals.

Too little damping would be disadvantageous. Many readers will have used simple balances which go on swinging long after a weight has been added, so that waiting for them to settle down wastes a lot of time. More advanced balances are damped so that they do not swing to and fro but move directly to the equilibrium position. The optimum amount of damping is *just* enough to prevent swinging: more than this will slow down the response.

The damping of the otolith organs of a teleost (*Acerina*) has been investigated. The head of the fish was clamped to a photographic plate, and head and plate together were vibrated with known amplitude in an X-ray beam. This gave a sharp radiograph of the skull of the fish (since the head was rigidly attached to the plate) with the otoliths blurred. From the amount of blur it was possible to show that at least the largest otolith in each ear had very nearly the optimum amount of damping. It would reach equilibrium after a change in acceleration or position in about 25 ms. The other otoliths are much smaller but about as stiff and must reach equilibrium even faster if they, too, are critically damped. They can be expected to be less sensitive to small accelerations but might be more sensitive to certain frequencies of vibration.

Lampreys have a single long otolith in each ear, consisting of a mass of calcium carbonate crystals embedded in a cupula. Its attachment to the wall of the ear is partly vertical and partly horizontal and the hair cells are oriented in a variety of ways (Fig. 3-9). The hair cells on the vertical (middle) part of the attachment all have the cilium at the top or the bottom of the tuft and so are presumably sensitive to vertical displacements of the otolith caused by vertical accelerations. Those along the median edges of the horizontal (anterior and posterior) parts of the attachment have the cilium on the anterior or posterior side of the tuft. This presumably makes them sensitive to fore-and-aft displacements of the otolith caused by longitudinal acceleration or by pitching movements. The remaining hair cells have the cilium on the lateral side (or, in a restricted area, on the median side) and are presumably sensitive to transverse acceleration and to rolling. It has been confirmed by recording action potentials from axons in certain parts of the auditory nerve that lamprey ears are indeed sensitive to tilting. It was not possible to demonstrate anatomically where these axons came from, but it is presumed that they were from the otolith organ. Responses to vibration have also been recorded. The dorsal macula is a small neuromast without an otolith in the dorsal part of the ear (Fig. 3-9).

There is evidence that a similar neuromast in rays is sensitive to vibrations but the mechanism does not seem to have been explained.

Hagfishes, like lampreys, have only one otolith in each ear but other fish have three, apparently homologous with the three parts of the lamprey organ marked M. utriculi, M. sacculi and M. lagenae in Fig. 3-9.

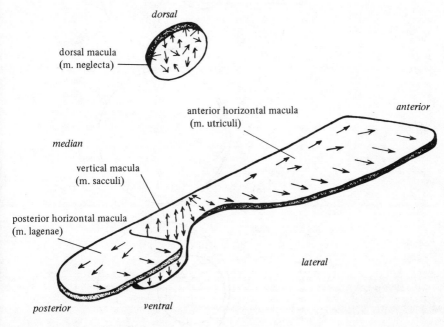

Fig. 3-9. A diagram of the right otolith organ and dorsal macula of a lamprey, showing the orientation of the hair cells. The arrowheads indicate the position of the kinocilium in the hair cell tufts. From O. Lowenstein, M. P. Osborne & R. A. Thornhill (1968). *Proc. R. Soc. Ser.* B, **170**, 113–34.

It is assumed that most readers will be familiar with the basic structure of vertebrate eyes. They may not however have considered the special problem of underwater vision. Air has a refractive index of 1. The refractive indices of water and of the fluids which fill the eye are about 1.33. Therefore light entering an eye from air can be strongly converged by refraction at the convex surface of the cornea, and quite a weak lens may suffice to focus it on the retina. A cornea of radius of curvature r would have a focal length $4r$. In water, virtually no refraction occurs at the corneal surface and focussing must depend entirely on the lens. A strong lens is needed.

Fish lenses are nearly always approximately spherical. Samples taken from their cores have a refractve index of about 1.53 (about the same as crown glass). If the lens had this refractive index throughout, its focal

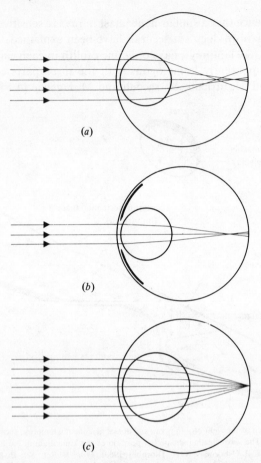

(a)

(b)

(c)

Fig. 3-10. Diagrams of fish eyes showing (*a*) the spherical aberration which would occur if the lens were homogeneous and of wide aperture, (*b*) how this can be remedied by stopping down, and (*c*) how it can be remedied by a non-homogeneous lens.

length would be four times its radius: images of distant objects would be formed four radii from the centre of the lens. The images would be poor, because of spherical aberration (Fig. 3-10*a*). They could be improved by making the pupil small, so that only light travelling reasonably near the centre of the lens reached the retina (Fig. 3-10*b*). This would reduce the sensitivity of the eye, since less light would get into it. It would also tend to make the image rather coarse: diffraction effects make it impossible for light of wavelength λ passing through an aperture of diameter d to form distinct images of objects separated by angles less than about λ/d radians. However, the acuity of vertebrate eyes depends on the spacing of sensory cells on the retina as well as on diffraction effects.

Fish lenses actually have focal lengths of about 2.5 radii, not 4, and those that have been examined have no detectable spherical aberration (Fig. 3-10c). These remarkable and desirable properties result from the lens not being homogeneous. Its refractive index is about 1.53 at the centre but less in the outer layers.

When distant objects are focussed on the retina, nearby ones are not: accommodation is required to focus objects at different distances. Mammals accommodate by altering the shape and so the focal length of the lens. The lens is elastic and is suspended by taut suspensory ligaments which tend to flatten it. Contraction of the ciliary muscle slackens the suspensory ligaments and allows the lens to bulge. Other vertebrates have a wide variety of mechanisms of accommodation. Lampreys have a muscle that flattens the cornea, pushing the lens towards the retina and bringing more distant objects into focus. Selachian and teleost eyes have muscles within them which move the lens.

WATER AND IONS

The freezing point of the blood plasma of various lampreys and ammocoetes has been measured, so that the osmotic concentration could be calculated. The concentrations found have been around 250 mosmol l^{-1}, which is many times the concentration of fresh water but only about a quarter of the concentration of seawater. Ammocoetes and lampreys in fresh water have plasma more concentrated than the water, so ions must tend to diffuse out from the blood and water must tend to diffuse in. Lampreys in the sea face the reverse situation: ions tend to diffuse in and water to diffuse out.

Consider first an ammocoete or lamprey in fresh water. To maintain the composition of its blood it must get rid of the water that diffuses in and replace the ions that diffuse out. It gets rid of the water as urine. It is not easy to collect and measure the urine produced by an ammocoete, so an indirect method of measuring it has been devised. When the animal is in fresh water the rate at which water diffuses in is matched by the rate of urine production. If it is transferred to a solution of the same osmotic concentration as its own blood water will stop diffusing in immediately, but it will presumably take a little time to halt urine production and during this time the animal will lose weight. It can be assumed that the initial rate of loss of weight after transfer is the rate at which urine is produced in fresh water. It has been found in this way that ammocoetes of *Lampetra planeri* produce about 200 cm³ urine kg body weight $^{-1}$ day^{-1}, and adults of *L. fluviatilis* (in fresh water) 160 cm³ kg^{-1} day^{-1}. Even higher rates have been found when the urine of lampreys has been collected (for instance in a balloon tied round the tail), but these may have been unnaturally high: fish are apt to produce urine faster than usual when they are handled or interfered with.

The urine is much more dilute than the blood plasma, with an osmotic concentration of only about 40 mosmol l^{-1}, but since so much urine is produced the quantity of salts lost in this way is considerable. This loss is in addition to the loss by diffusion. The losses are apparently made good by uptake of ions through the gills. This has been demonstrated in experiments with ammocoetes which were put in water containing radioactive isotopes of salts. It was shown by autoradiography that ions were taken up rapidly by the gills, but hardly at all by the skin. $^{22}Na^+$, $^{42}K^+$ and $^{36}Cl^-$ were taken up irrespective of what other ions were present. Anions other than chloride were taken up only in the presence of sodium or potassium ions. Cations other than sodium and potassium were taken up only in the presence of chloride ions. It was concluded that there are mechanisms for taking up sodium, potassium and chloride ions and that other ions enter with them to preserve electrostatic balance. There are cells in the gill filaments, between the lamellae, which are believed to be responsible for ion uptake. They are packed with mitochondria. Adult lampreys living in fresh water have similar cells.

In the sea, lampreys must get rid of excess salts which diffuse into their bodies, and must replace the water which is lost. They are hard to catch in the sea and they unfortunately lose the ability to survive in seawater when they return to fresh water to breed. The experiments which have been done to try to find out how lampreys regulate the composition of their blood in the sea have therefore been done on lampreys caught on their way up-river, when they can no longer survive in pure seawater but can still survive in half-strength seawater. This has nearly twice the concentration of the blood plasma so it presents the lamprey with the same problem as seawater, in a less intense form.

The lampreys (*L. fluviatilis*) were anaesthetized. The water was shaken out of their gill cavities and any water in the gut or urine in the kidney ducts was squeezed out by massage. They were weighed. The papilla through which urine is excreted was closed by a ligature and the anus was plugged. The lampreys were allowed to recover from the anaesthetic, and then put into half-strength seawater containing a little of the indicator phenol red. After about 24 hours they were removed and reweighed. The urine and gut contents were removed, measured and analysed. The net change in the weight of the body (excluding urine and gut contents) was noted.

Most of the animals lost 6% or more of their weight but two kept their weight more or less constant. These two had retained better than the others the ability to regulate their water content in a concentrated environment. They produced no urine, and none of the others produced more than 6 cm^3 urine (kg body weight)$^{-1}$ day^{-1}. This is far less than the lampreys produced in fresh water. It was much too little to fill the kidney ducts in the course of the experiment, so the ligature on the urinary papilla presumably did not interfere with urine production.

The purpose of the phenol red was to find out how much water the lampreys drank and absorbed from the gut. Phenol red itself is apparently not absorbed, and it could not be excreted through the plugged anus. Hence the phenol red from all the water the lampreys swallowed remained in the gut. Its concentration in the gut fluid at the end of the experiment was measured by colorimetry, after treating the fluid with alkali to bring out the red colour. (Phenol red is an indicator, and is colourless in neutral and acid solutions.) If it was found, for instance, that the gut of a lamprey contained 1 cm^3 fluid with three times the phenol red concentration of the aquarium water, it could be inferred that the lamprey had swallowed 3 cm^3 water and absorbed 2 cm^3. In this way it was shown that the two lampreys which maintained their weight absorbed 35 and 82 cm^3 water kg^{-1} day^{-1}.

Table 3-1 gives more information about one of these lampreys (the other was similar). It shows that, though the blood plasma and the gut fluid had about the same total osmotic concentration, the concentrations of individual ions in them were very different. The gut fluid had very much higher concentrations of magnesum and sulphate ions than either the plasma or the environment. Apparently, these ions are selectively excluded from the body. The phenol red in the gut fluid showed that this particular lamprey had swallowed 3.1 times as much water as was left in the gut. If none of the magnesium and sulphate had been absorbed one would expect their concentrations in the gut fluid to be 3.1 times as high as in the environment: one could expect 84 mEq l^{-1} magnesium and 43 mEq l^{-1} sulphate. The observed concentrations are even higher than this: it appears that magnesium and sulphate may actually be excreted into the gut contents. Other ions (mainly sodium and chloride) must have been absorbed with the water, otherwise the total osmotic concentration of the gut fluid would be 3.1 times that of the environment. These ions must presumably have been excreted again, but they cannot have been excreted in urine since no urine was produced.

Some lampreys caught in upstream migration have in their gills mitochondrion-rich cells which are much larger than the ion-uptake cells described on p. 62. Of the fourteen experimental lampreys, only the two which were able to maintain their weight had them. It is believed that these cells excrete the sodium and chloride ions which are absorbed from the gut or which diffuse into other parts of the body.

TABLE 3-1. *Data concerning one of the lampreys which maintained its weight in 50% seawater, in the experiment described in the text*

This data is from Pickering, A. D. and Morris, R. (1970). *J. exp. Biol.* **53**, 231–44

	Osmotic concentration (mosmol l^{-1})	Concentration (mEq l^{-1})			
		Na$^+$	Mg^{2+}	Cl$^-$	SO$_4^{2-}$
Environment (50% seawater)	540	227	27	262	14
Blood plasma	300	155	4	132	3*
Gut fluid	280	64	93	143	63

* Measurement from other animals, in fresh water.

Though the two lampreys which maintained their weight did not produce urine, most of the others did. Though this urine was much more concentrated than the urine produced in fresh water, it never had a higher osmotic concentration than the blood plasma. Lampreys (and fishes in general) seem unable to produce urine more concentrated than the plasma. This means that they cannot gain water by absorbing seawater and passing its salts as urine. The kidneys may nevertheless play a part in ionic regulation in the sea, by excreting magnesium and sulphate. The concentrations of these ions in the urine of the experimental lampreys in 50% seawater averaged 56 and 34 mEq l^{-1} respectively, far higher than their concentrations in the plasma.

Teleost fishes have blood plasma of fairly similar osmotic concentration to lampreys, and regulate their salt and water contents in essentially the same way, both in fresh water and in the sea. Marine teleosts produce small quantities of urine containing high concentrations of magnesium and sulphate. Hagfishes and sharks are quite different from lampreys in the composition of their plasma, as will be explained (pp. 74 and 115).

It has been shown that the kidneys serve to get rid of water when the lamprey is in fresh water and probably to get rid of magnesium and sulphate when it is in the sea. It remains to describe them and explain how they work.

A kidney consists of a large number of similar nephrons, packed together to form a more or less compact organ. Each nephron is a long slender, tangled tubule. In the mud-puppy (*Necturus*, an amphibian), for instance, the nephrons are around 25 mm long. At one end the nephron leads into a branching collecting duct which collects urine from it and from many other nephrons and carries it to the kidney duct (this is known as the ureter in reptiles, birds and mammals, but not in fishes and amphibians, where its embryological origin is different). The other end of the nephron ends blindly, normally as a Bowman's capsule (Fig. 3-11). It is as though the end of the nephron had been blown up into a bulb, and the end of the bulb had then been pushed in to give it a wineglass-like shape. A group of blood capillaries (the glomerulus) fills the bowl of the wineglass. The blood in these is separated from the fluid in the nephron only by the thin

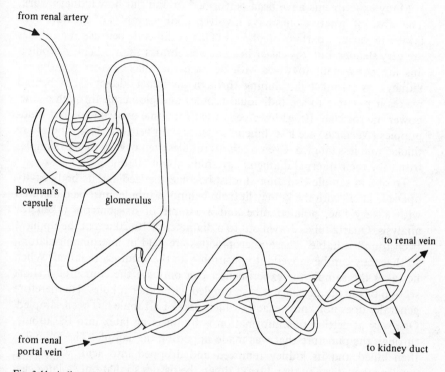

Fig. 3-11. A diagram showing the structure of a typical nephron.

capillary walls and the thin wall of the capsule. The blood leaving the glomerulus travels in an arteriole which breaks up again into a second set of capillaries, investing the tubule. Other capillaries around the tubule are supplied by the renal portal vein which brings venous blood from the posterior part of the body.

Though there is considerable uniformity of kidney structure among the vertebrates, there are also substantial differences. Some species of lampreys do not have a separate Bowman's capsule for each tubule but have a small number of capsules with many tubules opening from each. Some marine teleosts have no capsules or glomeruli. There are differences between groups of vertebrates in the structure of the cells which form the tubule;

the tubule always has several histologically distinct segments but the cells of corresponding parts of the tubules of different vertebrates are often quite different. The loop of Henle is a special feature of mammal and bird nephrons which will be described in Chapter 12. Mammals have no renal portal vein and it is doubtful whether the so-called renal portal vein of birds forms capillaries in the kidney or merely passes through it.

Many experiments have been performed to find out how kidneys work. The most informative ones have involved taking samples of the fluid contained in various parts of nephrons. This is difficult, because the tubules are very slender, but it is easier in some vertebrates than others. Amphibians are convenient to work with because they have rather translucent kidneys. A strong light shining through an intact kidney from behind makes it possible to see individual tubules and glomeruli through a low-power microscope. Frogs have been used for some experiments but mudpuppies (*Necturus*) are less difficult to work with because their tubules are thicker and less tangled. Even *Necturus* tubules are not easy to take samples from, for their internal diameters are only about 50 μm.

To obtain samples, an animal must be anaesthetized and its body cavity opened. Then, with the kidney lit from behind, a tubule must be punctured with a very fine, pointed tube and a sample of its contents taken for analysis. Quartz tubes drawn out to a diameter of 10–20 μm at the tip have been found suitable. These micropipettes are held in micromanipulators, so that they can be moved into position with the accuracy required. When he takes the sample the experimenter can only see the course of a short section of the tubule, since it is so tangled with other tubules. He therefore generally does not know exactly which part of a tubule has been sampled. One way of getting this information is by injecting latex into the tubule, through the puncture that was made in getting the sample. The animal is then killed and its kidney removed and dropped into acid, which first hardens the latex and then breaks down the tissues so that only a latex cast of the tubule remains intact. A blob of latex that seeped out before it hardened always marks the point of puncture. The experimenter can thus see the course of the whole tubule, and the position on it of the point where the sample was taken.

The methods of analysis that are used have to be suitable for dealing with very small samples. Electron-probe microanalysis has been used to measure the concentrations of elements in the samples in recent investigations, but other methods had to be used in earlier investigations before this method was available. Samples taken from Bowman's capsule are always found on analysis to be very similar indeed to blood plasma. They have very closely the same concentrations of sodium, potassium, chloride, glucose, urea, and indeed of every reasonably small molecule for which they have been analysed. The only substantial difference between these samples and blood plasma is that they contain very little protein. It is as though the plasma were being filtered through small pores in the walls of the glomerular capillaries and of the capsule, small molecules passing through and large ones being left behind in the blood. In one series of experiments the blood flowing through frog kidneys was replaced by a stream of saline solution of similar ionic composition containing egg albumin (which has molecules of radius about 2 nm,) and horse serum albumin (radius 3.3 nm). When the fluid in Bowman's capsules was sampled and analysed it was found to contain almost as high a concentration of egg albumin as the saline solution, but no serum albumin. The egg albumin molecules apparently passed quite freely through the pores of the filter but the serum albumin ones were too large. This suggests that the serum is filtered through pores of radius about 3 nm. However, these pores have not been identified in electron micrographs, and experiments with rats suggest a process more complicated than simple filtration: dextran molecules pass through more easily than protein molecules of similar size.

Samples taken from other parts of nephrons show that when fluid leaves Bowman's capsule and travels down the tubule, its composition changes gradually. Fig. 3-12 shows as an example results of analyses for

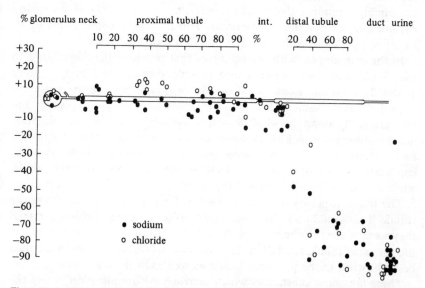

Fig. 3-12. A diagram showing the concentrations of sodium and chloride in samples of fluid taken from different parts of *Necturus* nephrons. Concentrations are expressed by showing the percentage difference from the concentration in the blood plasma. From P. A. Bott (1962). *Am. J. Physiol.* **203**, 662–6.

sodium and chloride. The concentrations of these ions remain constant all along the proximal part of the tubule, at about the values found in the blood plasma. They fall along the distal part of the tubule, so that their concentrations in the urine which finally leaves the body may be only small fractions of their concentrations in the plasma. Thus the proximal and distal parts of the tubule differ in function; they also differ in the structure of their cells. Analyses for glucose show that its concentration is about the same in Bowman's capsule as in the plasma but declines to a very low value as the fluid travels along the proximal tubule and remains at this low level all along the distal tubule.

These results indicate that glucose is reabsorbed from the fluid in the proximal tubule, as are sodium and chloride in the distal tubule. A much more complete picture of events along the tubule has been obtained in experiments with inulin. This is a polysaccharide with a molecule small enough to pass through the glomerular filter. Animals seem to be unable to metabolize it, and apparently neither reabsorb it from the tubules nor secrete it into them. Various experiments have been performed to establish this; for instance, evidence that reabsorption does not occur has been obtained by injecting radioactive inulin into proximal tubules in *Necturus*, and testing the blood leaving the kidney for radioactivity. No transfer of inulin from the nephron to the blood could be detected. One can thus be reasonably confident that inulin enters the nephron only at Bowman's capsule and leaves it only in the urine. This makes it possible to use inulin to investigate movements of water in and out of the tubule.

In the experiments with mud-puppies that provided the data shown in Fig. 3.12, inulin was injected into the bloodstream and samples were analysed for inulin as well as for ions. It was found that the concentration of inulin was the same in Bowman's capsules as in the blood plasma, but rose gradually along the whole length of the tubules so that it could be as much as four times as high in the urine as in the plasma. This is explained by the supposition that up to three-quarters of the water which enters the capsule from the glomerulus is reabsorbed through the walls of the tubule, while all the inulin is left behind in the urine.

The inulin experiments indicate that water is reabsorbed all along the tubule. This would make the concentrations of sodium and chloride rise, if they were not reabsorbed as well. Their concentrations remain constant in the proximal tubule and fall in the distal one, so sodium and chloride must be reabsorbed in the proximal tubule as well as in the distal one.

There are some substances which increase in concentration, along the length of the tubule, more rapidly than inulin. They must be added to the urine (after the filtration process) by secretion through the wall of the tubule. They include *para*-amino hippuric acid, which is a waste product formed from benzoic acid in the kidney.

Inulin is not the only substance that can be used as an indicator of re-absorption of water by the kidney. Creatinine behaves in the same way in some vertebrates, though there is evidence that it is secreted by others. It has the advantage over inulin that it is a natural waste product and is present in the blood without having to be injected. Urine collected from lampreys living in fresh water was found to contain creatinine at concentrations which averaged 1.5 times the concentration in the plasma. This implies that if creatinine is neither secreted nor reabsorbed (which may or may not be the case in lampreys) about one-third of the water entering the capsule must have been being reabsorbed. In salt water, lampreys produce far less urine, but it is not clear to what extent this is due to slower filtration of plasma and to what extent to faster reabsorption of water. We have seen that the concentrations of magnesium and sulphate in the urine of lampreys kept in 50% seawater averaged about twelve times as high as in the plasma. This could be achieved by filtering plasma quite rapidly and reabsorbing 11/12 of the water without the magnesium and sulphate. Alternatively it could be achieved by filtering plasma more slowly and secreting these ions into it from the tubule wall.

It can now be seen how the kidney fulfils its function of getting rid of some constituents of the blood (notably certain ions, metabolic waste products and excess water) while retaining others. Filtration at the glomerulus prevents large molecules from escaping. Small, wanted molecules, such as glucose, and certain ions are recovered by reabsorption. Reabsorption of water, and secretion into the tubule, are both processes which make it possible to excrete unwanted materials at concentrations higher than their concentrations in the blood.

Metabolism of protein produces nitrogenous waste. In terrestrial vertebrates this is produced largely as urea and water, which are excreted by the kidneys. In fish it is mainly ammonia, and excretion of nitrogenous waste is not an important function of the kidneys; most of the ammonia diffuses out of the body at the gills. This has been demonstrated by fixing a fish in a small, divided tank of water, with its body passing through a tight-fitting hole in the partition. Ammonia appears in the water around its head very much faster than in the water around the posterior parts of the body.

HAGFISHES

Class Agnatha, order Myxiniformes

The hagfishes are a small group of bottom-living marine fish. Like lampreys, they have a rather eel-like shape, no jaws, no pectoral or pelvic fins, gill pouches with small external openings and no scales. In a great many other respects they are strikingly different from lampreys.

The North Atlantic species *Myxine glutinosa* lives in muddy places. Specimens kept in aquaria with mud provided spend a lot of time buried in the mud with only the snout protruding. Though *Myxine* usually lives at depths of more than 100 m there is a Norwegian fjord where it is plentiful in water only 30 m deep, shallow enough for frogmen to observe it. The bottom there has lots of hillocks, each with a hole leading down from its

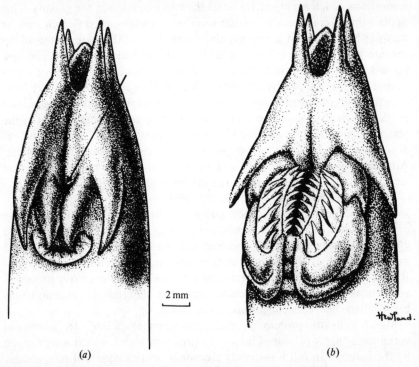

2 mm

(*a*)   (*b*)

Fig. 3-13. Ventral views of the head of *Myxine* with the toothplates (*a*) withdrawn and (*b*) everted. From A. Brodal & R. Fänge (eds.) (1963). *Biology of Myxine*, Oslo University Press, Oslo.

apex. It is believed that these are hagfish burrows, but hagfishes have been seen entering them on only a few occasions. When the divers brought net bags of dead fish hagfish assembled within a few minutes, presumably attracted by odour.

Not all hagfishes live in muddy places: *Eptatretus* is caught on rocky bottoms, off California.

Fish caught on lines at the bottom in areas where hagfishes are plentiful are often attacked by them. The hagfishes cut a hole through the skin of the fish and then eat out its viscera and flesh, leaving the rest of the skin intact. By the time the fisherman hauls his catch to the surface it may be little more than a skin containing a skeleton and a group of well-fed hagfishes. It is thought probable that hagfishes generally attack dead and dying fish rather than healthy ones: a healthy fish could surely escape since hagfishes (unlike lampreys) have no sucker, and no other means of holding on to prey. Polychaete worms and other invertebrates have been found in hagfish guts, so the diet is not exclusively (and perhaps not even mainly) fish.

Though hagfishes have no jaws they have a remarkable apparatus that is capable of cutting through the skin of a fish. It is a flexible plate of cartilage carrying rows of horny (keratinous) teeth. It is normally kept inside the mouth cavity, folded in half along the middle so that the points of the teeth face inwards (Fig. 3-14*a*, *b*). It can, however, be pulled forward, on to the animal's chin as it were, by a protractor muscle that uses the end of a cartilage as a pulley. When this happens the plate opens out flat (Figs. 3-13, 3-14*d*). Feeding involves alternating movements of the toothplates: they move forward and open, then back and close. The teeth rasp at prey and (since they point posteriorly) tend to pull chunks of prey into the mouth. The hagfish cannot hold on to its prey as it rasps; specimens eating dead fish in an aquarium kept themselves in position by swimming movements.

Like lampreys, hagfishes have only one nostril. It opens on the snout rather than the top of the head and it does not end in a blind sac but opens into the mouth cavity (Fig. 3-14*a*). There are five to fifteen pairs of gill pouches, which open directly from the pharynx as in ammocoetes, not from a separate water tube as in adult lampreys. Since hagfishes take solid food, which is relatively unlikely to escape through the gills, there is no need for a separate water tube. The gill pouches have individual external openings in some hagfishes but *Myxine* has only one external opening on each side of the body (Fig. 3-14*e*). It can be shown by releasing dye into the water that water for respiration is not taken in through the mouth, but through the nostril. Thus breathing makes water flow over the sensory epithelium of the nose. (The same effect is achieved in lampreys by a different method, p. 50.) Respiratory water leaves by the external gill openings. The respiratory current can be stopped by blocking the nostril, but this does not kill hagfishes kept in well aerated tanks. Specimens have survived a period of twelve days with plugged nostrils, presumably obtaining the oxygen they needed by diffusion through the skin. This is probably possible because resting hagfishes (*Eptatretus*) at 10 °C use only about 0.2 cm^3 oxygen kg body weight^{-1} min^{-1}. (Resting teleosts of similar weight at the same temperature use oxygen up to five times as fast.) Hagfishes are generally found at substantial depths in the sea, where dissolved oxygen concentrations are apt to be very low. The gills may not be necessary for survival in a well aerated aquarium, but are probably essential in natural habitats.

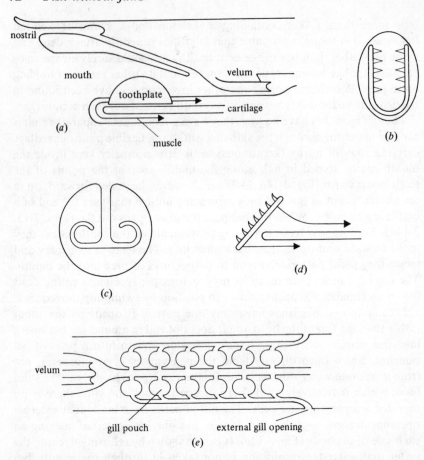

Fig. 3-14. Diagrams of the head of *Myxine*. (*a*) is a median vertical section of the head with the toothplates withdrawn and (*b*) shows in transverse section how the toothplates are folded. (*c*) is a transverse section through the velum, (*d*) shows the toothplates everted, and (*e*) is a horizontal section showing the gill pouches.

The structure known as the velum seems to play a major part in pumping water over the gills. Its position, at the junction of the nasal tube with the mouth cavity, is shown in Fig. 3-14*a*. It is shaped in transverse section like an upside-down T (Fig. 3-14*c*). It is stiffened by rods of cartilage and operated by muscles which alternately roll up and unroll its side pieces. The sequence of movements shown in Fig. 3-15 has been observed in dissected living animals. The scrolls unroll slowly, not touching the wall of the pharynx (*a* to *c*). They roll up much faster and at this stage (*d*) are in contact with the wall of the pharynx, so that the water above them is driven back over the gills.

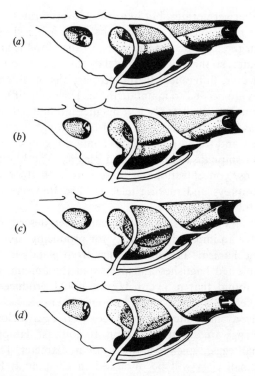

Fig. 3-15. Four stages in the cycle of movement of the velum of *Myxine*. From R. Strahan (1958) *Acta zool., Stockh.* **39**, 227–40.

Hagfishes seem to be able to keep water flowing over the gills by movements of the velum alone, but some observations by X-ray cinematography indicate that movements of the gill pouches are generally involved as well. A solution of Hypaque, which is opaque to X-rays, was introduced into the mouth cavity through the mouth or nostril. When it was pumped into the gill cavities, these became visible in radiographs. X-ray cine films generally showed them expanding and contracting, presumably participating actively in pumping. There are striated muscle fibres in the walls of the gill cavities.

Pressure transducers have been used to record pressures from various blood vessels in *Myxine*. It was found that blood leaves the heart at pressures of up to only about 10 cmH$_2$O (1 kN m^{-2}). This is well below the pressures that are usual in other vertebrates (see p. 115). There are other pumps that help the heart; there are heart-like structures in various parts of the venous system, and the contractions of the gill cavities also help to propel the blood.

Unlike all other vertebrates, but like many invertebrates, hagfishes have about the same total concentration of salts in their blood as in seawater. This means that the osmotic concentration of the blood is about the same as that of seawater, so there is little tendency for water to diffuse in or out. There are, however, differences in the proportions of different ions, and active processes are needed to maintain these. The main differences are that the plasma contains considerably less magnesium and sulphate than sea-water. Urine has been collected from hagfishes through cannulae sewn into the kidney ducts and found generally to contain higher concentrations of magnesium and sulphate than the blood plasma. The kidney apparently maintains the low concentrations of these ions in the blood by excreting them, as in lampreys and maring teleosts. The Bowman's capsules are large, and the tubules short.

The peculiar ears of hagfishes have already been described (p. 59).

Hagfishes have an impressive capacity for producing slime, presumably as a defence mechanism. The slime comes from glands in the skin. One investigator molested hagfishes and measured the amount of slime they produced. He found that a 33 cm *Myxine* could produce slime which, when it had taken up water, occupied about 500 cm^3.

Hagfishes lay a small number of large, oval eggs inclosed in horny shells. *Myxine* eggs are about 2 cm long. In contrast, lampreys lay great numbers of small eggs, typically about 1 mm in diameter. The advantages of large and small eggs will be discussed in Chapter 4. Here, we will merely note a difference in mode of development, which will be already familiar to readers who have studied the embryology of frogs and chickens. Lamprey eggs, frog eggs and other small vertebrate eggs are divided into halves and then quarters by the first two cell divisions; the whole egg is involved in cell division. Hagfish eggs, chick eggs and other large verte-brate eggs develop differently: cell division is confined to one end of the egg so that the embryo rests on a mass of undivided yolk.

BONE AND IVORY

Though lampreys and hagfishes have no bone or ivory the extinct groups which are included with them in the class Agnatha did, so it is appropriate to consider these very important materials now.

The main constituents of bone are roughly equal volumes of collagen fibres and inorganic crystals. (Since the inorganic materials are much denser than the collagen they make up considerably more than half the weight of bone.) The crystals are tiny, commonly about 20 nm long. They seem to be firmly attached to the collagen fibres. They contain calcium, phosphate and hydroxyl ions in about the same proportions as hydroxy-apatite ($3 \ Ca_3(PO_4)_2 . Ca(OH)_2$).

Table 3.2 shows some of the properties of mammal bone and collagen. The properties of fish bone and collagen are probably very similar. The data on collagen is from experiments on collagen obtained from bone by dissolving out the inorganic material, but tests on untreated ligaments and tendons (consisting of collagen fibres and little else) have given similar results. The Young's modulus of collagen is about 10^9 N m^{-2}. That of the crystals in bone has not been measured but it seems likely (by analogy with other crystals) that it is about 10^{11} N m^{-2}. The Young's modulus of bone lies between these values. Thus the crystals give bone a much higher modulus than collagen by itself would have; they make bone a great deal stiffer and less extensible than collagen. This effect is not surprising. More remarkable is the contribution which the crystals seem to make to the strength of bone. Bone has a higher tensile strength than collagen, although only half its volume is occupied by collagen. One might not expect very short, and presumably brittle, crystals to strengthen bone, but similar effects are well known to engineers and are exploited in man-made composite materials. For instance, the filled rubbers that are used for making tyres are mixtures of rubber and fine soot (carbon black). Though soot seems so unpromising a reinforcing material, rubber containing 50% carbon black can be as much as sixteen times as strong as pure rubber, as well as having a higher Young's modulus.

TABLE 3.2. *Mechanical properties of bone,
and of collagen obtained from bone*

Data from Ascenzi, A., Bonucci, E. and Checcucci, A. (1966) in F. G. Evans (ed.), *Studies on the anatomy and function of bone and joints.* Springer, Berlin

	Bone	Collagen
Young's modulus (N m^{-2})	10^{10}	10^9
Tensile strength (N m^{-2})	1.3×10^8	9×10^7

The strength of a homogeneous material like steel is the same, whatever direction it is stressed in, but timber is much stronger when stressed along the grain than when stressed across the grain. Similarly, the strength of bone is generally not the same in all directions because the collagen fibres generally run predominantly in certain directions. Three main patterns are found.

Woven-fibred bone. The fibres are tangled, running in all directions.

Surface bone (Fig. 3-16*a*). The bone consists of a series of lamellae (layers), like plywood. Within each lamella, fibres tend to run parallel. Typically the fibres in successive lamellae run at right angles to each other just as the grain runs at right angles in successive layers in plywood.

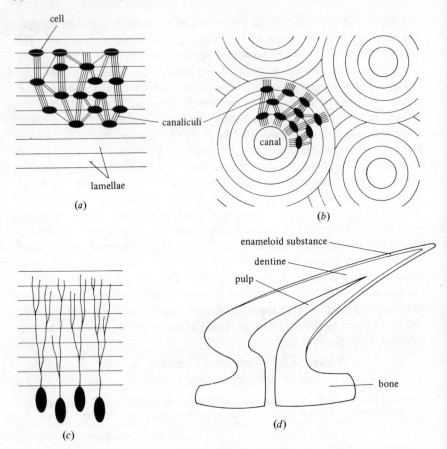

Fig. 3-16. Diagrams showing the structure of (*a*) surface bone, (*b*) osteon bone, (*c*) dentine, and (*d*) a placoid scale.

Osteon bone (Fig. 3-16*b*).   The bone again consists of lamellae, with the fibres about at right angles in successive lamellae. Instead of being more or less flat the lamellae are cylindrical. Each of the units, called osteons, consists of a central canal containing blood vessels with lamellae arranged concentrically around it. The fibres in successive lamellae are longitudinal and circular, or left-handed and right-handed helices.

   Dentine, or ivory, is very similar to bone in the properties considered so far. It has the same main constituents. Woven-fibred and surface dentine occur, and also denteons, which are the dentine equivalent of osteons. The difference between bone and dentine concerns their manner of growth. In both, the collagen and crystals are laid down extracellularly. The cells that lay them down become embedded in bone, but not in dentine. The cells remain scattered throughout bone, in cavities connected by short, fine canals (canaliculi, Fig. 3-16). The cell bodies retreat as dentine grows, so that they are confined to soft tissue adjacent to the dentine (in the pulp cavities of teeth, for instance) or to a layer of bone underlying the dentine. Canals containing processes of the cells extend into the dentine (Fig. 3-16c). These canals tend to be much longer than the canaliculi of bone, for they penetrate the whole thickness of the dentine. As the dentine grows and the cells retreat, the canals elongate correspondingly.

Though the distinction between bone and dentine may seem clear enough, it is not always clear whether a particular tissue should be identified as bone or as dentine. There are intermediate hard tissues in which most of the cells retreat but a few become enclosed. There are mixtures: denteons may fill the spaces in spongy bone. The cells in bone sometimes disappear immediately after they have been enclosed (and this happens in many teleost fishes). There is a form of dentine which lacks the usual canals for cell processes.

Skin consists of two layers, an outer epidermis of epithelial cells and an inner dermis of connective tissue. The lining of the mouth cavity has the same structure. Some bones develop from cartilage in positions deep in the body (see p. 133), but hard tissues also develop in the dermis, forming scales or dermal bones. Dentine tends to be formed in the outer parts of the dermis and bone in the deeper parts, so scales and dermal bones tend to consist of an outer layer of dentine and an inner layer of bone.

Fig. 3-17 (*a*, *b* and *c*) shows the structure of dermal bone from members of the very early group of fossil fish known as the subclass Heterostraci. These fish are described later in the chapter. They include the earliest known vertebrate fossils. The specimens illustrated all have an outer covering of knobs of dentine (odontodes) on an inner layer of bone permeated by relatively stout canals, which presumably housed blood vessels. These canals must also have housed the bodies of the dentine cells, for the fine canals in the dentine converge on them. There are no cell cavities in the bone.

The outermost surface of dentine is often covered by a layer of material containing a higher proportion of inorganic crystals. The enamel of mammal teeth is an example. It is harder than dentine, but more brittle. Similar substances found in fish are often called enameloid substances rather than enamel, because of evidence that they may not be formed in the same manner as is the enamel of mammals. It is not always easy to distinguish enameloid substance from dentine in fossils, but it seems to have been present in some Heterostraci and absent from others.

Fig. 3-17 (*d*) shows a scale from a much more advanced fish, one of the

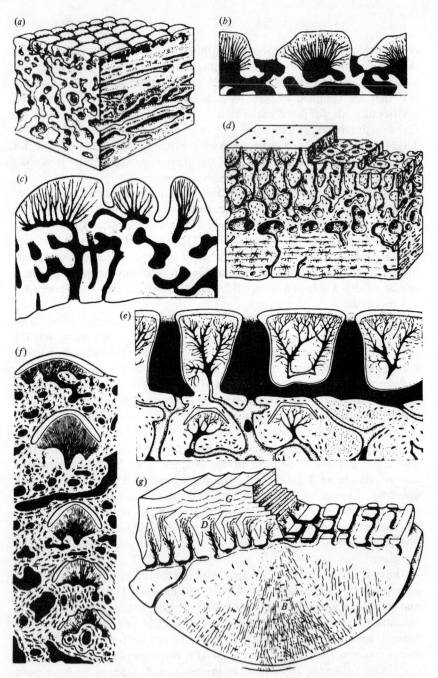

Fig. 3-17. Diagrams, and drawings of sections of, (*a*), (*b*) dermal bone of *Psammosteus* (a Devonian heterostracan); (*c*) dermal bone of *Eriptychius* (an Ordovician heterostracan); (*d*) a scale of *Megalichthys* (a Carboniferous crossopterygian); (*e*) a scale of *Porolepis* (a Devonian crossopterygian); (*f*) dermal bone from another Devonian crossopterygian and (*g*) a scale of *Cheirolepis* (a Devonian palaeoniscoid). From T. Ørvig (1967). In A. E. W. Miles (ed.) *Structure and chemical organization of teeth*. Academic Press, New York.

Crossopterygii described in Chapter 7. It is remarkably like the specimens from Heterostraci: the main difference is that there are cell cavities
in the bone. There are two layers of bone, a spongy layer of woven-fibred
bone and a much more compact layer of surface bone. The outer surface of
the dentine is covered by enameloid substance, as in some Heterostraci.
The scale thus consists of four layers, which are (starting at the outer surface) enameloid substance, dentine, woven-fibred bone and surface bone.
This type of scale is called cosmoid.

Fig. 3-17 (*e*, *f*) also shows crossopterygian specimens, but these do not
have this simple arrangement of four layers. In (*e*) there is a row of odontodes at the surface and a deeper row embedded in bone. In (*f*) there is a
series of five odontodes, one above the other, all but the topmost embedded
in bone. The deepest odontodes were presumably formed first and were
probably in the outer part of the dermis at the time of their formation. As
the dermis thickened they came to lie deeper and deeper in it; bone was
formed around them and new odontodes were formed above them, close
to the new outer surface of the dermis.

The structure of a rather different type of scale is shown in Fig. 3-17 (*g*).
It is from one of the Palaeoniscoidei, a group of fish that included ancestors
of the teleosts and which is described in Chapter 5. It has an outer layer of
enameloid substance (*G*), a middle layer of dentine (*D*) and an inner layer
of surface bone. Notice that the enameloid substance is not a single thin
layer as in (*d*), but is quite thick and consists of a series of superimposed
layers. This type of scale is called ganoid. It is possessed by two genera of
modern fish (*Polypterus* and *Calamoichthys*) that are regarded as surviving
palaeoniscoids by many zoologists, as well as in fossils.

The structure of ganoid scales reflects their manner of growth. As each
ring of dentine is added to the circumference of the scale a new layer of
bone is laid down on its inner surface and a new layer of enameloid substance on its outer surface.

The scales (called placoid scales) and teeth of sharks are believed to have
evolved from groups of odontodes. They consist of dentine with an outer
coating of enameloid substance and a base of bone (Fig. 3-16*d*). Cells are
enclosed in the bone when it is first formed, but subsequently disappear.
The dentine cell bodies lie in the soft tissue (pulp) at the centre of the scale
or tooth. The teeth of other vertebrates are similar structures.

OSTRACODERMS

Class Agnatha, orders Heterostraci, Osteostraci, Anaspida and Thelodonti

There are four extinct orders of fish that, like lampreys and hagfishes, have
no jaws. They are known collectively as ostracoderms. This name is derived

from a Greek word meaning 'with skin like earthenware' and refers to the fact that most ostracoderms (unlike lampreys and hagfishes) were covered by thick scales or plates of bone. Most ostracoderms were quite small: few fossils longer than about 30 cm have been found. Ostracoderms have been found in Ordovician, Silurian and Devonian rocks. The earliest are also the earliest known vertebrates, and are members of the order Heterostraci. A section of a fragment of dermal bone from one is shown in Fig. 3-17 (*c*). None of these Ordovician fossils is complete enough to give a clear impression of the appearance of the intact fish, so the example of the Heterostraci shown in Fig. 3-18*a* is a later one, from the Devonian period. This particular species has its head covered partly by large plates of dermal bone and partly by an irregular mosaic of small ones. The Ordovician species have the whole head covered by a mosaic of small plates, but many others have only large plates. The trunk is covered by overlapping scales.

The number of bones in a heterostracan carapace, or in the skull of a higher vertebrate, must tend to be a compromise between strength and the need to grow. A flat plate of bone can grow simply by thickening and the addition of material at the edges. An arched plate must grow in a more complicated way if its radius of curvature is to be increased to match the increasing radius of the head: old bone must be removed as well as new bone being added. If numerous small plates cover the head they can be more or less flat, and can grow in the simple way. If there are only a few large plates the carapace or skull may be stronger but the plates must be arched, so that their growth requires more destruction and reconstruction of bone and (presumably) more energy. A sutureless carapace could grow as tubular leg-bones grow, by addition of material on the outside and removal from the inside.

Most Heterostraci had heads that were broader than they were high, though few were as flat as *Drepanaspis* (Fig. 3-18*a*). The flattened shape suggests that they habitually rested on the bottom. Their eyes must have been small, fitting the small holes in the orbital plates. The lower margin of the mouth was formed by a row of narrow plates of bone lying side by side. These may have been joined by an extensible membrane so that they could be lowered, forming a scoop, which might perhaps have been used for scooping up the layer of detritus that tends to accumulate on the surface of mud. Detritus seems a likely food for jawless, bottom-living fishes, particularly if, as is probable, they evolved from ancestors with filter-feeding habits like those of *Branchiostoma* and ammocoetes. Various modern fishes including grey mullets (*Mugil*) feed on detritus.

Though we have a great many specimens of the carapaces that enclosed the heads of Heterostraci, no remains of internal skeletons have been found except for traces in some Ordovician fossils. Presumably there was a

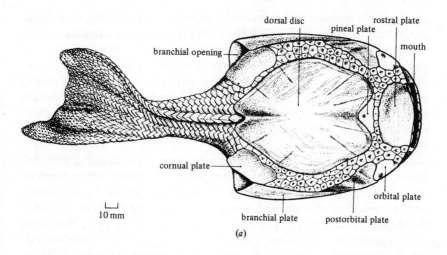

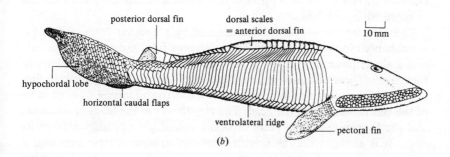

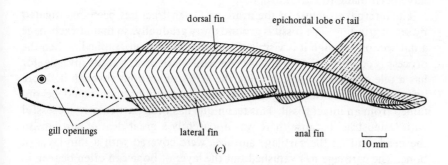

Fig. 3-18. (*a*) *Drepanaspis*, a member of the Heterostraci, (*b*) *Hemicyclaspis*, a member of the Osteostraci, and (*c*) *Jamoytius*, a member of the Anaspida. From J. A. Moy-Thomas (1971). *Palaeozoic fishes*, 2nd edit., revised by R. S. Miles. Chapman & Hall, London.

cartilaginous cranium and some sort of gill skeleton, and perhaps cartilaginous vertebrae. Such information as we have on the internal structure of the Heterostraci has been obtained by studying the inner surfaces of the head plates. Depressions on these indicate the presence of a long brain, at least two pairs of semicircular canals and a pair of nasal sacs. No nostrils perforate the carapace, so the nasal sacs may have opened into the mouth cavity. More lateral depressions suggest the presence of about seven pairs of gill chambers, which must all have been connected to the single pair of external gill openings.

The Osteostraci is another order of early Agnatha with rather flattened heads encased in a bony carapace (Fig. 3-18b). A typical species (such as the one illustrated) had a trunk that was roughly triangular in section, with a dorsal line of scales and a pair of ventrolateral ridges emphasizing the corners of the triangle. The flat belly would help to make the fish stable when it rested on the bottom. The tail was of the type called heterocercal, like the shark tails that will be described and explained in Chapter 4. Many Osteostraci had a pair of flippers in the position occupied by pectoral fins in more advanced fishes.

The carapace consisted of bone covered by a peculiar form of dentine. It probably did not develop until the fish was fully grown. Its main part was a single piece, but it was broken up into a mosaic of small plates in four areas. Three of these were on the dorsal surface (one is visible in Fig. 3-19b) and their function is uncertain. The fourth and largest covered most of the ventral surface of the head. The mouth was at the front of this area, and there were ten or so pairs of small, round gill openings around its edge. It is thought that this ventral mosaic area may have provided a sufficiently flexible floor to the mouth and gill cavities, to allow breathing by enlargement and contraction of the cavities. Similar movements may have been made to suck in food.

The internal structure of the heads of Osteostraci has been investigated by serial grinding. The fossil is ground away gradually, so that at each stage a flat section through it is visible. Each section is photographed. When the process is complete the fossil has been reduced to dust, but the investigator has a pile of photographs showing the series of sections through it. More information can often be got from the photographs than could be obtained from an intact fossil. This technique has been particularly successful with Osteostraci because there was apparently a great deal of cartilage in the head, and all the cartilage surfaces were covered with a thin layer of bone. The cartilage has vanished but the layer of bone can often be seen in the sections. This layer even covered the walls of all channels through the cartilage, so the paths of nerves and blood vessels can be traced.

A vertical longitudinal section is shown in Fig. 3-19. Notice the shape of the brain cavity and its connections with two median holes in the roof of the carapace. The cavity would fit a brain and nasal sac very like those of the lamprey illustrated in Fig. 3-5. The two holes must have housed a median (lamprey-like) nostril and the pineal organ. Other sections show that there were only two semicircular canals in each ear.

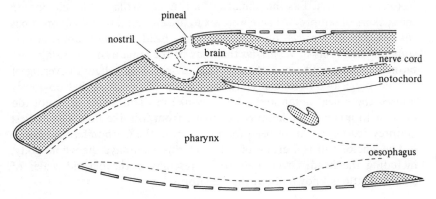

Fig. 3-19. A diagrammatic median section of the head of an osteostracan.

Though the Osteostraci are so different from lampreys in shape and in being armoured, there are striking similarities. Like lampreys, they had no jaws, they had numerous small, round gill openings, they had only two semicircular canals in each ear and they seem to have had the same peculiar sort of olfactory organ with a single dorsal nostril. These similarities suggest evolutionary relationship.

The Anaspida is another order of ostracoderms. Most members had thick scales on the trunk and a mosaic of small dermal bones all over the head, but *Jamoytius* (Fig. 3-18c) had no bones on the head, and scales so thin and flexible that some have been folded almost double in the squashing which accompanied fossilization, without being broken. The chevrons along the side of the body in Fig. 3-18c are scales. Anaspids were slender fish, and seem to have been somewhat flattened from side to side like most modern fish that spend a lot of time swimming. Their tails were like shark tails turned upside-down: the rear end of the body extends into the dorsal lobe of the tail in Osteostraci and sharks, but into the ventral lobe in Anaspida. It will be explained in Chapter 4 that shark tails produce upward forces but that the upside-down shape of anaspid tails does not necessarily imply downward forces. *Jamoytius* had median dorsal and anal fins and a long lateral fin on each side of the body. The anterior part of the lateral fin corresponds in position to the pectoral fin of a shark or of a primitive teleost such as a trout, and the posterior part to the pelvic fin.

The dorsal and lateral fins correspond in position to the dorsal and ventro-lateral ridges of Osteostraci (Fig. 3-18*b*); indeed, most Anaspida did not have a dorsal fin but merely a line of big scales.

The mouths of Anaspida seem to have been elliptical, taller than wide. There seem to have been no jaws, and a dark ring on fossils of *Jamoytius* has been interpreted as the remains of a ring of cartilage, like the ones in the suckers of lampreys. There were six to fifteen small, round gill openings (like those of lampreys) on each side of the head and there are stains on some of the fossils of *Jamoytius* that may be the remains of a cartilaginous branchial basket. In some of the fossils of species that had a covering of dermal bone, two holes can be found on top of the head that obviously housed the pineal organ and a lamprey-like median nostril. It is not too difficult to imagine the lampreys evolving from fish like this. The earliest lamprey fossils that have been found are from the Carboniferous period.

The Thelodonti is a group of ostracoderms than have the whole body, including the head, covered by scales resembling the placoid scales of sharks. Little is known about their anatomy.

FURTHER READING

General works on fishes

Alexander, R. McN. (1970). *Functional design in fishes*, 2nd edit. Hutchinson, London.

Hoar, W. S. & D. J. Randall (eds.) (1969–72). *Fish physiology* (6 vols.). Academic Press, New York.

Moy-Thomas, J. A. (1971). *Palaeozoic fishes*, 2nd edit., revised by R. S. Miles. Chapman & Hall, London.

Satchell, G. H. (1971). *Circulation in fishes*. Cambridge University Press, London.

Wheeler, A. (1969). *The fishes of the British Isles and North-West Europe*. Macmillan, London.

Lampreys

Hardisty, M. W. & I. C. Potter (eds.) (1971–2). *The biology of lampreys* (2 vols.). Academic Press, New York.

Feeding and respiration

Gradwell, N. (1972). Hydrostatic pressures and movements of the lamprey, *Petromyzon*, during suction, olfaction and gill ventilation. *Can. J. Zool.* **50**, 1215–23.

Hill, B. J. & I. C. Potter (1970). Oxygen consumption in ammocoetes of the lamprey *Ichthyomyzon hubbsi* Raney. *J. exp. Biol.* **53**, 47–57.

Potter, I. C., B. J. Hill & S. Gentleman (1970). Survival and behaviour of ammocoetes at low oxygen tensions. *J. exp. Biol.* **53**, 59–73.

Roberts, T. D. M. (1950). The respiratory movements of the lamprey (*Lampetra fluviatilis*). *Proc. R. Soc. Edinb. Ser. B*, **64**, 235–51.

Sense organs

Dijkgraaf, S. (1963). The functioning and significance of the lateral-line organs. *Biol. Rev.* **38**, 51–105.

Groen, J. J., O. Lowenstein & A. J. H. Vendrik (1952). The mechanical analysis of the responses from the end-organs of the horizontal semicircular canal in the isolated elasmobranch labyrinth. *J. Physiol., Lond.* **117**, 329–46.

Nicol, J. A. C. (1963). Some aspects of photoreception and vision in fishes. *Adv. mar. Biol.*, **1**, 171–208.

Tansley, K. (1965). *Vision in vertebrates.* Chapman & Hall, London.

Water and ions

Pickering, A. D. & R. Morris (1970). *Osmoregulation of Lampetra fluviatilis* L. and *Petromyzon marinus* (Cyclostomata) in hyperosmotic solutions. *J. exp. Biol.* **53**, 231–43.

Riegel, J. A. (1972). *Comparative physiology of renal excretion.* Oliver & Boyd, Edinburgh.

Hagfishes

Brodal, A. & R. Fänge (eds.) (1963). *Biology of Myxine.* Oslo University Press, Oslo.

Bone and Ivory

Bourne, G. H. (ed.) (1972). *The biochemistry and physiology of bone*, 2nd edit. Academic Press, New York.

Currey, J. D. (1970). *Animal skeletons.* Edward Arnold, London.

Ørvig, T. (1967). Phylogeny of tooth tissues: evolution of some calcified tissues in early vertebrates. In Miles, A. E. W. (ed.) *Structural and chemical organization of teeth*, vol. 1, pp. 45–109. Academic Press, New York.

4

Sharks and some other fish

This chapter is about sharks and related fish (class Selachii) and about some other fish that have jaws but are not included in the great class Osteichthyes, which is the subject of Chapters 5, 6 and 7. These others belong to the class Holocephali (which includes the rabbit-fishes) and the two extinct classes Acanthodii and Placodermi.

Many sharks are large, and the whale shark *Rhincodon typus* is the largest known fish. Specimens over 17 m long, weighing 40 tons, have been caught. Few selachians are really small: the smallest is probably *Euprotomicrus*, which grows to an adult length of about 20 cm.

SWIMMING

Fish swimming raises very difficult problems in hydrodynamics, and much of our understanding of it results from the work of mathematicians rather than zoologists. Sir James Lighthill's mathematics has been particularly illuminating. Abstruse hydrodynamics would obviously be out of place in this book, but a little elementary hydrodynamics can be extremely helpful. The next few paragraphs outline principles which will be useful here, and again when bird flight is discussed.

When a solid body moves through a fluid the pattern of flow around it depends on its shape: the fluid flows differently around a sphere and around a flat plate. The pattern also depends on the attitude of the body: it is different for plates moving broadside-on and edge-on. So much is obvious. The pattern also depends on the size and speed of the body and on the properties of the fluid. The effects of these latter factors can be combined by describing them as the effects of Reynolds number. For bodies moving in water, the Reynolds number is 10^6 (length in m) $\times$ (velocity in m s^{-1}). A large (1 m) fish swimming fast, at 10 m s^{-1}, has a Reynolds number of 10^7. A small (5 cm) fish swimming slowly, at 5 cm s^{-1}, has a Reynolds number of 2500. For bodies moving in air, the Reynolds number is 7×10^4 (length in m) $\times$ (velocity in m s^{-1}). If two bodies of the same shape, moving through fluids in the same attitude, have the same Reynolds number, the patterns of flow around them are identical.

When a body moves through a fluid, the fluid exerts a force on it in the direction opposite to the motion. This force is called drag. Consider a body moving at velocity U through fluid of density ρ. Let its frontal area be A. This is the area of the view of the body, seen along the direction of motion. For many bodies it equals the maximum cross-sectional area, when they are moving along their long axes. The drag is given by the equation

$$\text{Drag} = \tfrac{1}{2}\rho U^2 A C_D. \tag{4.1}$$

The quantity C_D, known as the drag coefficient, is a constant for bodies of the same shape, attitude and Reynolds number. In the range of Reynolds numbers that we are concerned with, between about 10^3 and 10^7, it is 0.2–0.5 for spheres. It is higher for parachute shapes but lower, about 0.04–0.1, for streamlined shapes. These are torpedo-like shapes, rounded at the nose and tapering gently at the rear. The taper at the rear is particularly important. The ideal shape, giving least drag for given volume and velocity, is a streamlined one with the length about 4.5 times the maximum diameter. If the fins were removed from the shark illustrated in Fig. 4-1, its body would correspond closely to this ideal.

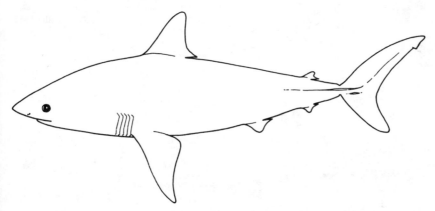

Fig. 4-1. Porbeagle shark, *Lamna nasus* (length up to 3 m). A drawing by Valeri Du Heaume from A. Wheeler (1969). *The fishes of the British Isles and North-West Europe.* Macmillan, London.

When a symmetrical body moves along its axis of symmetry, drag is the only force that the fluid exerts on it. Another component of force, known as lift, acts on asymmetrical bodies and on symmetrical ones moving at an angle to the axis. Lift acts at right angles to the direction of motion but not necessarily upwards (in spite of its name). Aerofoils and hydrofoils are bodies designed to produce large lift forces without too much drag. The wings of aeroplanes are the most familiar example. The lift forces on them must be big enough to support the weight of the aircraft but the drag should be as small as possible (and is in practice only a small fraction of the lift). The bigger the drag the more thrust the engines must exert. Fig. 4-2a shows a section through an aerofoil or hydrofoil with lift and drag acting on it.

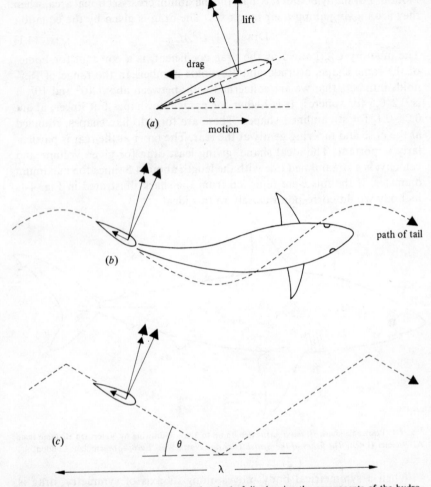

Fig. 4-2. (*a*) A diagrammatic section through a hydrofoil, showing the components of the hydrodynamic force which acts on it; (*b*) a diagram showing how fish such as *Lamna* use their tails as hydrofoils, for propulsion; and (*c*) a simplified diagram which is discussed in the text.

Areas and Reynolds numbers have to be considered when forces on hydrofoils are being calculated. The area used is not the frontal area but the plan area S (i.e. the area of a full-scale plan of the hydrofoil). The Reynolds number is calculated in the same way as for other bodies, and the length used is the chord of the hydrofoil (i.e. the distance from its front edge to its rear edge). For a hydrofoil moving at velocity U through fluid of density ρ

$$\text{Lift} = \tfrac{1}{2}\rho U^2 S C_L. \qquad (4.2)$$

C_L is a quantity known as the lift coefficient. It depends not only on the shape of the hydrofoil and its Reynolds number but also the angle of attack (α, Fig. 4-2a) at which it is set to its direction of motion. It is low or zero, depending on the shape of the cross-section of the hydrofoil, when $\alpha = 0$. It increases as α increases until it reaches a maximum, generally when $\alpha \simeq 20°$. Further increase in α results in a fall in C_L because of a drastic change in the pattern of flow: the hydrofoil stalls. The maximum lift coefficient which can be obtained with well designed hydrofoils is about 1.0 at a Reynolds number of 10^3, and 1.5 at a Reynolds number of 10^7.

An increase in angle of attack increases drag as well as lift. Provided the hydrofoil is not stalled

$$\text{Drag} \simeq \tfrac{1}{2}\rho U^2 S C_{D0} + (2kL^2/\pi R \rho U^2 S), \tag{4.3}$$

where L is the lift and C_{D0} is the drag coefficient at the angle of attack at which there is no lift. k depends on how the hydrofoil tapers towards its tips but is generally about equal to 1. R is the aspect ratio of the hydrofoil, that is its span divided by its average chord. Notice that most lift can be obtained for least drag, if the aspect ratio is large. The shape of the cross-section is also important. The best hydrofoils for high lift and low drag are those with a streamlined cross-section (as in Fig. 4-2a), provided the Reynolds number is higher than about 10^5. At lower Reynolds numbers a thin, arched (cambered) plate is better.

This hydrodynamics can now be applied to fishes. Different fish have different shapes and correspondingly different swimming actions. The easiest to explain is the action of sharks such as *Lamna* (Fig. 4-1) and some similarly shaped teleosts such as the tunnies (Scombridae). These have streamlined bodies of near-ideal proportions. Their tail fins are hydrofoils, with the span vertical. As the fish swims forward the tail moves from side to side, following a wavy path (Fig. 4-2b). It is held at an angle of attack to its path, so that lift acts on it as shown. As it moves towards the fish's right the resultant of lift and drag acts forward and to the left. As it moves to the left the resultant acts forward and to the right. In a complete cycle of tail movements the components of force to left and right cancel out but the forward components remain to propel the fish, overcoming the drag on the body.

What factors are likely to affect the efficiency of swimming? Suppose for simplicity that the tail moves in a zigzag so that its path is always at an angle θ to the path of the body (Fig. 4-2c). The lift L has a forward component $L \sin \theta$ and the drag D on the tail has a backward component $D \cos \theta$. The net thrust is thus ($L \sin \theta - D \cos \theta$). This must equal the drag on the body. In each cycle of tail movements the body moves forward λ, so the useful work done propelling it is $\lambda(L \sin \theta - D \cos \theta)$. In addition, work is done against the drag on the tail. In each cycle of movements the tail moves a distance $\lambda \sec \theta$, so this work is $\lambda D \sec \theta$. The total work is thus $\lambda(L \sin \theta - D \cos \theta + D \sec \theta)$ and the efficiency is

$$\frac{\lambda(L \sin \theta - D \cos \theta)}{\lambda(L \sin \theta - D \cos \theta + D \sec \theta)}.$$

Dividing numerator and denominator by $\lambda \sin \theta$ gives

$$\text{efficiency} = \frac{L - D \cot \theta}{L - D\,[\cos^2 \theta - 1)/\sin \theta \cos \theta]}.$$

$$= \frac{L - D \cot \theta}{L + D \tan \theta}.$$

This must give too high an estimate of efficiency, because the tendency of the tail to wag the fish has been ignored; additional work will be wasted, moving the body of the fish from side to side. However, this crude analysis is enough to show that for high efficiency, L/D should be as large as possible. A tail with a large aspect ratio will help to achieve this, and the tails of sharks such as *Lamna* do indeed have reasonably large aspect ratios, of about three to five. Tails of higher aspect ratio, around eight, are possessed by tunnies.

A few measurements of shark swimming speeds have been obtained by recording the rates at which hooked sharks pulled out a line. One such record is of a *Carcharinus* about 1 m long which reached a top speed of 4 m s^{-1}. The mean chord of this shark's tail fin must have been about 0.07 m, so the Reynolds number of the tail, at 4 m s^{-1}, must have been about 3×10^5. This is high enough for a streamlined section to be advantageous, and the tail is in fact streamlined in horizontal section both in typical sharks and in tunnies.

The above is a reasonable description of swimming by typical sharks and by tunnies. It is not a good description of swimming by dogfish (Figs. 4-3 and 4-20), still less by lampreys or eels. These fish do not have hydrofoil-like tail fins and their swimming action does not even approximate to simple side-to-side movement of the tail. The body is thrown into waves, which travel posteriorly along it, driving water backwards and the fish forwards.

What conditions promote high efficiency, in swimming by undulation? Fig. 4-4a shows two successive positions, separated by a small time interval δt, of a fish swimming to the right. The fish is swimming at velocity U so its snout moves forward a distance $U \delta t$. Waves are travelling backwards along the body at velocity V relative to the body, or $(V - U)$ relative to

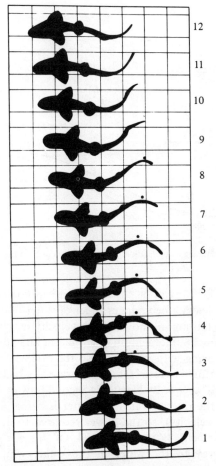

Fig. 4-3. Tracings from a film of a dogfish (*Squalus*) swimming. The interval between successive tracings is 0.1 s. The grid squares have 3 inch (7.6 cm) sides. From J. Gray (1933). *J. exp. Biol.* **10**, 88–104.

the water. Thus each crest in the wave pattern moves a distance $(V - U)\,\delta t$ in the interval. The tail is moving transversely with velocity W.

Consider the particle of water which occupies position X_1 at the beginning of the time interval and X_2 at the end, when it is about to be left behind by the fish. It is still moving transversely with velocity w, when it is left behind. It is apparent from the similar triangles in Fig. 4-4b that

$$w\,\delta t = W\,\delta t(V - U)/(U + V - U)$$
$$w = W(V - U)/V. \tag{4.5}$$

We need to know the momentum and kinetic energy of the moving water

(a)

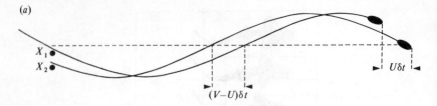

(b)

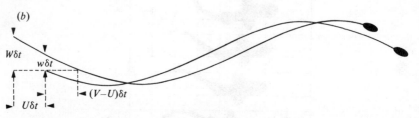

Fig. 4-4. Diagrams of swimming by undulation, which are explained in the text.

left behind by the fish. It can be shown that it is a fair approximation to think of the fish as leaving behind a cylinder of water of diameter equal to the height s of the tail fin, all moving with velocity w. Unit length of this cylinder has mass $\pi \rho s^2/4$, momentum $\pi \rho s^2 w/4$ and kinetic energy $\pi \rho s^2 w^2/8$. The cylinder is growing with velocity U so the rate at which the water is being given momentum is $\pi \rho s^2 w U/4$ and the rate at which it is being given kinetic energy is $\pi \rho s^2 w^2 U/8$. By Newton's Second Law of Motion the force on the tail equals the rate at which the water is being given momentum. Power is (force × velocity), so the power required to drive the tail from side to side with velocity W is $\pi \rho s^2 w U W/4$. Of this $\pi \rho s^2 w^2 U/8$ is wasted, giving kinetic energy to the water in the wake, while the remainder $(\pi \rho \, s^2 w U/4)\,(W - \tfrac{1}{2}w)$ serves to propel the fish. The efficiency of propulsion is this power actually used in propulsion, divided by the total power required to drive the tail.

From the above expressions

$$\text{Efficiency} = (W - \tfrac{1}{2}w)/W,$$

and from this and equation 4.5

$$\text{Efficiency} = (V + U)/2V. \tag{4.6}$$

The small dogfish shown in Fig. 4-3 is swimming with velocity $U = 14$ cm s^{-1}. The waves are travelling backwards relative to the water with velocity $(V - U) = 8$ cm s^{-1}. Hence the efficiency can be estimated as 0.8. This is an overestimate because equation 4.6, like equation 4.4 ignores the tendency of the tail to wag the fish.

U must always be smaller than V if the waves are travelling backwards relative to the water. Equation 4.6 shows that efficiency will be highest if U is as nearly equal to V as possible. How can this be achieved? If the body of the fish has drag coefficient C_D and frontal area A the power required to propel it (drag × velocity) is $\frac{1}{2}\rho U^3 A C_D$. The power available has already been estimated, so

$$\tfrac{1}{2}\rho U^3 A C_D = (\pi \rho s^2 w U/4)(W - \tfrac{1}{2}w).$$

Substituting for w from equation 4.5 and rearranging

$$(V^2 - U^2)/V^2 = 4U^2 A C_D/\pi s^2 W^2. \tag{4.7}$$

For the efficiency to be high, $(V^2 - U^2)$ must be small. Equation 4.7 shows that large s (a tall tail fin) will help to achieve this.

Notice that the transverse motion of only the extreme rear end of the body has been considered. It is not necessary or even desirable that the wave motion should have the same amplitude all along the body. Fish which swim by undulation swim like the dogfish (Fig. 4-3), with the amplitude increasing towards the tail.

There is no sharp distinction between the two swimming methods which have been described, between swimming with a hydrofoil tail and swimming by undulation. *Lamna* obviously belongs to one category and the dogfish to the other, but many fish lie between these extremes. They undulate their bodies but the amplitude is only appreciable near the posterior end, so the movement of waves along the body is not at all obvious. They have fairly tall tail fins, but the shapes of these fins do not immediately suggest hydrofoils.

The rate at which oxygen is used in swimming has been measured in experiments with various teleosts but apparently not, so far, with selachians. Typical apparatus is shown in Fig. 4-5. It is a water tunnel designed on the same principles as the wind tunnels used by aircraft engineers. The loop of pipes (6–12) is filled with water which is driven clockwise around it by the centrifugal pump (5) powered by the electric motor (1). The fish is confined to the test section (11) by nylon mesh at (10) and (12). It is trained to swim against the current, at just such a speed as neither to gain nor lose ground. Its situation is then the same as if the water were stationary and it were swimming at the same relative speed *provided* the water in the test section flows smoothly and at the same velocity at all points in the cross-section of the tube. The grids (8) and reduction cone (9) are designed to achieve this as nearly as possible. An oxygen electrode inserted at (6) was used to measure the oxygen concentration in the water. Measurements taken at ten-minute intervals were used to calculate the rate at which the fish was using oxygen. The experiment was stopped and the water replaced with fresh, aerated water before the oxygen concentration

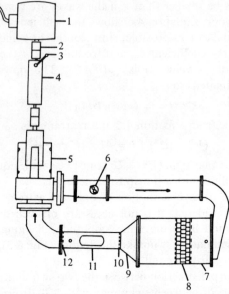

Fig. 4-5. Apparatus for measuring the oxygen consumption of swimming fish. The labels are explained in the text. From G. J. Farmer & F. W. H. Beamish (1969). *J. Fish. Res. Bd Canada*, **26**, 2807–21.

fell too far. In experiments with this apparatus it was found that 80 g. *Tilapia* at 25 °C could maintain for 30 min or more speeds up to 0.6 m s^{-1}. At this speed they used about 5 cm^3 oxygen (kg body weight)$^{-1}$ min $^{-1}$, which is about four times the resting rate. Similar results have been obtained with other small teleosts. *Tilapia* could swim for short periods considerably faster than 0.6 m s^{-1} but short bursts of high speed are achieved by anaerobic respiration, as will be explained.

The muscles which drive the tail from side to side to propel the fish make up more than 50% of the weight of most fish. It is mainly them that we eat, when we eat fish. The muscles on the left of the vertebral column bend the fish towards the left, and the muscles on the right to the right. Two main types of muscle are involved. The bulk of the muscle is white but there is a proportion of red fibres, generally as a lateral strip (Fig. 4-6). In the dogfish *Scyliorhinus* about 18% of the fibres are red. The two types of muscle remain different in colour after cooking: the distinction can be seen particularly plainly in herring (*Clupea*, a teleost).

Dr Quentin Bone has shown that the two types of muscle have distinct functions. He did this by experiments on dogfish which had had their brains destroyed. The rest of the body was kept alive by pumping aerated water through a tube into the mouth, so that it flowed over the gills.

Dogfish prepared in this way still make swimming movements, if fixed by the head with the tail free in a tank of water. The movements continue at a slow, more or less steady rate, but if the tail is pinched the fish swims faster or makes a strong, sustained bend. Bone stuck electrodes into the red and white parts of the swimming muscle and recorded potentials from them. During slow swimming movements potentials were detected in the red muscle, but not in the white. (Dogfish red muscle does not seem to propagate action potentials, and it is likely that these potentials were action potentials in the nerves rather than the muscles.) When the fish was pinched and so stimulated to make more vigorous movements, muscle action potentials were detected in the white muscle. It is not certain whether the red muscle was active as well because the potentials in the white muscle are much larger than those detected (during slow swimming) in the red. They are large enough to affect the electrode in the thin layer of red muscle and mask any activity in the red muscle itself. These experiments are interpreted as evidence that the red muscle alone is used for slow and sustained swimming but that the much greater bulk of white muscle comes into use for violent movements and bursts of speed.

Bone also investigated the effects of exercise on the stores of food materials in the red and white muscles. Some fish were kept unexercised for one or two days before they were killed for analysis. Others were killed in a state of exhaustion, after they had been made to swim vigorously for up to ten minutes. Yet others had their brains destroyed and were set swimming in the manner that has been described. They were kept swimming slowly for up to 50 h, and then killed. The results of the experiments are shown in Table 4-1. Neither exhaustion nor prolonged slow swimming had any appreciable effect on the glycogen content of the red

TABLE 4-1. *Mean glycogen and fat contents of samples of red and white muscle from the myomeres of dogfish.*

The fish were unexercised, exhausted or made to swim slowly for many hours, prior to killing and sampling. Data from Q. Bone (1966). *J. mar. biol. Ass. U.K.* **46**, 321–49.

	Glycogen content (%)		Fat content %	
	Red	White	Red	White
Unexercised	1.9	1.1	3.2	0.8
Exhausted	1.8	0.5*	4.9	0.6
Slow swimming	1.6	1.1	1.7*	0.5

* Significantly different from the values for unexercised fish.

muscles, or on the fat content of the white ones. Exhaustion (after less than ten minutes exercise) halved the glycogen content of the white muscle but had no significant effect on the red. (The fat content of the red muscles seems to be lower in unexercised fish than in exhausted ones. This may be a real difference due to the fish being starved while they were kept unexercised, but very few exhausted fish were analysed for fat and the difference is not statistically significant.) Prolonged slow swimming had no effect on the white muscle but led to a reduction in the fat content of the red muscle. It seems that violent exercise is powered by white muscle fuelled by glycogen, but slow swimming by red muscle fuelled by fat.

The fat is probably metabolized aerobically, and the glycogen anaerobically. Prolonged swimming is most appropriately powered by aerobic metabolism since an oxygen debt cannot be allowed to accumulate indefinitely. The fat used in the red muscle is presumably oxidized completely to carbon dioxide and water. The red fibres have an appropriately rich blood supply and they are slender, so the distances the oxygen has to diffuse are short. They contain myoglobin (a haemoglobin-like compound) which gives them their red colour and is involved in the transport of oxygen within the cells. They have plenty of mitochondria, the organelles which house the enzymes of the Krebs cycle and so are involved in aerobic, but not in anaerobic, metabolism.

In contrast, the white fibres probably get their energy anaerobically, by converting glycogen to lactic acid. This conversion releases energy without using oxygen. However, lactic acid accumulates and oxygen is eventually required to dispose of it; some of the lactic acid can be oxydized and the energy so released used to convert the rest back to glycogen. There is a limit to the concentration of lactic acid which can be tolerated, so violent activity quickly leads to exhaustion. However, until the limit has been reached the power output of the muscles can be greater than aerobic metabolism would allow, because it is not limited by the rate at which oxygen can be supplied by the gills and blood system. Anaerobic metabolism makes bursts of very fast swimming possible, provided they are separated by time for recovery.

The white fibres have a much less good blood supply than the red ones, they are thicker, they lack myoglobin and they contain relatively few mitochondria. Thus they lack the adaptations for aerobic metabolism which are found in the white muscle.

Power is the rate at which work is done, so the power output of a muscle is the force it is exerting multiplied by the rate at which it is shortening. The maximum force that can be exerted is less when the muscle is shortening fast than when it is shortening slowly. Thus the power output that can be achieved diminishes at very high rates of contraction. It seems to be a general rule that vertebrate striated muscles can exert most power when contracting at about 0.3 of the maximum possible rate. This rule has been established by experiments with other muscles, but it presumably applies to fish swimming muscles and seems to explain one feature of their arrangement.

The red muscle fibres of selachians run parallel to the long axis of the body. All are about the same distance from the vertebral column (Fig. 4-6) and all must shorten by about the same fraction when the fish bends.

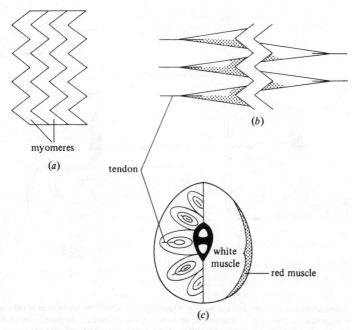

Fig. 4-6. Lateral views of (*a*) part of the trunk of a selachian, skinned, and (*b*) a single myomere. (*c*) A transverse section of a selachian, showing the myomeres.

Some of the white fibres are close to the median plane and some far from it. If they too ran parallel to the long axis of the body they would have to shorten by very different fractions. The ones close to the median plane would hardly shorten at all and their rate of contraction (expressed as lengths per second) would be relatively low. The ones far from the median plane would shorten more and faster. If all the white fibres had similar properties there would be no speed of swimming at which all were contracting at the optimum rate for maximum power output. Appropriately, most of the white fibres are not parallel to the long axis but are inclined at angles of up to about 37°.

The actual arrangement is rather complicated. The myomeres have the shape shown in Fig. 4-6. Note how they are drawn out into long forward- and backward-pointing cones. The cone of one myomere fits into the hollow of the next, and transverse sections show concentric rings of interlocking myomeres. The septa between the myomeres are thin sheets of connective tissue but stout tendons project from them at the apices of the cones. The white fibres converge on these tendons, as shown schematically in Fig. 4-7a. A mathematical analysis of the pattern led to the conclusion

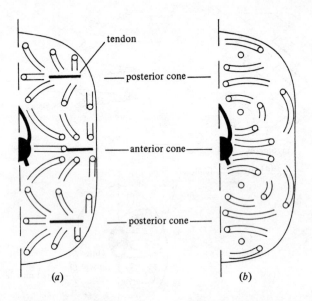

tendon

posterior cone

anterior cone

posterior cone

(a)   (b)

Fig. 4-7. Diagrams showing the arrangement of white muscle fibres in the myomeres of (a) a selachian and (b) a typical teleost. Each diagram represents a thick transverse section, viewed from the posterior side. From R. McN. Alexander (1969). *J. mar. biol. Ass. U.K.* **49**, 263–90.

that it makes all the white fibres shorten by about the same fraction and at about the same rate, when the fish bends. Swimming movements like those shown in Fig. 4-3 involve about 20% changes in the lengths of the red fibres and 15% changes in the lengths of the white ones. The white fibres may well all be contracting at the optimum rate for maximum power output, when the fish is swimming at its top speed.

BUOYANCY AND FINS

Nearly all selachians live in the sea, in water of specific gravity about 1.026. Many of the tissues in their bodies are denser: the muscle has a specific gravity of about 1.06, the cartilage 1.1 and the scales probably about 2. Consequently most selachians are denser than the water they live in, and must produce vertical hydrodynamic forces as well as horizontal ones when they are swimming horizontally. Their density can be determined by applying Archimedes' Principle, by weighing in air and then with the body immersed in water. Dogfish (*Scyliorhinus*) are typical with a specific gravity of about 1.075 (density 1075 kg m^{-3}). A 1 kg dogfish has a volume of about 1/1075 m^3 so the upthrust exerted on it by seawater of density 1026 kg m^{-3} is only 1026/1075 = 0.954 kg wgt. The difference of 46 g wgt between this and its weight must be balanced by upward hydro-dynamic forces, when the fish swims. It is balanced partly by lift acting on the pectoral fins, just as the weight of an aeroplane is balanced mainly by lift on the wings, and partly by the action of the tail. The pectoral fins are held at an appropriate, small, angle of attack. The contribution of the tail will be explained shortly.

Fig. 4-8*a* shows the vertical components of the forces which act on a swimming dogfish. The Archimedes' upthrust acts at the centre of buoyancy of the fish (i.e. at its geometrical centre). The centre of gravity where the weight acts is about 1.7 mm posterior to this (in a 1 kg fish) because the tail end is denser than the head. The upthrust of 954 g wgt thus exerts a

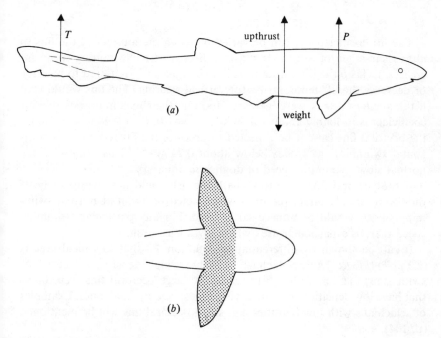

Fig. 4-8. (*a*) The vertical components of the forces which act on a swimming dogfish; (*b*) The dogfish viewed from above, showing the area referred to in the text.

moment of $1.7 \times 954 = 1620$ g wgt mm about the centre of gravity. The upward forces P on the pectoral fins and T on the tail probably act about 100 mm and 370 mm, respectively, from the centre of gravity. At equilibrium the upward forces must balance the downward force

$$954 + P + T = 1000,$$

and the moments of the forces about the centre of gravity must balance

$$100P + 1620 = 370T.$$

The solution of these simultaneous equations is $P = 33$, $T = 13$. When the fish swims horizontally, at whatever speed, vertical hydrodynamic forces of 33 g wgt (0.32 N) and 13 g wgt (0.13 N) are required on the pectoral fins and tail, respectively.

The pectoral fins act as hydrofoils and the lift they produce is given by equation 4.2 (p. 88). We have just seen that the lift required is 0.32 N. The density of seawater is 1026 kg m^{-3}. The area S is not the area of the fins alone, but their area together with the plan area of the strip of trunk between them, as indicated in Fig. 4-8b. This is how the areas of aircraft wings are reckoned, in calculations of this sort. The area S would be about 0.010 m^2 for our 1 kg dogfish. Putting these values in equation 4.2

$$0.32 = \tfrac{1}{2} \times 1026 \times 0.010 U^2 C_L,$$
$$U^2 = 0.06/C_L.$$

The lift coefficient C_L can be varied by using the muscles of the fins to alter their angle of attack. It must be decreased as swimming speed increases, to keep the lift constant. It is also adjusted as required to increase or decrease the lift temporarily, for vertical steering. The fins would stall if the angle of attack were increased too far: their maximum possible lift coefficient is probably about 1.0. When $C_L = 1.0$, $U = 0.24$ m s^{-1}. Hence the pectoral fins cannot be expected to produce the lift required for horizontal swimming, at speeds below about 0.24 m s^{-1}. This is close to the normal slow swimming speed of dogfish in aquaria.

If the pectoral fins were too small, the fish could not swim slowly. If they were unduly large, the drag which acted on them at normal swimming speeds would be unnecessarily large. For any particular swimming speed there is a particular area which gives least drag.

It can be shown by differentiating equation 4.3 that this ideal area is $(2P/\rho U^2)(k/\pi R C_{D0})^{1/2}$. Selachians that have high densities (P large) or swim slowly (U small) are best served by large pectoral fins: selachians that have low densities or swim fast are best served by small ones. Examples of selachians with low densities and small pectoral fins will be mentioned (p. 104).

The tails of dogfish and other sharks are heterocercal: the posterior end of the vertebral column turns upwards and extends into the dorsal lobe

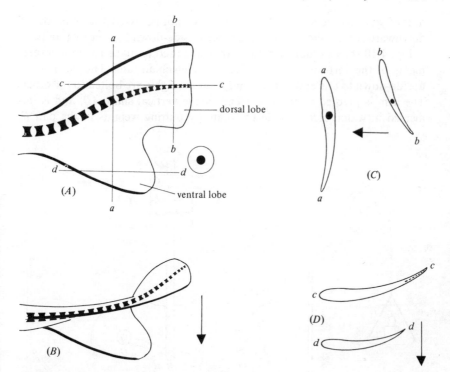

Fig. 4-9. The tail of a Port Jackson shark, moving to the left in swimming. The tail is seen (*A*) from the side, (*B*) from above, (*C*) in vertical section and (*D*) in horizontal section. From J. R. Simons (1970). *J. exp. Biol.* **52**, 95–107.

of the tail. Fig. 4-9 shows the heterocercal tail of the Port Jackson shark, and the way it bends in swimming. The edges indicated by thick black lines are stiffer than the rest of the tail. The tail is moving in the direction indicated by the arrow, with the stiffened edges leading and the more flexible parts lagging behind. Because it bends in this way the ventral lobe of the tail (seen in section *aa*, Fig. 4-9*C*) tends to deflect water upwards while the more posterior, dorsal lobe (*bb*) tends to deflect water downwards. Consequently the force on the ventral lobe has a downward component, and the force on the dorsal lobe has an upward component. The upward force on the dorsal lobe is greater than the downward force on the ventral lobe, so the tail as a whole produces the required upward force T.

Note that this upward force is the difference between upward and downward forces on different parts of the tail. If the ventral lobe were larger or if the relative stiffness of the lobes were slightly different the overall effect of the tail might be to produce downward force. A heterocercal tail need not necessarily produce an upward force. Similarly the hypocercal tails of some ostracoderms (Fig. 3-18*c*) need not necessarily have produced downward forces, though they look like upside-down heterocercal tails.

Fig. 4-10 shows crude apparatus that was used to make rough measurements of the vertical forces produced by selachian tails. The tail is fixed upside-down to one end of the arm *FG*, and weights are bolted to the other. The arm is pivoted to the lower end of the vertical shaft *AB*, and is immersed in water. *AB* is made to rotate by putting weights in the pan *E*,

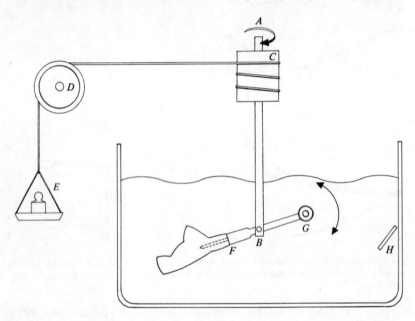

Fig. 4-10. Apparatus used to measure the vertical component of the force which acts on a selachian tail when it is moved sideways through water. From R. McN. Alexander (1965). *J. exp. Biol.* **43**, 131–8.

so that the tail moves sideways-on through the water and produces a downward force. The faster the rotation, the greater the force. The speed is adjusted to balance the force against the weights at *G*. A series of measurements was made using different weights so that a graph could be drawn of the vertical force against the speed of tail movement. It was found that fresh dogfish tails produced the force *T* needed for horizontal swimming only when moving at a speed equivalent to swimming at about 0.8 body-lengths per second. When a dogfish swims horizontally at any other speed it must presumably use the muscles of the caudal fin to adjust the stiffness of its lobes, so as to obtain the required vertical force.

Selachian fins, including the caudal fin, have the basic structure shown in Fig. 4-11*a*, *b*. They are segmented structures, with muscles which

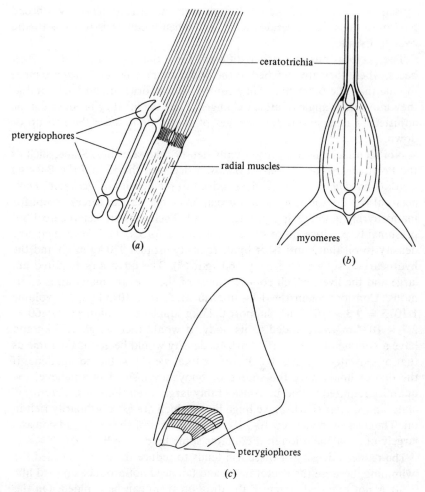

Fig. 4-11. Diagrams of selachian fins. (*a*), (*b*) are a dissection and a transverse section through a typical dorsal fin, respectively. (*c*) shows the skeleton of a pectoral fin.

develop in the embryo by separating off the myotomes. The increase in length of the fin as the embryo grows generally fails to keep pace with the growth of the body, so that a dorsal fin derived from fourteen segments (for instance) may only extend over six myotomes in the adult fish. Each segment of the fin contains two or more rods of cartilage (pterygiophores) connected end-to-end. The distal pterygiophores lie between sheets of packed ceratotrichia, which are very thin rods of a form of collagen known as elastoidin. The ceratotrichia extend to the edge of the fin, and stiffen it. There are a great many of them in each segment, but there is only one muscle on each side of the fin in each segment. It is attached by a broad tendon to the sheet of ceratotrichia, and when it contracts it bends the fin over to its side.

The pectoral fins are rather different from the rest (Fig. 4-11c). Their bases, where they are attached to the body, are narrow, and they are more mobile than the other fins. They can be swung forward and back, and as they move their angle of attack changes. Thus the lift they produce can be adjusted to compensate for changes of speed, or to steer the fish up or down.

Not all selachians are substantially denser than seawater. Some, such as the Portuguese shark (*Centroscymnus*, a deep-sea shark) and the Basking shark (*Cetorhinus*), are very close indeed to the density of seawater. Their pectoral fins do not have to produce appreciable vertical forces except for manoeuvring, and are appropriately small. Their tails are heterocercal but presumably produce more or less horizontal forces. They owe their low density to oil that consists of lipids (density around 920 kg m^{-3}) and the hydrocarbon squalene (density 860 kg m^{-3}). The quantities required are large and the liver, which contains most of the buoyant material, is enormous. Consider again the 1 kg dogfish of density 1075 kg m^{-3}, volume $1/1075 = 9.3 \times 10^{-4}$ m^3. Suppose 0.23 kg squalene of volume $0.23/860 = 2.7 \times 10^{-4}$ m^3 were added to its body. It would then weigh 1.23 kg and have a volume of 1.20×10^{-3} m^3. Its density would be about the same as that of seawater, but nearly 20% of its total weight would be squalene. If the denser lipids were the source of buoyancy instead of squalene, the quantity required would be considerably larger even than this. *Centroscymnus* and similar sharks have huge livers which are extraordinarily rich in oil. The liver oil makes up 16–24% of the weight of the body and consists largely of squalene, though it contains some lipid as well.

The reduced density of these fish tends to reduce the energy needed for swimming, because the pectoral fins and tail need not produce upward lift. If lift is not required, some of the drag on them can be avoided. On the other hand the buoyant material increases the bulk of the fish and so tends to increase the drag on the trunk. Is the net effect a decrease or an increase in the energy needed for swimming?

Consider yet again the 1 kg dogfish. Its pectoral and caudal fins would have chords of a few centimetres and it would swim mainly at speeds around 0.25 m s^{-1}. Hence the Reynolds numbers of the fins would be of the order of 10^4. The drag on hydrofoils of moderate aspect ratio at Reynolds numbers of this order seems always to be at least 0.1 of the lift. The total upward force that is needed is 0.45 N (0.32 N from the pectoral fins and 0.13 N from the tail), and producing it would increase the drag on the fish by at least 0.045 N.

Now consider the same fish with its density decreased to that of sea-water by the addition of $2.7 \times 10^{-4}\,\text{m}^3$ squalene. If the extra material were distributed uniformly all along the length of the fish (about 0.7 m), the cross-sectional area would be increased uniformly by $2.7 \times 10^{-4}/0.7 = 4 \times 10^{-4}\,\text{m}^2$. It is more likely that the increase in cross-sectional area would be largest in the thickest part of the body, so as to keep the shape of the fish streamlined. The increase in frontal area would then probably be around $10^{-3}\,\text{m}^2$. The drag coefficient of the body would probably be around 0.1. Hence the drag which acted on the fish when it swam at $0.25\,\text{m s}^{-1}$ would be increased by $\frac{1}{2}\rho A U^2 C_D = \frac{1}{2} \times 1026 \times 10^{-3}(0.25)^2 \times 0.1 = 0.003\,\text{N}$. When it swam faster at $1\,\text{m s}^{-1}$ the increase would be 0.05 N. The squalene would have made possible a reduction of drag at low speeds, but not at high ones.

This calculation involves too many guesses and assumptions for the estimates of drag to be at all reliable. However, one conclusion seems justified. Squalene is most effective in reducing drag (and so saving energy) at low swimming speeds.

GILLS AND JAWS

Selachians have their gills arranged in essentially the same way as ammo-coetes (Fig. 3-2). The filaments and secondary lamellae have the same basic structure and arrangement. There are parabranchial cavities between the gills and the narrow external gill slits. Only the gill skeleton is strikingly different. It is not a continuous network of cartilage but consists of distinct bars of cartilage jointed together (Fig. 4-12a). The main cartilages which constitute the branchial arches do not lie external to the gills, but internal to them. They are known as the pharyngo-, epi-, cerato- and hypobranchials. Slender gill rays extend laterally from them in the partitions between the gill slits. Further cartilages stiffen the edges of the gill openings. This is so different from the gill skeleton of ammocoetes that it is not at all obvious how one type could have evolved from the other.

The most anterior of the gill arches is the hyoid arch, which is rather different from the rest. It has typical gill filaments only on its posterior face. Its skeleton has the usual ventral elements (the cerato- and hypohyals) but there is only one dorsal element. This is the hyomandibular cartilage, and corresponds to the epibranchials. It articulates with the cranium. Immediately anterior to it is the small round opening known as the spiracle, which is apparently a vestigial gill slit. It was probably a full-length slit at an early stage in evolution, but its lower parts have been obliterated. Anterior to the hyoid arch are the jaws, an upper palatoquadrate cartilage and a lower Meckel's cartilage. They are attached to the hyoid arch by

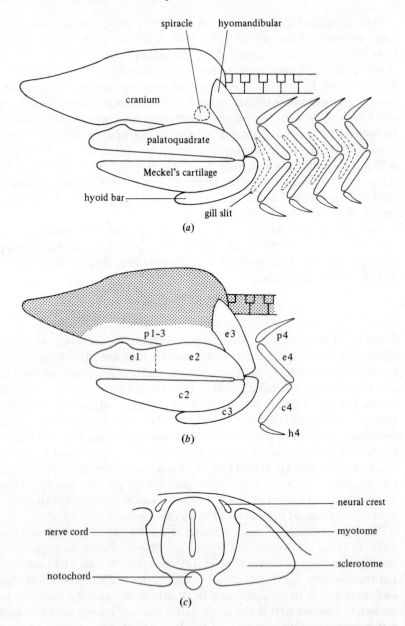

Fig. 4-12. (*a*) A diagram showing the main cartilages of a selachian skull. (*b*) A diagram showing how the jaws are believed to have evolved from parts of two gill arches anterior to the hyoid arch. p, e, c, h, pharyngo-, epi-, cerato- and hypobranchial elements. (*c*) A diagrammatic transverse section of a vertebrate embryo.

ligaments. The left and right palatoquadrates meet in the middle of the upper jaw and the two Meckel's cartilages in the middle of the lower one.

The jaws are believed to be modified gill arches, like the hyoid arch. The evidence is largely embryological. Each somite in a vertebrate embryo differentiates into two main parts, the myotome and the sclerotome (Fig. 4-12c). The myotome becomes muscle and the sclerotome cartilage which in some vertebrates later becomes bone. As the nerve cord is formed by rolling up of the neural plate, a group of cells known as the neural crest separates from alongside it. Different parts of the neural crest have different fates, forming spinal ganglia, pigment cells and cartilage. The skulls of vertebrates appear initially as cartilage, whether they are to remain cartilaginous as in selachians or to become largely bony. It can be shown by careful anatomical study of embryos, supplemented by experiments in which parts of the embryo are stained or damaged, that this cartilage develops from two sources. The main part of the cranium develops from sclerotome, as do the vertebrae. The gill arches, the jaws and the remainder of the cranium develop from neural crest. There is evidence in some fossil fishes (members of the Crossopterygii) of an additional rudimentary gill opening, above the palatoquadrate cartilage and anterior to the spiracle. This has led to the suggestion that there were originally two gill arches anterior to the hyoid arch. These are supposed to have formed the jaws, as indicated in Fig. 4-12b. The parts of the cranium which develop from neural crest are supposed to be dorsal elements of these two arches and of the hyoid arch, which have become incorporated in the cranium.

The breathing movements of dogfish can be observed when they are resting or swimming slowly in an aquarium. The anterior part of the body gets deeper and wider, then shallower and narrower, as the mouth cavity and parabranchial cavities enlarge and contract. The gill slits, the mouth and the spiracle open and close. The water movements involved can be made visible by introducing a little milk from a pipette. Milk released near the mouth is drawn in through the mouth and expelled through the last three of the five pairs of gill slits. Milk released near a spiracle is drawn in through the spiracle and expelled through the first three gill slits of the same side of the body.

The pressures involved in breathing were investigated by Professor George Hughes. A dogfish was anaesthetized by putting it in seawater mixed with anaesthetic, and fixed in a specially designed holder. This was put into a tank of seawater containing only a little anaesthetic so that the fish was less deeply anaesthetized. It made apparently normal breathing movements. Pressure transducers were used to record pressures through slender tubes which were inserted into the mouth, or by way of a gill slit into a parabranchial cavity. The opening and closing of mouth and gill slits were recorded simultaneously by displacement transducers. Typical results are shown in Fig. 4-13. Both the mouth and the gill slits are open in the interval between breaths. At this stage the mouth cavity and para-branchial cavities are expanded. The mouth and spiracle close and the gill slits remain open, as the cavities contract and water is driven out. The pressure is now positive (above ambient) both in the mouth cavity ('oro-branchial pressure') and in the parabranchial cavities. The mouth and spiracles open again and the gill slits close briefly as the cavities expand and water is drawn in. The pressure is negative (below ambient) in the mouth and parabranchial cavities. Fig. 4-13 and other records show that, whether the pressure is positive or negative, it is nearly always greater in the mouth cavity than in the parabranchial cavities. Water must be kept flowing from the mouth to the parabranchial cavities both while the cavities are

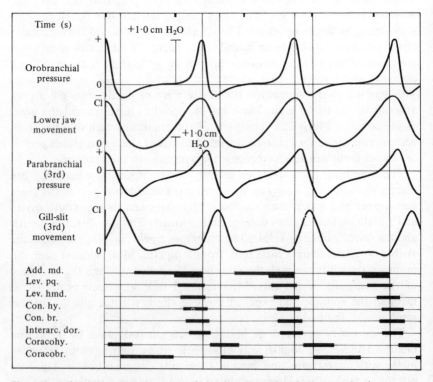

Fig. 4-13. A diagram based on the results of experiments described in the text, showing pressure changes, movements and periods of activity of various muscles during three respiratory cycles of a dogfish. The muscles, indicated by abbreviations, are the adductor mandibulae, levatores palatoquad-rati and hyomandibulae, constrictores hyoideus and branchiales, interarcualis dorsalis, coraco-hyoideus and coracobranchiales. From G. M. Hughes & C. M. Ballintijn (1965). *J. exp. Biol.* **43**, 363–83.

expanding and the pressures are negative, and while they are contracting and the pressures are positive. It was explained in Chapter 3 that one-way flow like this is believed to occur in the ammocoete larva, but it was not possible to cite pressure records to demonstrate it. Ideally the pressure difference and rate of flow across the gills should be kept constant, but the pressure difference falls to zero in the intervals between breaths.

Some sharks do not make breathing movements as they swim, but simply swim with their mouths open. This keeps water flowing more steadily over the gills than the fluctuating pressures of breathing can do. It thus reduces the energy needed for breathing, but it does not eliminate it: the work required to drive the water over the gills is done by the swimming muscles.

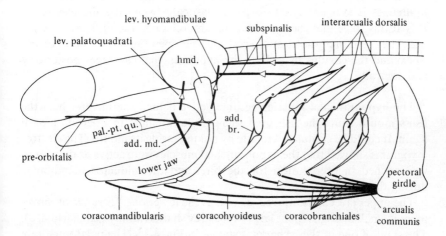

Fig. 4-14. A diagram showing the main muscles of a dogfish head, excluding the constrictor muscles. add. br., adductor branchialis; add. md., adductor mandibulae; hmd., hyomandibular cartilage; pal.-pt.qu., palatoquadrate cartilage. From G. M. Hughes & C. M. Ballintijn (1965). *J. exp. Biol.* **43**, 363–83.

The main muscles of a dogfish head are shown in Fig. 4-14. Their sequence of contraction in respiration has been investigated by electromyography. A dogfish was fixed in a holder in the same way as for the pressure recordings (and indeed pressures were recorded at the same time as the electromyographs). Pairs of electrodes were inserted through small cuts in the skin, into the muscles which were to be investigated. Recordings were made from up to four muscles at a time. The results are summarized in Fig. 4-13. The hypobranchial muscles (coracomandibular, coracohyoid and coracobranchial) are inactive in normal gentle respiration: the activity indicated in the figure occurs only in heavy breathing. In quiet breathing the pattern of muscle activity is very simple, with three main phases:

(i) The adductor mandibulae contracts, closing the mouth. The spiracle is also closed.

(ii) The adductor mandibulae remains active, holding the mouth closed, while many other muscles contract and compress the mouth cavity and parabranchial cavities. Water is driven out through the gill slits. The active muscles include the constrictors which run vertically in the partitions between the gills. The more anterior muscles contract slightly before the more posterior ones.

(iii) All these muscles relax. The mouth and spiracle open, the mouth cavity and parabranchial cavities expand and water is drawn in. This does not seem to require any muscle activity in gentle breathing; the movements are apparently produced by elastic recoil. Though the gill skeleton is jointed, it is stiff enough to expand elastically when the muscles relax. It is at this stage that the hypobranchial muscles contract in heavy breathing. They reinforce the elastic recoil, pulling open the mouth and helping to expand the cavities by pulling the hypobranchial cartilages posteriorly and ventrally.

The hypobranchial muscles became active if the experimenter held the fish's mouth shut: the fish used them in its attempts to open its mouth again. It could be stimulated to bite by poking inside its mouth. The electromyograms showed that the hypobranchial muscles were active as the mouth was opened in preparation for the bite, and the adductor mandibulae during the bite itself.

The upper jaw can be moved, as well as the lower one. Its range of movement is small in dogfish, but large in the grey sharks (family Carcharinidae). The skull of one of these sharks is shown in Fig. 4-15, The palatoquadrate cartilages have projections known as orbital processes, which fit into grooves in either side of the cranium. In (*a*) the palatoquadrate is retracted so that it lies flush with the undersurface of the snout. In (*b*) it has slid forward, down the grooves, and projects considerably below the snout. This protrusion is limited by ligaments (the ethmopalatine ligaments) that connect the orbital processes to the cranium. The ligament is not illustrated in Fig. 4-15*a*, but would be slack. It is taut in Fig. 4-15*b*. Protrusion is probably brought about largely by contraction of the levator palatoquadrati and levator hyomandibulae, which in grey sharks run at quite different angles from those shown for the dogfish in Fig. 4-14. They run at such angles as to tend to pull the palatoquadrates and hyomandibulae forwards.

The feeding behaviour of various grey sharks has been observed and filmed in a pen at a marine laboratory. When sharks take small food in mid-water they simply swim at it with their mouths open. When picking

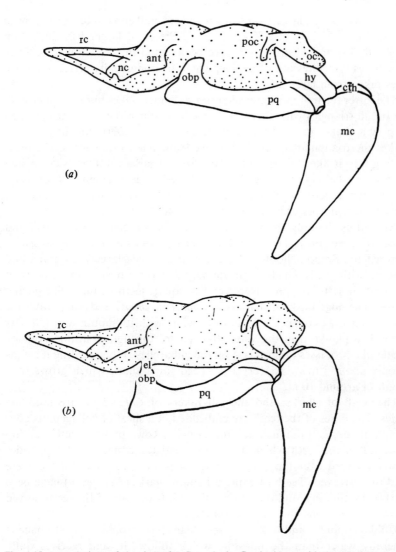

Fig. 4-15. Lateral views of the skull of a Grey shark, *Carcharhinus*, with the mouth open. The palatoquadrate cartilage is protruded in (*b*). *el*, ethmopalatine ligament; *hy*, hyomandibula; *mc*, Meckels' cartilage; *obp*, orbital process; *pq*, palatoquadrate. From S. A. Moss (1972). *J. Zool., Lond.* **167**, 423–36.

up food from the bottom they protrude the upper jaw: this makes the food easier to get at than if the jaw were kept flush with the underside of the head. If the jaw were kept permanently protruded the fish would be less well streamlined than with the jaws flush.

Grey sharks feed largely on fish which are small enough to be swallowed whole, but they also attack larger prey. Pieces of larger fish (including other sharks) and of porpoises have been found in their stomachs. They sometimes take bites out of people, when the opportunity arises. How are bites taken from large prey?

The lower teeth of most grey sharks are slender spikes but the upper ones are broad triangles with sharp cutting edges. When the shark has gripped its prey it shakes its head violently. Because of the inertia of the prey this makes the triangular upper teeth move from side to side across the prey, cutting into it like the teeth of a saw. Some sharks roll from side to side instead of shaking their heads, but the effect is the same. Protrusion enables the upper jaw to extend down into the kerf so that the cut can go considerably deeper than the height of the teeth.

As well as the grey sharks, many other sharks feed on whole fish and pieces of larger prey. Their teeth have sharp points and cutting edges. In contrast the Smooth hound (*Mustelus canis*) feeds largely on crabs and lobsters with shells which might damage sharp teeth but can be crushed by blunt ones. Its jaws are covered by flat, square teeth arranged like paving stones. The huge Basking shark (*Cetorhinus maximus*) feeds on planktonic animals such as copepods. It simply swims with its mouth and gill slits open. The plankton in the water that passes through is strained out by bristle-like processes (gill rakers) on the gill arches. When feeding it swims at only about 1 m s^{-1}, a very low speed for a shark which grows to a length of around 10 m.

The teeth of sharks, and the gill rakers of *Cetorhinus*, are modified scales. The bases of the teeth are embedded in a sheet of collagenous connective tissue, continuous with the dermis. Rows of new teeth develop posterior to the teeth which are in use, and the connective tissue slides slowly forward over the jaw to bring them into position. Old teeth are shed or destroyed. Teeth of captive Lemon sharks (*Negaprion*) have been marked by clipping, and their fate followed. It was found the teeth moved moved forward one row every eight days or so.

Unlike lampreys and hagfishes, selachians have stomachs. A stomach is generally wider than the intestine which follows it, and receives quite different secretions. Cells in the wall of a stomach secrete hydrochloric acid and the precursors of enzymes of the pepsin type, which attack proteins by breaking them into shorter peptide chains. Pepsins act fastest in acid conditions. By contrast, the many enzymes which are active in the intestine work fastest in neutral or mildly alkaline conditions. Some of them are secreted by the wall of the intestine and some by the pancreas, and bile from the gall bladder is mixed with them. These secretions are mildly alkaline, and reduce the acidity of the materials which enter the intestine from the stomach. Products of digestion are absorbed in the intestine, but not in the stomach. Details of the digestive processes can be found in the textbooks of physiology.

The intestines of selachians contain a helical partition, the spiral valve. Food travelling through must take a helical path involving, in at least one species, as many as twenty complete turns.

BLOOD AND BLOOD VESSELS

The heart of a shark is illustrated in Fig. 4-16. The cardinal veins from either side of the body and the hepatic vein from the liver open into the small sinus venosus at the back of the heart. It opens into the large, dorsally placed atrium, which opens in turn into the ventricle which lies ventral to

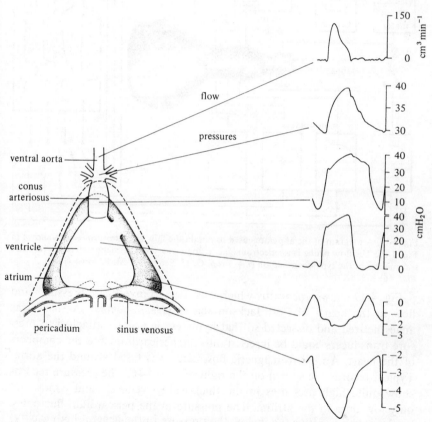

Fig. 4-16. A ventral view of the heart of the Port Jackson shark, *Heterodontus*, and records of blood flow and pressure changes at the positions indicated, during a single heart beat. From G. H. Satchell (1971). *Circulation in fishes*. Cambridge University Press, London.

it. Finally the conus arteriosus joins the ventricle to the ventral aorta, which carries the blood anteriorly towards the gills. All the chambers of the heart, from the sinus to the conus, have muscular walls, but the walls of the ventricle and conus are much thicker than those of the sinus and atrium. Flap valves at the openings between the chambers prevent blood flowing in the wrong direction. There are four or five valves, each consisting of a ring of flaps, in the conus (see Fig. 4-17). The whole heart is enclosed in a chamber called the pericardium, walled partly by the cartilage of the pectoral girdle and partly by sheets of collagenous connective tissue.

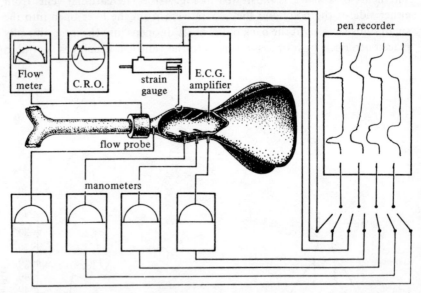

Fig. 4-17. A diagram of the apparatus used to obtain the flow and pressure records shown in Fig. 4-16. The flow probe is an electromagnetic flowmeter and the manometers are pressure transducers of the type described on p. 4. From G. H. Satchell & M. P. Jones (1967). *J. exp. Biol.* **46**, 373–82.

Fig. 4-17 shows apparatus which was used to investigate pressures and flow in the hearts of Port Jackson sharks (*Heterodontus*). A shark was anaesthetized and dissected so that hypodermic needles attached to pressure transducers could be inserted into the pericardium and the chambers of the heart. An electromagnetic flowmeter was fitted around the aorta. Typical records are shown on the right of Fig. 4-16. The pressure records show quite high pressures in the thick-walled ventricle and conus, but only low ones in the atrium. The pressure in the pericardium fluctuates, but is always negative (i.e. below the pressure in the general body cavity) owing to the elasticity of the pericardium walls. Two main phases of the heartbeat can be distinguished:

(i) The atrium contracts, emptying itself into the ventricle. The pressure is slightly positive in the atrium, and about zero in the ventricle.

(ii) The ventricle contracts, driving blood out into the conus and aorta at pressures of 30–40 cmH$_2$O (3–4 kN m^{-2}). This reduces the volume of the heart and so the pressure in the pericardium. The atrium is relaxing and the negative pressure in the pericardium makes it expand, drawing in blood from the sinus venosus and veins.

The conus arteriosus continues to contract after the ventricle has started relaxing, so that high pressure is maintained in it for longer. The pressure in the ventral aorta is remarkably steady, fluctuating only between 30 and 40 cmH$_2$O. This is partly due to the prolonged contraction of the conus, and partly to the elastic properties of the ventral aorta, which is inflated by blood as the ventricle and conus contract, and then recoils elastically. The recoil keeps the pressure quite high and should keep blood flowing through the gills until the ventricle contracts again. However, the flowmeter records show the flow falling abruptly to zero at the end of the contraction of the ventricle. Perhaps the elastic part of the aorta is distal to the point where the flowmeter was fitted. Steady flow of blood through the gills, if it occurs, will give the same advantages as steady flow of water over them (p. 44).

The ventral aorta owes its elastic properties to a high proportion of elastin in its walls. Analysis of the aortas of five sharks of various species showed that elastin constituted an average of 31% of the dry weight of the wall of the ventral aorta, but only 9% in the dorsal aorta. The corresponding figures for collagen were 46% and 69%. There is also smooth muscle in the walls of the arteries: it must account for most of the fraction of the dry weight that is neither elastin nor collagen.

Elastin is a protein with properties like rubber: it can be stretched a long way, and snaps back when released. Collagen is a fibrous protein which cannot be stretched anything like as easily or as far. The Young's modulus of collagen is about a thousand times that of elastin, and the properties of arteries would be dominated by the collagen fibres if they were tight. In fact the collagen fibres seem to be arranged loosely, so that the elastin can play an important part. The ventral aorta in particular probably has to be distended considerably, to tighten its collagen fibres.

The blood plasma of hagfishes has an osmotic concentration close to that of seawater (about 1 osmol l^{-1}) but that of lampreys and teleosts has much lower concentrations around 300 mosmol l^{-1}. In all these fish, most of the osmotic concentration is due to salts dissolved in the plasma and only a little to other materials. In marine selachians the total osmotic concentration is generally a little above that of seawater but only about half of it is due to inorganic salts: most of the rest is due to urea and tri-methylamine oxide, which are present in remarkably high concentrations. The other extracellular fluids are similar in composition to the plasma.

Since the osmotic pressure of the body fluids is a little above that of seawater, there is some tendency for water to move into the body, particularly at the gills. Inorganic ions must also tend to diffuse in while urea and trimethylamine oxide tend to diffuse out.

As well as the kidneys there is an organ called the rectal gland, which helps to regulate the composition of the body fluids. It is a small finger-shaped gland which protrudes from the posterior end of the intestine and discharges through a duct into it. Its function was investigated in a series of experiments with Spiny dogfish, *Squalus acanthias*. Cannulae were fastened into its duct, and also into the urinary opening. The dogfish were fastened to a board in some of the experiments but in others they were free to swim in an aquarium, trailing the cannulae behind them. The rectal gland secretion was collected through one cannula and the urine through the other. They flowed at variable rates, but the average for each was 0.5 cm^3 kg body weight^{-1} h^{-1}. Their compositions were quite different, as Table 4-2 shows. The urine contained about the same concentrations of sodium and chloride as the plasma, but much more magnesium (and sulphate and phosphate) and much less urea and trimethylamine oxide. The latter compounds must be reabsorbed from the kidney tubules, so that their rate of loss does not exceed the rate at which they are produced as waste products of protein metabolism. Divalent ions, on the other hand, must be secreted into the tubules. Experiments with inulin (see p. 68) have shown that the glomerular filtration rate is only about four times the rate of urine production, while the concentration of magnesium (for instance) is more than ten times as high in the urine as in the plasma.

TABLE 4-2.   *Composition of various fluids from* Squalus acanthias

Data from J. Burger (1967) in P. W. Gilbert, R. F. Mathewson & D. P. Rall (eds.) *Sharks, skates and rays.* © Johns Hopkins Press, Baltimore

	Concentrations (mEq l^{-1} or mM l^{-1})					Osmotic concentration (mosmol)
	Na$^+$	Cl$^-$	Mg^{2+}	Urea	TMAO*	
Seawater	440	490	50	0	0	930
Blood plasma	250	240	>1†	350	70	1000
Urine	240	240	40	100	10	800
Rectal gland secretion	500	500	0	18	—	1000

* Trimethylamine oxide.
† Few measurements have been made, but values of 3–4 mEq l^{-1} seem to be typical.

The secretion of the rectal gland (Table 4-2) consists of little but water and sodium chloride. It contains very little urea or magnesium, but the sodium and chloride concentrations are about twice as high as in the plasma and much the same as in seawater.

There are cells in the gills of selachians which are believed to be capable of excreting salts, but they seem to excrete less than the kidneys and rectal gland.

Though the great majority of selachians live in the sea there are a few which live in fresh water. There are sharks in Lake Nicaragua and in the San Juan River which joins it to the sea, in the Zambezi and in the Ganges. Men have been attacked by the sharks in all these places. These sharks are all very like the marine Bull shark, *Carcharinus leucas*. The Nicaraguan and Zambezi sharks seem actually to belong to this species, but the Ganges shark seems to be distinct. The sawfish *Pristis* and the stingrays *Dasyatis* etc. have species which enter or even live permanently in fresh water and are found in many tropical rivers. The body fluids of many of the fresh-water selachians have been analysed. They contain rather lower concentrations of salts than marine species, and much lower concentrations of urea. For instance, the plasma of marine *Carcharinus leucas* has been found to contain 223 mEq Na^+ and 333 mM urea, and that of the Lake Nicaragua shark 200 and 132, respectively.

Though salts tend to diffuse into marine selachians they must tend to diffuse out of freshwater ones. There is no need for a rectal gland to excrete salts, and the rectal glands of Lake Nicaragua sharks have a degenerate structure which suggests that they are functionless. *Pristis* urine has been collected and found to be much more dilute than the blood plasma: its osmotic concentration was 55 mosmol, and that of the plasma 550 mosmol.

REPRODUCTION

The males of most fish do not fertilize the eggs until they have been laid, but selachians copulate and the eggs are fertilized within the female. Males have the posterior ends of their pelvic fins modified to form organs known as claspers (Fig. 4-18), which serve the function of a penis. Each is cigar-shaped, but with a deep groove along one side, and stiffened by a long stem cartilage. They normally lie flat against the body, pointing posteriorly, but they can be erected individually by muscle action. Other muscles can spread the small terminal cartilages apart, so that the end of the clasper is greatly widened. Both erection and spreading can be demonstrated on a shark fastened upside-down out of water, by stimulating the muscles electrically. When the clasper is erect the basal end of the groove is against the cloaca and semen can be extruded into it. The clasper grooves have often been found to contain sperm. Associated with the claspers are two siphon sacs, which lie under the skin of the belly and open into the clasper grooves. Each can be pumped full of water by repeated erection of the corresponding clasper. They have muscles in their walls which presumably empty them again, squirting water along the clasper grooves.

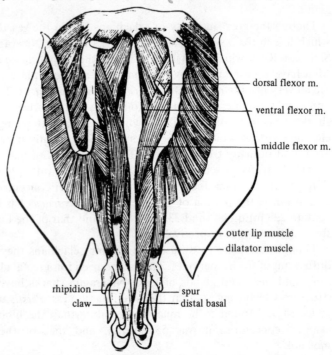

dorsal flexor m.

ventral flexor m.

middle flexor m.

outer lip muscle
dilatator muscle

rhipidion
claw
spur
distal basal

Fig. 4-18. Dissection from the dorsal aspect of the pelvic fins and claspers of *Squalus*. The claspers are at the bottom of the drawing, i.e. at the posterior end of the fins. The rhipidion, claw, spur and distal basal are the terminal cartilages referred to in the text. Contraction of the dilatator muscle swings the spur in one direction, and the other terminal cartilages in the other. From P. W. Gilbert & G. W. Heath (1972). *Comp. Biochem. Physiol.* **42A**, 97–119.

When dogfish (*Scyliorhinus*) copulate, the male wraps his body around the female in such a way that he can insert a clasper into her cloaca. Only one is inserted. It is presumed that the terminal cartilages are spread after insertion to anchor it, and that the sperm are flushed along the groove by water from the siphon sac. Less slender sharks probably could not wind around their females like dogfish. Lemon sharks (*Negaprion*) are moderately stout and have been seen copulating, swimming side by side with their tails beating in unison.

Internal fertilization raises two alternative possibilities. The egg may be enclosed in a tough egg case. Since fertilization is internal it can occur before the shell (which would not easily be penetrated by sperm) is formed. Alternatively the egg can be retained inside the body until development has proceeded for some time, before it is laid. In either case the egg is protected in the interval between fertilization and the time when the embryo becomes an active young fish, able to feed and to take action to avoid predators.

The larger the egg the larger and less vulnerable can the young fish be expected to be, when it emerges from the egg case or from its mother. Selachian eggs are large like hagfish eggs. Dogfish (*Scyliorhinus*) eggs are about 13 mm in diameter, and most selachian eggs are larger. In contrast teleost eggs are small, like lamprey eggs. Salmon (*Salmo salar*) eggs are only 6 mm in diameter and most teleost eggs are considerably smaller.

The volume of eggs that can be produced at one time by a fish of given size seems to be limited to about 20% of the volume of the body. Hence an increase in the size of eggs implies a proportionate reduction in their number. If the volume of each egg is increased by a factor n the number of eggs must generally be diminished by the same factor. In this case, the increase in size will only be advantageous if it increases by the factor n the chance each egg has of producing a breeding adult.

Dogfish (*Scyliorhinus*) and various rays (species of *Raia*) lay each egg in the keratin envelope known as a mermaid's purse. This is more or less rectangular, with tendrils at the corners. Dogfish egg cases are often found near the low tide mark, with their tendrils entangled in seaweed. The case is initially closed, and is filled with a fluid which contains salts and urea in much the same concentrations as the blood plasma. Later pores open in it, allowing seawater to circulate through. The circulation is maintained by movements of the embryo's tail. Once the pores have opened the embryo must have to use energy maintaining the difference in composition between its body fluids and the water, but it is probably essential that they should open, to enable the embryo to obtain enough oxygen for the later stages of development. Until they open, oxygen can only enter by diffusion through the wall of the egg case. Once they have opened water from outside can reach the gills. Dogfish embryos have filamentous extensions of the gills, which protrude from the sides of the body, while they are in the egg case.

Table 4-3 shows that dogfish eggs increase in weight between being laid and hatching. However, the increase is due to uptake of water from the environment. The amount of organic matter they contain decreases. This is inevitable, since energy is required for development and the foodstuffs which are used cannot be replaced.

TABLE 4-3.   *The mean initial weight and organic content of eggs of various selachians, and the mean weight and organic content of the embryo at the time of hatching or of birth*

Data from J. Needham (1950) *Biochemistry and morphogenesis*. Cambridge University Press, London.

	Weight (g)		Organic content (g)	
	Initial	Final	Initial	Final
Scyliorhinus canicula	1.3	2.7	0.61	0.48
Squalus acanthias	23	41	8.6	5.2
Mustelus canis	5.5	189	2.8	32
Dasyatis violacea	2	118	0.9	16

The Spiny dogfish (*Squalus acanthias*) does not lay its eggs, but retains the embryos in its uterus for almost two years until they are ready to swim and feed. At first they are surrounded by a fluid resembling blood plasma but later by one much more like seawater: presumably seawater leaks in through the cloaca. Their respiration and excretion apparently depend on diffusion of oxygen and waste products through the uterine fluid. Table 4-3 shows that the eggs lose organic matter and take up water as they develop, like the eggs of *Scyliorhinus*.

The Smooth hound (*Mustelus canis*) also keeps its embryos in the uterus, but not for so long. They are born after 10–11 months, having grown much faster. The new-born young is not only very much heavier than the egg, but it contains about ten times as much organic matter (Table 4-3). Foodstuffs are taken up from the mother in the course of development. This is made possible by a placenta.

At first the embryo *Mustelus* depends on the foodstuffs in the yolk. A yolk sac with blood vessels in it grows round the yolk. Blood circulates through, carrying foodstuffs back to the embryo. As the yolk is used, part of the yolk sac becomes firmly attached to the wall of the uterus so that its blood vessels are very close to maternal blood vessels. Foodstuffs, waste products and gases can then diffuse between the blood of the embryo and that of its mother. The arrangement is just like that of the yolk-sac placenta of mammals (Fig. 12-2) and it apparently functions in the same way. The fluid in the uterus remains similar in composition to blood plasma throughout pregnancy.

The stingray *Dasyatis* also keeps its eggs in the uterus, and they too gain organic matter as they develop (Table 4-3). However, there is no placenta. The fluid in the uterus is a milky suspension containing 13% organic matter (largely fat). Apparently foodstuffs are secreted into the uterus by the mother, and absorbed by the embryos.

SENSE ORGANS

The lateral line neuromasts of lampreys and hagfishes all protrude from the surface of the body, but selachians and teleosts have neuromasts enclosed in sunken canals, as well as superficial ones. Fig. 4-19 shows how these are arranged. Neuromasts and openings to the external water alternate along the length of the canal. A difference in pressure two successive pores causes water to move along the canal between them, deflecting the cupula.

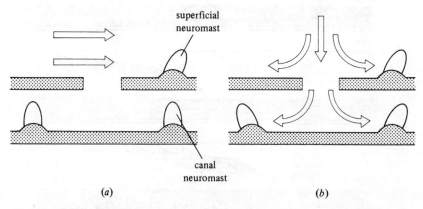

Fig. 4-19. A diagram showing superficial and canal neuromasts and how they are affected by water movements (*a*) parallel to and (*b*) at right angles to the body surface.

Enclosure in a canal affects the response of the neuromast to stimuli. Superficial neuromasts are very sensitive to water movements parallel to the surface of the body (Fig. 4-19*a*). Canal neuromasts are not, because the pressure differences between successive pores are then small. However, both types of neuromast are sensitive to local water movements at right angles to the body (Fig. 4-19*b*). Water striking the body at right angles is deflected along it, so the superficial neuromasts are stimulated. There is a region of high pressure where the water strikes the body, so water will tend to move along the canals. When a fish is swimming, or resting on the bottom in a current, water is constantly flowing longitudinally over it. The effects of this flow on the superficial neuromast must tend to mask the effects of gentler water movements produced by wriggling prey or approaching predators. Canal neuromasts can be expected to be less sensitive to it, and so better able to detect movements due to prey and predators, if these have components at right angles to the surface of the body. The main disadvantage of canal organs is that they must tend to be less sensitive than superficial ones to vibrations above a certain frequency, because of the inertia of the water in the canal.

Action potentials in nerves and muscles involve changes of electrical potential at their surfaces. Any aquatic animal which produces action potentials must therefore set up electric fields in the water around itself. These fields are, however, extremely weak. Fig. 4-20 illustrates a series of

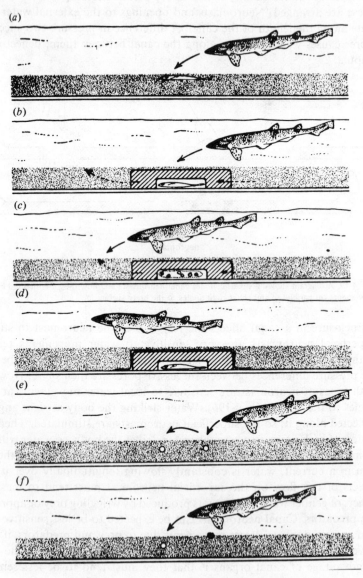

Fig. 4-20. Diagrams illustrating the experiments on the electric sense of dogfish, which are described in the text. From A. J. Kalmijn (1971). *J. exp. Biol.* **55**, 371–84.

experiments which demonstrated that dogfish (*Scyliorhinus*) can detect the electric fields generated by potential prey. The dogfish were kept in shallow water in a child's inflatable paddling pool, with sand on the bottom. Young plaice (*Pleuronectes*) or other objects were introduced into the pool while the dogfish were lying inactive on the bottom, as well fed dogfish do. When the object had been hidden under the sand the dogfish were stimulated to search for food by putting a little fish juice into the water. They were then observed.

Young plaice which were put in the tank buried themselves in the sand with only parts of the head exposed (Fig. 4-20*a*). A dogfish passing within 15 cm of a plaice would generally uncover and capture it: it would suck off the sand covering the plaice (blowing out the sand through its gill slits) and then seize the plaice and shake it until it tore into pieces small enough to swallow. It could of course have found the plaice by sight, smell or water movements. The next experiment (Fig. 4-20*b*) was designed to eliminate these possibilities. The plaice was put in a chamber made of agar jelly, of about the same electrical conductivity as seawater. The broken arrows in the diagram indicate tubes used to pass aerated seawater through the cavity, so that the plaice was not asphyxiated. The dogfish responded as before if they swam near the chamber and tried to uncover the plaice. They showed no interest if the chamber was empty. Fig. 4-20*c* shows an experiment in which the living plaice was replaced by pieces of dead fish. The dogfish did not uncover the chamber but tried to find food at the end of the outlet tube carrying water from it. They were presumably responding to the odour of the pieces of fish. Fig. 4-20*d* shows that there was no response to a live plaice in the chamber if it was covered by a very thin sheet of a plastic which is a good electrical insulator. More direct evidence that the plaice were detected by an electric sense was obtained by burying electrodes in the sand (Fig. 4-20*e*). The dogfish showed no interest in the electrodes if no current was flowing, but when the electrodes were used to produce an electric field simulating the field around a living plaice the dogfish responded as though to a real plaice and uncovered them. They responded to the buried electrodes in preference to a piece of dead fish on the surface of the sand (Fig. 4-20*f*).

Other experiments showed that the electric sense could be abolished, by cutting the nerves to the structures known as the ampullae of Lorenzini. These are jelly-filled tubes in the heads of selachians. They are often many centimetres long (and more than a metre in large rays) but are generally only a millimetre or two in diameter. One end opens at the surface of the head but the other is blind, and at this blind end is a group of sensory cells, served by about six axons from the seventh cranial nerve. Electrical recordings have been made from these axons. When conditions are constant,

action potentials occur at a steady rate. The rate can be modified by prod-
ding, by changes of temperature or salinity, and by electrical potentials.
It is particularly sensitive to electrical potential gradients along the length
of the ampullae. The ampullae seem well adapted for detecting such
gradients. Their walls have high electrical resistance but the jelly has a low
resistance: it contains ions in about the same concentrations as seawater
and so has much higher conductivity than the surrounding tissues. Consequently, when there is a potential difference between the ends of the ampulla
the electric current tends to be channelled along the ampulla. The current
through the sensory cells must be higher than if the jelly had the same
conductivity as the tissues. Different ampullae run in different directions,
and so are sensitive to potential gradients in different directions.

Another sense used by selachians in finding food is the sense of smell.
Its use was apparent in the experiments on the electric sense in which fish
juice was used to stimulate the dogfish to search for food, and in the situation illustrated in Fig. 4-20c. The manner in which it is used was investigated in experiments with sharks in large pens in shallow tidal water.
The odours used were glutamic acid and trimethylamine, both substances
which might be released by potential food. The odours were released into
the water outside the pen, so as to be carried through by the tidal current.
The response of the sharks was recorded by time-lapse photography.
Lemon sharks (*Negaprion*) simply swam upstream when the odour was
introduced, to the point where the current entering the pen was fastest.
This did not take them towards the point of release of the odour in cases
when this was in the slower part of the current. Apparently the odour
started the sharks searching for food, but they took their direction from
the current rather than from the odour itself. Nurse sharks (*Ginglymostoma*)
behaved differently. They swam along a sinuous path, to the point where
the odour entering the pen was strongest. They apparently used gradients
of odour intensity to locate the source of an odour.

The organs of smell are a pair of large sacs which occupy most of the
space in the snout, anterior to the eyes. The nostrils are on the underside
of the head. A single nostril on each side of the head of the embryo becomes
elongated and constricted as development proceeds, so that the adult has
effectively two openings on each side. One of these openings is generally
just outside the mouth, or even inside it, so that breathing probably tends
to make water flow in at one nostril and out at the other.

ACANTHODIANS

Class Acanthodii

The remaining sections of this chapter are concerned with the evolution of the selachians and of some other groups of jawed fish. The acanthodians are considered first because they include the earliest known fossil fish with jaws. They seem to have arisen in the Silurian and become extinct in the Permian. They resemble modern jawed fish and differ from Agnatha not only in having jaws, but also in the structure of the gill skeleton, in having separate pectoral and pelvic fins and in having three semicircular canals in each ear.

Most acanthodians were small, less than 20 cm long, but a few much bigger ones are known. They were covered by small, squarish scales which did not overlap and were very similar in internal structure to the scale of *Cheirolepis* (Fig. 3-17*g*).

The acanthodians had heterocercal tails (Fig. 4-21*a*). Each fin (except the caudal one) had a spine in its front edge, and primitive acanthodians had in addition a row of spines on each side of the body, between the pectoral and pelvic fins. Notice that these spines follow the line of the lateral fins of *Jamoytius* and the ventrolateral ridge of *Hemicyclaspis* (Fig. 3-18). The spines seem to be modified scales, like the one in front of the dorsal fin of *Hemicyclaspis*. In the more primitive acanthodians (Fig. 4-21*a*) they are broad and superficial, simply sheathing the front edge of the fin. In more advanced ones they are slender and their bases were embedded in the muscles of the trunk.

The jaws and the main elements of the gill skeleton were bone, but are otherwise very like those of sharks and dogfishes. However, the palato-quadrates do not meet at the front of the mouth and they articulate with the cranium (also consisting of bone) in such a way that it seems most un-likely that they could have been protrusible. The hyomandibular articulated with the cranium as in sharks and fitted into a groove on the median surface of the palatoquadrate. In advanced acanthodians a gill cover attached to the rear edge of the hyoid arch covered all the gill slits, like the operculum of teleosts. In more primitive ones such as *Climatius* (Fig. 4-21*a*) the hyoidean gill cover is smaller and did not cover the more posterior gill slits: each had its own cover more or less as in sharks.

Acanthodians had rather large eyes set well forward in the head, leaving little room in the snout anterior to them. The nasal capsules must have been small as in teleosts, not large as in sharks. None of the acanthodians seem to have had dorsoventrally flattened bodies, nor do they show other signs of bottom-living habits. It seems likely that they spent at least a good deal of their time swimming in mid-water. A few had quite large pointed teeth firmly fixed to their jaws, and probably ate other fish. Many others had no teeth and presumably ate smaller food. Some had long gill rakers which would have prevented small items from escaping through the gills. The remains of small invertebrates (probably ostracods) have been found

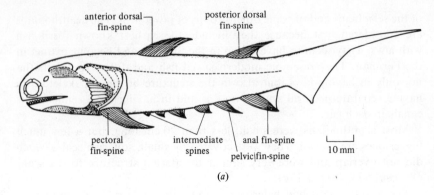

anterior dorsal
fin-spine

posterior dorsal
fin-spine

pectoral
fin-spine

intermediate
spines

anal fin-spine

pelvic fin-spine

10 mm

(a)

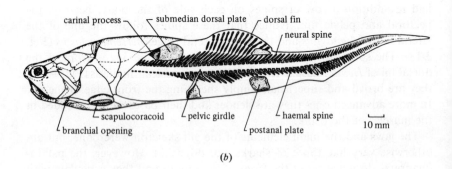

carinal process

submedian dorsal plate

dorsal fin

neural spine

scapulocoracoid

pelvic girdle

haemal spine

branchial opening

postanal plate

10 mm

(b)

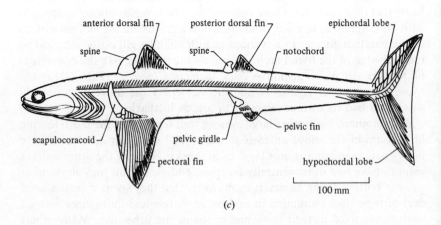

anterior dorsal fin

posterior dorsal fin

epichordal lobe

spine

spine

notochord

scapulocoracoid

pelvic fin

pelvic girdle

hypochordal lobe

pectoral fin

100 mm

(c)

Fig. 4-21. (*a*) *Climatius*, an acanthodian; (*b*) *Coccosteus*, a placoderm; (*c*) *Cladoselache*, a Devonian shark. From J. A. Moy-Thomas (1971) *Palaeozoic fishes*, 2nd edit., revised by R. S. Miles. Chapman & Hall, London.

inside fossils of two species: they are presumably stomach contents, preserved with the fish. These acanthodians probably fed on plankton much
as anchovies (*Engraulis*, which also have long gill rakers) do.

The acanthodians have been described in this chapter because they
include the earliest known fish with jaws, but they are probably not at all
closely related to sharks. There is a good deal of evidence suggesting that
they are more closely related to the Osteichthyes which are described in the
next few chapters.

PLACODERMS

Class Placodermi

The placoderms are a group of fish almost entirely restricted to the Devonian period, but a few have been found in late Silurian and early Carboniferous rocks. In the Devonian they seem to have been very plentiful. They
vary widely in size and structure but most seem to have been to some extent
dorsoventrally flattened, which suggests bottom-living habits. Most had
most of the head covered by dermal bones, and the shoulder region covered
by a separate ring of dermal bones (Fig. 4-21*b*). The head shield and the
trunk shield seem each to have been more or less rigid, but they were hinged
together so that the head could rock up and down on the trunk. They articulated with each other at two points, on either side of the body and in
line with the joint between the cranium and the first vertebra.

The palatoquadrates are known to have been fixed rigidly to the inside
of the head shield in one group of placoderms, and to the cranium in
another. It was very probably true of placoderms in general that they could
not be moved relative to the rest of the skull. The gills apparently discharged through a single opening on each side of the body, between the
head shield and the trunk shield. The only way in which the volumes of
the mouth and gill cavities could be changed substantially was probably
by nodding the head: tilting the head up about its hinge with the trunk
shield would enlarge the ventrally situated gill cavities. Placoderms probably nodded their heads to pump water over their gills, and to suck in food.
Study of *Coccosteus* (Fig. 4-21*b*) indicates that the range of movement at
the hinge was probably at least 10°. It can be estimated from the dimensions
of the fish that movement through 10° would have altered the volume of
the head by about 4% of total body volume. Most teleosts change the
volumes of their heads, in heavy breathing and in feeding, by 5–8% of the
body volume.

Placoderms have toothplates (rather than typical teeth) on the palatoquadrate, on the underside of the cranium and on the lower jaw. In some

these plates had sharp cutting edges. In others they were flatter and were presumably used for crushing molluscs and crustaceans.

Most placoderms had heterocercal tails and the fins in general were very like those of early sharks, in internal structure as well as in external appearance. It is widely believed that the sharks and placoderms are quite closely related.

RABBIT-FISHES

Class Holocephali

The rabbit-fishes are a small group of marine fishes, 60 cm to 2 m long, *Chimaera* is the best known example. They live near the bottom, mainly at depths of 100 to 1500 m. They have long slender tails and big pectoral fins, and use both in swimming. At least one species has enough oil in its body to make its density approximately the same as seawater. They feed mainly on bottom-living invertebrates, taking small bites from them by means of the ridges of very hard, calcified tissue which they have on their jaws. Prey such as ophiuroids and crabs are found in the gut in pieces not more than 2 cm long. The palatoquadrate cartilages do not articulate with the cranium as in sharks, but are fused to it as a single block of cartilage.

There is a Devonian fossil called *Ctenurella* which shows a remarkable mixture of placoderm and rabbit-fish features. It has head and trunk shields which, though small, consist of bones arranged as in many placoderms. There is (indirect) evidence that the palatoquadrate cartilages were fused to the cranium, and there are other similarities to rabbit-fish. *Ctenurella* is regarded as a placoderm and it may not be particularly closely related to the rabbit-fishes, but it shows how rabbit-fishes may have evolved from placoderms.

SHARKS AND RAYS

Class Selachii

The fossil record of selachians is not as good as might be wished: fossil teeth and fin spines are plentiful enough but the skeleton is seldom preserved. Ordinary cartilage normally disappears without trace in fossils, and even cartilage which has been reinforced by calcification (as in many shark skeletons) seems to survive much less well than bone. However, enough skeletons have been found to give a reasonably clear indication of the evolution of the class.

The earliest known selachians are from the Devonian period and are not too different from modern sharks. *Cladoselache* (Fig. 4-21*c*) is an example, which may not be typical in all respects. Its general appearance must have been shark-like but its jaws are longer than in most modern sharks and the mouth reached the tip of the snout. The palatoquadrate articulated with the cranium behind the eye and was probably also attached to it below the eye. It seems clear that the palatoquadrate could not have been protruded, but it could probably have been swung laterally to widen the mouth cavity. The pectoral and pelvic fins were triangular, apparently attached to the trunk all along the base. The pectoral fins did not have narrow bases as in modern sharks, so their range of movement must have been limited. There were spines in front of the dorsal fins. All the fins had pterygiophores reaching close to the fin margins, instead of having most of the fin stiffened only by ceratotrichia.

Though these are points of difference between early selachians and most modern ones, there are some modern selachians that retain primitive features. For instance, the Frilled shark (*Chlamydoselachus*) has jaws very like those of *Cladoselache*. The Spiny dogfish (*Squalus*) has dorsal fin spines. The pectoral fins of rays (Raiiformes) have pterygiophores reaching close to their margins.

The rays are, of course, very strikingly different from sharks. The tail is slender but the pectoral fins are enormous and often meet in front of the head. When we eat skate we discard the reduced trunk muscles and eat only the muscles of these fins (the 'wings'). The muscles, like the pterygiophores, extend almost to the margins of the fins. Swimming is achieved by beating the pectoral fins up and down in such a way that waves travel posteriorly along them, increasing in amplitude as they go. This undulation propels the fish, in essentially the same way as the horizontal undulation of the body in slender fish.

The rays must have evolved from more shark-like selachians. The guitar-fishes (*Rhinobatus*) show how this could have happened. They are intermediate in shape between dogfish and rays, with rather flattened bodies and rather large pectoral fins. They swim at low speeds by tail movements alone, but undulate their pectoral fins as well when they swim fast.

A flattened body is likely to be advantageous to a fish which spends a lot of time resting on the bottom, particularly in shallow coastal water where tidal currents and bright sunlight are encountered. Flattening makes the fish less easily displaced by currents, and helps to eliminate the shadows that can make round-bodied fish, resting on a sunlit bottom, very conspicuous. Where the edges of fins rest on the bottom, fringing shadows may be completely eliminated. The early stages of the evolution of rays may have depended on these advantages of flattening and large fins. Rays often bury themselves almost completely in sand leaving little but the eyes and spiracles exposed: this makes them very inconspicuous.

The spiracles of sharks are more dorsally placed than the gill slits. In the evolution of rays, the pectoral fins have extended forwards between the spiracles and the gill slits, so that the spiracles open on the dorsal surface and the gill slits on the ventral. When a ray is resting quietly on the bottom of an aquarium it breathes in only through the spiracles, and not through

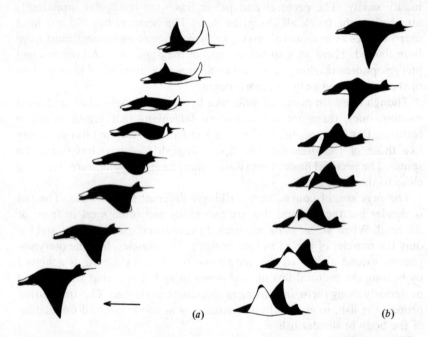

Fig. 4-22. Tracings from an underwater film of a Manta ray (*Mobula diabolis*) swimming. (*a*) shows the downstroke and (*b*) the upstroke. The arrow shows the direction of swimming. The movements shown (one complete beat) occupied 3 s. From W. Klausewitz (1964). *Zool. Anz.* **173**, 111–20.

the mouth. When rays are buried they probably keep their mouths shut: they could hardly breathe in through their mouths without taking in sand or mud.

The eagle rays (Myliobatidae) and the manta rays (Mobulidae) have tails which are even more reduced than those of other rays, and pectoral fins which are more like wings (Fig. 4-22). These fins have a higher aspect ratio than in ordinary rays, and they do not meet in front of the head. They are flapped up and down like the wings of a bird as the ray swims. Comparison of Fig. 4-22 with Fig. 11-8 shows that the similarity to birds is very close indeed. If these rays are denser than water (as they seem to be) the wing movements must produce an upward force as well as a forward one, just as in birds.

FURTHER READING

General

Budker, P. (1971). *The life of sharks*. Weidenfeld & Nicolson, London.
Gilbert, P. W., R. F. Mathewson & D. P. Rall (eds.) (1967). *Sharks, skates, and rays*. John Hopkins Press, Baltimore.
Lineaweaver, T. H. & R. H. Backus (1970). *The natural history of sharks*, André Deutsch, London.
See also the list of general works on fishes at the end of Chapter 3.

Swimming

Alexander, R. McN. (1969). The orientation of muscle fibres in the myomeres of fishes. *J. mar. biol. Ass. U.K.* **49**, 263–90.
Bone, Q. (1966). On the function of the two types of myotomal muscle fibre in elasmobranch fish. *J. mar. biol. Ass. U.K.* **46**, 321–49.
Lighthill, M. J. (1969). Hydromechanics of aquatic animal propulsion. *Ann. Rev. Fluid Mech.* **1**, 413–46.

Buoyancy and fins

Alexander, R. McN. (1965). The lift produced by the heterocercal tails of Selachii. *J. exp. Biol.* **43**, 131–8.
Alexander, R. McN. (1972). The energetics of vertical migration by fishes. *Symp. Soc. exp. Biol.* **26**, 273–94.
Bone, Q. & B. L. Roberts (1969). The density of elasmobranchs. *J. mar. Biol. Ass. U.K.* **49**, 913–38.
Corner, E. D. S., E. J. Denton & G. R. Forster (1969). On the buoyancy of some deep-sea sharks. *Proc. R. Soc.* B, **171**, 415–29.
Simons, J. R. (1970). The direction of the thrust produced by the heterocercal tails of two dissimilar elasmobranchs: the Port Jackson shark, *Heterodontus portusjacksoni* (Meyer) and the Piked dogfish, *Squalus megalops* (Macleay). *J. exp. Biol.* **52**, 95–107.

Gills and jaws

Hughes, G. M. & C. M. Ballintijn (1965). The muscular basis of the respiratory pumps in the dogfish (*Scyliorhinus canicula*). *J. exp. Biol.* **43**, 363–383.
Moss, S. A. (1972). The feeding mechanism of sharks of the family Carcharhinidae. *J. Zool., Lond.* **167**, 423–36.

Blood and blood vessels

Satchell, G. H. & M. P. Jones (1967). The function of the conus arteriosus in the Port Jackson shark, *Heterodontus portusjacksoni J. exp. Biol.* **46**, 373–82.
Sudak, F. N. (1965). Some factors contributing to the development of sub-atmospheric pressure in the heart chambers and pericardial cavity of *Mustelus canis* (Mitchill). *Comp. Biochem. Physiol.* **15**, 199–215.

Reproduction

Needham, J. (1950). *Biochemistry and morphogenesis*. Cambridge University Press, London.

Price, K. S. & F. C. Daiber (1967). Osmotic environments during fetal development of dogfish, *Mustelus canis* (Mitchill) and *Squalus acanthias* Linnaeus, and some comparisons with skates and rays. *Physiol. Zool.* **40**, 248–60.

Sense organs

Kalmijn, A. J. (1971). The electric sense of sharks and rays. *J. exp. Biol.* **55**, 371–384.

Acanthodians

Miles, R. S. (1964). A reinterpretation of the visceral skeleton of *Acanthodes*. *Nature, Lond.* **204**, 457–9.

Placoderms

Miles, R. S. (1969). Features of placoderm diversification and the evolution of the arthrodire feeding mechanism. *Trans. R. Soc. Edinb.* **68**, 123–70.

5

Teleosts and their relatives

This is the first of three chapters about the fish of the class Osteichthyes, commonly referred to as the bony fishes. Most of these fish have skeletons which consist mainly of bone, and they are called bony fishes to distinguish them from the selachians which have cartilaginous skeletons. However, even selachians have bone in the bases of their scales and teeth (Fig. 3-16d), and many extinct fish which are not included in the Osteichthyes had a great deal of bone. The thick bony scales and carapaces of many ostracoderms and placoderms, and the bony skeletons of acanthodians, have already been described.

It will be helpful to distinguish between cartilage bone and dermal bone. Structures described as consisting of cartilage bone appear first in embryos as cartilage, which is later replaced by bone. They are derived ultimately from sclerotome or neural crest (p. 107), and are homologous with the cartilages of selachian skeletons. Dermal bones are formed firstly in the dermis of the skin or the lining of the mouth, without prior formation of cartilage. They are formed in the same way as scales and like them are sometimes coated with outer layers of dentine and enameloid substance (p. 77). There are no structures in selachians which are homologous with particular dermal bones, though the dermal bones and scales of bony fishes may be regarded as being in a general sort of way homologous with the placoid scales of selachians.

The lungs and swimbladders of Osteichthyes are much more distinctive features than their bone. Most modern Osteichthyes have a lung or a swimbladder but no other fish, living or extinct, are known to do so.

This chapter is about the teleosts and the more primitive Osteichthyes from which they evolved. The special features of some of the advanced teleosts are described in Chapter 6. Finally the lungfishes and crossopterygians, Osteichthyes which are not at all closely related to teleosts, are described in Chapter 7.

FEEDING AND RESPIRATION

An understanding of the processes of feeding and respiration requires knowledge of the structure of the skull. The skulls of bony fish consist partly of cartilage and cartilage bone, and partly of dermal bone. The total number of bones is generally very large (about 135 in the typical case of the herring, *Clupea*) but it is happily unnecessary in this book to describe more than a few of them individually.

Fig. 5-1 shows how cartilage bone and dermal bone combine in the structure of the skull. The diagram is based on teleosts but would be little different if it had been drawn to represent some other group of bony fish. The cranium consists largely of cartilage bone but is roofed by dermal

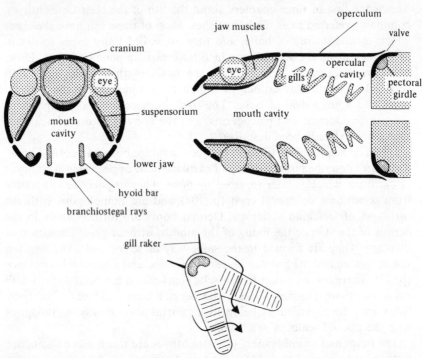

Fig. 5-1. Diagrammatic transverse and horizontal sections of the head of a typical teleost. Dermal bone is shown black and cartilage bone is hatched. An enlarged horizontal section through a gill is shown at the bottom.

bone and has a ventral covering of dermal bone where it forms the roof of the mouth cavity. The side walls of the mouth cavity are formed by the suspensoria, which each include cartilage bones homologous with the hyomandibular cartilage and parts of the palatoquadrate cartilage of selachians, and also dermal bones formed in the lining of the mouth. The lower jaw consists of a slender core of cartilage and cartilage bone homologous with Meckel's cartilage, invested by dermal bone. The hyoid bars consist of cartilage bones, homologous with the hyoid bars of sela-chians. The gills are supported by cartilage bones homologous with the corresponding cartilages in selachians. The opercula (gill covers), the branchiostegal rays ventral to them and the bony covering of the cheek consist of dermal bones. At an early stage in the development of the embryo the skull consists exclusively of cartilage and includes cartilages corresponding to all the main parts of the selachian skull. Later, bone develops in these cartilages, and dermal bones are also formed.

Fig. 5-2*a* is a diagram of the skull of a relatively primitive teleost such as a trout (*Salmo*). Fig. 5-2*b* shows the same skull with some of the dermal bones removed. Note that the suspensorium has two articulations with

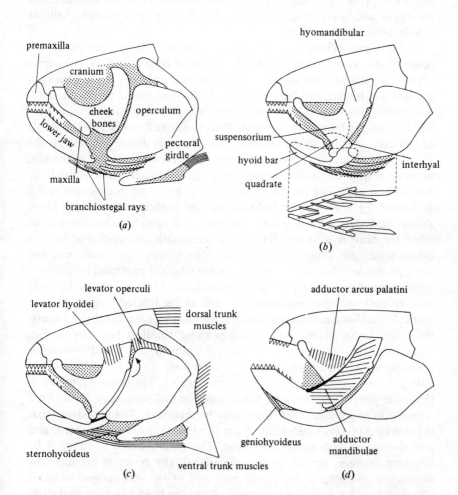

Fig. 5-2. Diagrams of the skull of a typical teleost such as a trout (*Salmo*). (*a*) shows the intact skull and (*b*) shows it after removal of the maxilla and cheek bones. A ventral view of the hyoid bars and branchiostegal rays is shown below (*b*). (*c*) and (*d*) show some of the principal muscles of the head.

the cranium, one anterior and one posterior to the eye. The latter articulation is formed by the hyomandibular. These articulations act as hinges, allowing the suspensorium to be swung laterally (widening the mouth cavity) or medially (narrowing it). The hyomandibular has a ball and socket joint with the operculum. The quadrate bone (one of the derivatives of the palatoquadrate cartilage) has a hinge joint with the lower jaw. The posterior end of the hyoid bar rests against the inner face of the operculum to which it is attached by ligaments, but is also connected through a small bone (the interhyal) to the hyomandibular. Thus the derivatives of the upper and lower parts of the hyoid arch form an unbroken chain, as in selachians.

If the suspensorium is swung laterally, widening the mouth cavity, the operculum is also carried laterally. Its posterior edge as well as its anterior edge move laterally because a stop on the ball and socket joint between the hyomandibular and the operculum limits movement there: the operculum can hinge outward about a vertical axis through the joint, but inward hinging is limited by the stop. Thus the mouth cavity and the whole of the opercular cavity are widened, and water is drawn in through the mouth. No water can enter under the posterior edge of the operculum because a strip of flexible tissue along this edge acts as a valve (Fig. 5-1*b*). None can enter the opercular cavity from below because the cavity is floored by a membrane stiffened by the branchiostegal rays. These slender bones are jointed to the hyoid bars, and open like the ribs of a fan when the head is widened. Narrowing the mouth and opercular cavities drives water out under the posterior edge of the operculum, but not through the mouth (unless it is wide open) which is protected by valves.

Since the joint between the hyomandibular and operculum is a ball and socket, it allows other movements as well as the limited hinging about a vertical axis that has been described. It also allows rotation about a transverse axis through the joint. Anticlockwise rotation (as seen in Fig. 5-2*c*) pulls the mouth open, because a ligament connects the anterior lower corner of the operculum to the lower jaw.

Fig. 5-3 is drawn from a film of a teleost feeding. The snout is tilted up as the lower jaw is lowered. The fish is more or less stationary as the food moves into its mouth, so the food must be sucked in. The film shows that this sucking involves widening of the head (as has been described), and also downward bulging of the throat. The mouth cavity is enlarged by lowering its floor, as well as by widening it. The bulge is produced by downward swinging of the hyoid bars and of the stiff tongue which is attached to them at the anterior end. After the food has been sucked in the mouth is closed and the water which has necessarily been drawn in with the food is driven out through the opercular openings. The movements of the maxilla which occur as the mouth opens and closes are considered on p. 168.

The parts played by the main muscles involved in producing these movements have been elucidated in electromyographic experiments by Dr J. W. M. Osse. He used the perch (*Perca*), which is an advanced teleost. It protrudes its premaxillae as it opens its mouth, as will be described in Chapter 6, but in other respects its feeding movements resemble those

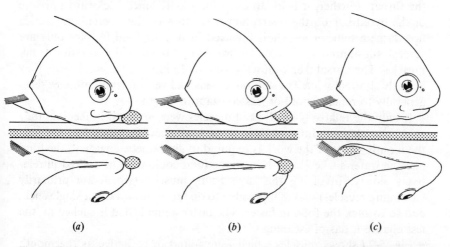

(a)   (b)   (c)

Fig. 5-3. Tracings of three frames of a film of an African knife-fish (*Papyrocranus afer*) feeding. The food is being taken from a glass platform with a mirror set at 45° to the horizontal below it, so the fish is seen in ventral as well as in lateral view. Successive frames are about 33 ms apart.

of primitive teleosts. Dr Osse was able to insert electrodes into as many as eight muscles in each fish. He inserted them under anaesthetic but when the fish recovered it was free to swim about in a small aquarium, trailing the electrode leads. It usually resumed feeding a few hours after recovery. It was fed on small fish or earthworms, and cinematograph film of feeding movements was taken simultaneously with the electromyographic records.

Fig. 5-2c shows muscles that Osse found were active when the mouth was opening and the mouth and opercular cavities were expanding. The levator operculi rocks the operculum, as indicated by the arrow, and so opens the mouth. The levator hyoidei helps to swing the suspensorium laterally, but is relatively unimportant: the other muscles that co-operate with it probably do most of the work. The ventral trunk muscles and the sternohyoideus prevent the tongue from moving forward (some of the trunk muscles hold the pectoral girdle firm against the pull of the sterno-hyoideus while others run forward with the sternohyoideus to insert directly on the tongue skeleton). The dorsal trunk muscles pull on the back of the cranium, bending the anterior joints of the vertebral column and raising the snout. This moves the posterior ends of the hyoid bars forward but their anterior ends, attached to the tongue, are fixed. The posterior ends can only move forward while the anterior ends stay fixed, if they also move laterally. The combined action of the dorsal and ventral trunk muscles and the sternohyoideus makes the hyoid bars splay apart, widening the head. It also tends to pull the tongue down, producing the bulge in the throat. Further, it helps to open the mouth since backward pressure of the hyoid bar on the operculum tends to rock the operculum. Notice how large a bulk of muscle is involved in sucking food in. Not only are several substantial head muscles involved, but also the anterior trunk muscles. The dorsal trunk muscles seem to do most of the work since they actually shorten, while the sternohyoideus and ventral trunk muscles seem generally to remain more or less constant in length.

If only the relatively small head muscles were involved in feeding, they could do relatively little work in a single contraction, sucking food into the mouth. Most of the work is required to give kinetic energy to the food and the water sucked in with it, so the food could only be sucked in relatively slowly. Involving the large trunk muscles (which are primarily swimming muscles) makes it possible to do far more work in a single suck, and so to suck the food in faster. The faster animal food is sucked in, the less chance it has of escaping.

Fig. 5-2*d* shows muscles which were found to be active as the mouth closed and the head cavities contracted. The adductor mandibulae closes the mouth. The geniohyoideus moves the hyoid bars back to their original position, narrowing the head. The adductor arcus palatini and another more posterior muscle run from the cranium to the median faces of the suspensorium and operculum, and help to pull them back to their original positions.

The breathing movements of teleosts are generally similar to their feeding movements, though much smaller in amplitude. Breathing movements have been investigated electromyographically in several species of teleost, including perch and the much more primitive trout. The trunk muscles and sternohyoideus seem generally to play no part in normal, gentle respiration, but the other muscles are involved in the same sequence as in feeding.

The pressures involved in respiration have been investigated in the same way as for selachians (p. 107), by inserting tubes attached to pressure transducers into the mouths and opercular cavities of teleosts held in suitable apparatus. In other experiments, teleosts have been allowed to swim freely in aquaria while connected to pressure-recording equipment by long flexible tubes. These experiments have shown that the pressures involved in gentle respiration are very similar to those recorded from selachians. If the fish is put in water of low oxygen content, or if a free-swimming fish is made to swim against a current in an annular tank, water has to be pumped faster through the gills and the pressures are larger, up to about ± 5 cmH$_2$O.

The methods used to record the pressures involved in respiration are not suitable for recording the pressures involved in feeding. Though teleosts continue to breathe when held in the apparatus used for the experiments with stationary fish, they cannot be expected to feed. Though the long flexible tubes used in the experiments with free-swimming fish transmit reasonably satisfactorily the relatively slow changes of pressure involved in respiration, the much faster changes involved in feeding would be seriously distorted. The method illustrated in Fig. 5-4a was therefore devised to record pressure changes from the mouth cavity during feeding. The tube from the pressure transducer is relatively short, and its walls are rigid. A ring of body wall cut from an earthworm is slipped over its end. Most fish soon learned to take this food from the tube. They could not do so without putting their mouths around the end of the tube. Records were therefore obtained of the pressure changes in the fishes' mouths, as they took the food. Films taken at the same time as the pressure records showed that the feeding movements were generally indistinguishable from those used in taking a piece of loose food from a horizontal surface. Occasionally, however, the fish did not suck the food off but grabbed it in its teeth and pulled. These occasions were identified from the films, and the records were discarded.

Records were obtained from ten species of teleost, and some consistent differences between species were found. However, the record of an orfe, *Idus*, shown as Fig. 5-4b, can be taken as typical. The pressure is initially steady before the fish put its mouth around the tube. There is a strong reduction of pressure as the food is sucked into the mouth, followed by a much smaller increase of pressure as the water, sucked in with the food, is driven out through the opercular openings. The pressure of -100 cmH$_2$O recorded as the food was sucked in represents much stronger suction than ever occurs in respiration. Similar pressures were often recorded even if the ring of worm fitted very loosely on the tube, but the orfe were apparently unable to suck much harder to dislodge tight-fitting pieces of worm. The pressure of about $+5$ cmH$_2$O recorded as the water was driven out, is within the range of pressures recorded during vigorous respiration. (No records seem to have been made of respiratory pressures of orfe, so the comparison is with records from other species.)

The film corresponding to the pressure record (Fig. 5-4b) shows that the mouth was wide open while the fish was sucking hard. The mouth

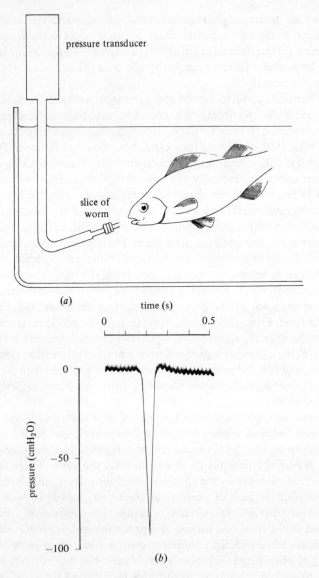

Fig. 5-4. (*a*) Apparatus used to record the pressure changes which occur during feeding in the mouth cavities of teleosts, and (*b*) a typical record made by an orfe (*Idus idus*).

cavity was not sealed off from the surrounding water, and the difference in pressure was drawing water very rapidly into the mouth.

When food has been caught, it has still to be swallowed. This is done with the help of the pharyngeal toothplates, which in typical teleosts, are plates of bone covered by teeth which are formed in the lining of the mouth and are attached to the underside of the cranium or to the gill skeleton. The teeth are generally small, sharp and backward-pointing. The lower toothplates can be moved forward and back by movements of the gill skeleton, so as to drag the food down the throat. Orfe and other cyprinid fish have peculiar pharyngeal teeth which are described on p. 189.

Large prey are swallowed whole, except by a very few species. Predatory teleosts can swallow other fish up to (generally) about 40% of their own length. Fish are swallowed head first. The inertia of the prey may be exploited in the early stages: the predator releases its grip and makes a sudden dart forward, getting a grip further along the prey. Later on the jaws are only used to hold the prey each time the lower pharyngeal teeth are released and moved forward; the prey is pulled in by the lower pharyngeal teeth. To illustrate the process, albino catfishes were fed to leaf-fishes (*Monocirrhus*) and recovered as soon as they had been swallowed. Toothmarks in their white, scaleless skin showed clearly where the jaw teeth and pharyngeal teeth had gripped.

In selachians and ammocoetes the gill slits are separated by the flaps that cover the parabranchial cavities (Fig. 3-2*a*). In teleosts the function of these flaps is taken over by the operculum. The gill filaments have the shape shown in Fig. 5-1, and water takes the path indicated between the secondary lamellae. The lamellae are very similar in structure to those of selachians and lampreys and the blood flows through them in the direction opposite to the flow of water. The gill rakers project from the gill arches like the teeth of a comb and prevent food that has been taken in at the mouth from escaping again through the gills.

The total area of the gills has been estimated for at least fifty species. For any one species, the area is generally about proportional to (body weight)$^{0.85}$, so comparisons between species are best made between specimens of similar weight. Striking differences between species are found, and can be related to differences in activity. The angler fish, *Lophius*, spends most of its time lying in a hollow on the bottom of the sea, waiting for prey. When small fish approach it attracts their attention by twitching the lure on its dorsal fin, and if they come near enough they are seized. It rarely chases prey (at least in aquaria) so it seems to lead a very inactive life. A 1.6 kg angler fish was found to have gills of total area 0.22 m². In contrast, the Skipjack tuna, *Katsuwonus*, is an active fish which swims perpetually, day and night. Specimens 44 cm long swim at about 66 cm s^{-1} for most of the time, and faster when feeding. A specimen of this length weighed 1.7 kg and was found to have a total gill area of 3.1 m², fourteen times as much as the angler fish. These are extreme examples, but they show that the range of gill areas for teleosts of given size is very wide.

It is easy to see that a fish which sustains a high level of activity for long periods must depend largely on aerobic metabolism and needs a large gill area. A generally sluggish fish may undertake short bursts of violent activity, using anaerobic metabolism in its white muscles, but if these bursts are infrequent a small gill area may suffice to repay the oxygen debts they produce as well as providing for resting metabolism. It is probably a positive advantage to the sluggish fish to have a small gill area because it is at the gills that most diffusion of water and ions occurs between the blood and the water in which the fish is swimming (see p. 61). The smaller the gill area, the less power has to be expended in maintaining the osmotic concentration of the blood. There is some evidence that the amounts of power expended may be quite substantial. The oxygen consumption of the teleost *Tilapia* has been measured swimming in water of various salinities in the water tunnel illustrated in Fig. 4-5. Oxygen consumption at any given swimming speed was lowest in water of the same osmotic concentration as the blood. It was about 25% higher in fresh water and about 40% higher in a solution almost as concentrated as seawater. Similar results have been obtained in experiments with trout (*Salmo*). (*Tilapia* and *Salmo* are unusual among teleosts in tolerating a wide range of salinities.) However, experiments on gills removed from eels indicate that the cost of osmotic regulation may be only about 2% of the resting metabolic rate.

BUOYANCY

Most teleosts have a sac of gas in the body cavity, which is known as the swimbladder. It necessarily reduces their density. Teleosts with no swimbladder, or with the swimbladder deflated, generally have specific gravities between 1.06 and 1.09. This is about the same range as for typical selachians which lack squalene; the bone of teleost skeletons is much denser than the cartilage of selachians (specific gravities about 2.0 and 1.1 respectively), but their high urea content must make selachian body fluids denser than the corresponding fluids in teleosts. Teleosts which have swimbladders generally have specific gravities very close indeed to that of fresh water (1.000) or seawater (about 1.026), whichever they live in.

A 1 kg teleost with no swimbladder, of specific gravity 1.080, would have a volume of $1000/1.080 = 926$ cm^3. A swimbladder containing 74 cm^3 gas would increase its volume to 1000 cm^3 without appreciably increasing its mass, and so give it a specific gravity of 1.000. Similarly a swimbladder containing 49 cm^3 gas would give it a specific gravity of 1.026.

Fig. 5-5 shows apparatus used to determine the densities of fish with swimbladders and the volumes of their swimbladders. An anaesthetized or dead fish is put into a flask that can be completely filled with water, excluding all bubbles. This water ends at a meniscus in the horizontal capillary tube. Pressure applied through the capillary compresses the gas in the swimbladder (since the body wall of the fish is flexible) but has only a very small effect on the volume of the tissues of the fish and on the water. It makes the meniscus move along the capillary and the distance it moves can be used (after correction for the compressibility of the water and the

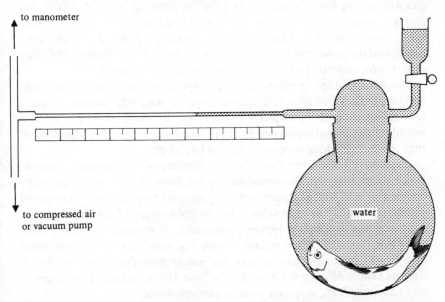

Fig. 5-5. Apparatus used to determine the densities of fish with swimbladders and the volumes of their swimbladders.

elasticity of the apparatus) to calculate the change in swimbladder volume. From this and Boyle's Law, the initial volume of the gas can be calculated.

The fish may initially float or sink in the apparatus. If it floats it can be made to sink by increasing the pressure. If it sinks it can be made to float by decreasing the pressure. The pressure is adjusted until the fish just sinks or just floats: that is, until its density is as nearly as possible the same as that of the water. The position of the meniscus shows how much the volume of the fish has had to be changed to bring this about. Hence the initial density of the fish can be calculated.

Such experiments have shown that many teleosts have densities within 0.5% of the density of the water they live in. The volumes of their swimbladders are generally around 70 cm³ (kg body weight)$^{-1}$ for freshwater fish and 50 cm³ kg^{-1} for marine ones.

Some sharks depend on hydrofoils to prevent them from sinking as they swim, but others contain low-density materials such as squalene which give them about the same density as seawater. It was calculated on p. 104 that the latter arrangement might save energy at least for slow-swimming species. A swimbladder should be an even more economical means of avoiding sinking, except in some circumstances that will be considered later in this chapter. Gas is so much less dense than squalene that 1 cm^3 gas will give as much buoyancy as 6 cm^3 squalene. It was calculated that enough squalene to give it the same density as seawater would increase the frontal area of a 1 kg dogfish by about 10^{-3} m^2. A swimbladder which had the same effect would increase the frontal area far less, and since drag is proportional to frontal area energy would be saved.

The swimbladder develops in the embryo as an outgrowth of the gut. Adults of the more primitive teleosts retain a duct (the pneumatic duct) connecting it to the oesophagus. Many of the more advanced teleosts including the Acanthopterygii (p. 194) lose the duct as they develop, so that the swimbladder becomes a closed bag of gas.

Gases have been taken from the swimbladders of fish caught at various depths, and analysed. Swimbladders of fish from shallow water generally contain around 80% nitrogen and 20% oxygen; they contain a mixture of gases very similar to atmospheric air. Swimbladders of fish from greater depths generally contain higher proportions of oxygen. However, trout and their relatives (Salmonoidei) commonly have very high proportions of nitrogen in their swimbladders, no matter how deep they are living. Ciscoes (*Leucichthys* spp.) from depths over 150 m in Lake Huron generally have over 95% nitrogen in their swimbladders.

Most swimbladders have extensible, flexible walls so that the pressure of gas inside matches that of the water around the fish. Some have less extensible walls and are kept inflated at a pressure a little above that of the water, but the pressure difference is small and can be ignored here.

The pressure in water is 1 atm at the surface and increases by 1 atm for every 10 m of depth, so the pressure at D m is $(1 + 0.1D)$ atm. This is also the pressure in the swimbladder of a fish at this depth. If the gas in the swimbladder contains x% oxygen, the partial pressure of oxygen in the swimbladder must be $(1 + 0.1D)x/100$ atm. It can be worked out in this way, from the analyses of gases from the swimbladders of fish caught at different depths, that fish living at D m generally have about $(0.2 + 0.1D)$ atm oxygen and 0.8 atm nitrogen in the swimbladders. This rule does not hold below depths of about 400 m, where higher partial pressures of nitrogen are usual. It does not, of course, apply to the salmonoid fish which usually have such high percentages of nitrogen, at all depths.

If the water were in equilibrium with the atmosphere, the partial pressures of dissolved oxygen and nitrogen in it would be about 0.2 atm and

0.8 atm, at all depths. About 0.8 atm nitrogen is indeed found at all depths, but the partial pressure of oxygen is always a good deal less than 0.2 atm at depths more than about 200 m. The partial pressures of gases in arterial blood are generally close to their values in the water, because of the exchange of gases which occurs in the gills. Hence the partial pressures of nitrogen in blood and swimbladder are generally both about 0.8 atm, but those of oxygen are generally 0.2 atm or less in the blood and (0.2 + 0.1 D) atm in the swimbladder. Oxygen must tend to diffuse from the swimbladder to the blood at a rate roughly proportional to the depth, since the difference in partial pressure is about 0.1 D atm. If the volume of the swimbladder is to be maintained, this loss must be made good. Teleosts with pneumatic ducts which live near the surface may do this by taking an occasional gulp of air at the surface. Others must use the process of secretion which will be described shortly.

Secretion of gas from a low partial pressure in the blood to a high partial pressure in the swimbladder is essentially a process of compression, and requires energy. If diffusion losses from the swimbladder can be kept small, energy will be saved. Most swimbladders have sparse blood supplies, except in the specialized regions of secretion and resorption. Many have walls made highly impermeable to diffusion of gases by inclusion of a layer of guanine crystals, quite tightly packed. Guanine is a purine. Its crystals have very low permeability to diffusion of oxygen, so diffusion is effectively limited to the narrow spaces between the crystals.

The conger eel (*Conger*) has this type of swimbladder wall. Professor E. J. Denton and his collaborators used the apparatus shown in Fig. 5-6 to investigate its permeability. The swimbladder is the sausage-shaped object, placed under the rings of the Perspex holder. Weights on the holder prevent it and the swimbladder from floating to the surface of the Ringer solution contained in the jar. The tube at the left hand end of the swimbladder enabled the experimenters to fill it with whatever gas they chose, and to remove gas when they wished for analysis. Another gas mixture was bubbled through the Ringer solution, saturating it with gases at known partial pressure. Gases diffused between the swimbladder and the solution, in accordance with the differences in partial pressure between them. For instance, if the swimbladder was filled with carbon dioxide, and a mixture of 95% oxygen with 5% carbon dioxide was bubbled through the solution, carbon dioxide diffused out of the swimbladder and oxygen diffused in. Carbon dioxide diffuses very much faster through water and tissues than oxygen does (for the same difference in partial pressure), so the swimbladder got smaller. The change in its volume was followed by taking advantage of Archimedes' Principle, by weighing swimbladder and holder, still immersed in the solution, from time to time. The results of experiments with

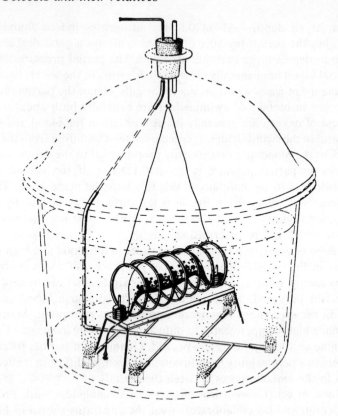

Fig. 5-6. Apparatus used to measure the rates of diffusion of gases into and out of a swimbladder. From E. J. Denton, J. D. Liddicoat & D. W. Taylor (1972). *J. mar. biol. Assoc. U.K.* **52**, 727–46.

various mixtures of gases made it possible to calculate that the permeability constants for carbon dioxide and for nitrogen diffusing through the swimbladder wall are each about 0.01 of the values that have been measured for frog connective tissue and muscle. The swimbladder is quite remarkably impermeable. No determinations were made for oxygen, presumably because its use in metabolism would have introduced substantial errors. (Errors due to metabolism in the calculations for carbon dioxide were much less serious, because carbon dioxide diffuses so much faster than oxygen.) It can be assumed that the constant for oxygen, like those for the other gases, is about 0.01 of the value in other tissues.

It can be calculated from the results of these experiments that when the difference in partial pressure of oxygen between the swimbladder and its surroundings is 1 atm, the swimbladder of a 2 kg conger would lose about 0.5 mm³ oxygen per minute. In a fish at a great depth it would lose oxygen faster: for instance, at 1000 m the partial pressure difference would be about 100 atm and the rate of loss would be about 50 mm^3 min^{-1}. Note that this 50 mm^3 is the volume of the lost gas *when measured at atmospheric pressure*. How much energy would be needed to replace this gas by secretion?

Consider a quantity of gas which at pressure p has volume V. It can be shown that the energy needed to compress it from a pressure p_1 to a pressure p_2 is $pV \ln (p_2/p_1)$. This formula can be applied to the swimbladder. The gas lost per minute has volume 50 mm^3 (5×10^{-8} m^3) at pressure 1 atm (10^5 N m^{-2}), so $pV = 5 \times 10^{-3}$ N m. The partial pressure of the dissolved oxygen in the water would, at 1000 m, be well below 0.2 atm: it would probably be about 0.01 atm. The partial pressure of oxygen in the swimbladder would be almost 100 atm. Hence $p_2/p_1 \simeq 100/0.01 = 10^4$ and $\ln (p_2/p_1) \simeq 9$. Thus the energy needed for secretion would be about $9 \times 5 \times 10^{-3} = 4.5 \times 10^{-2}$ J min^{-1}. The swimbladder being considered is from a 2 kg eel and metabolism using 1 cm^3 oxygen releases 20 J, so this is equivalent to about 1.1×10^{-3} cm^3 oxygen kg^{-1} min^{-1}. The process of secretion cannot be 100% efficient. It is not known what the efficiency is, but even if it is only 5% the metabolic cost to the eel of replacing the oxygen lost by diffusion would be only about 0.02 cm^3 oxygen kg^{-1} min^{-1}. This is about 1.5% of the resting metabolic rate of a 2 kg conger. It is too small to destroy the advantage of a swimbladder over other means of avoiding sinking, unless perhaps for an exceedingly sluggish fish. The conger swimbladder seems quite well suited to life at depths down to 1000 m, but it doubtful whether congers swim so deep.

The energy cost of replacing the gases which diffuse out of the swimbladder would be more serious for a smaller fish. Consider a 2 g fish, 10^{-3} times the weight of the conger. If it had the same proportions as the conger, lengths of its parts would be 10^{-1} times corresponding lengths in the conger and areas would be 10^{-2} times corresponding areas. The area of the swimbladder wall would be 10^{-2} times the area in the conger, so even if it was as thick as in the conger oxygen would diffuse out 10^{-2} times as fast. The rate of loss of oxygen *per unit body weight* would be ten times as high as for the conger and the metabolic cost of replacing the lost gases can be estimated as 0.2 cm^3 oxygen kg^{-1} min^{-1}. It seems doubtful whether a swimbladder would be advantageous to a very small fish living at a great depth. Eleven of sixteen teleost species which have been caught below 3500 m have swimbladders, but these species are not particularly small. Also, their swimbladder walls may be even less permeable to diffusing gases than that of the conger. The swimbladder walls of two species caught at 1400 m and 2800 m were found to contain ten times as much guanine, per unit area, as conger swimbladders.

Since pressure increases with depth, the swimbladder tends to be compressed when a fish swims deeper and to expand when the fish swims nearer the surface. Secretion can be used to compensate for increased depth, as well as to replace diffusion losses. This was first demonstrated in 1874. A wrasse (*Labrus*) which had previously been kept at the surface was put in a cage, lowered to a depth of 8–10 m and kept there for some days. It secreted so much gas into its swimbladder that when it was hauled to the surface the swimbladder was found to be almost double its original volume, (The pressure at 10 m is 2 atm, so twice as much gas is needed there as at the surface to fill the swimbladder.) The volume of the swimbladder returned to normal if the fish was kept in shallow water again. The wrasse has no pneumatic duct. A teleost with a duct would have allowed the excess gas to bubble out through it, as it was hauled to the surface.

The process of resorption is straightforward, since diffusion from the swimbladder is in any case inevitable. Many teleosts have means of speeding it up when required. The eel *Anguilla* has a pneumatic duct which could be used to pass gas to the mouth and out of the fish, but which serves as a gas-resorbing organ (Fig. 5-7*a*). Normally this duct is slender, and though it has plenty of blood vessels little blood flows through them. When gas is to be resorbed it swells and fills with gas from the main cavity of the swimbladder. Its blood vessels dilate, and it has been estimated from measurements of oxygen concentration in them and other vessels that as much as 20% of the blood pumped by the heart may travel through them. Advanced teleosts with no swimbladder duct commonly have a pocket in the swimbladder wall, known as the oval, where resorption occurs (Fig. 5-7*b*). It has plenty of blood vessels but these are generally constricted so that little blood flows through them, and the oval is closed off from the main cavity of the swimbladder by a sphincter. During resorption the blood vessels dilate and the sphincter opens.

Some teleosts have no obvious locality for gas secretion, which probably occurs over large areas of the swimbladder wall. Others have conspicuous red patches known as gas glands. Their function has been demonstrated by injecting cod (*Gadus morhua*) with yohimbine (a drug which stimulates gas secretion). The swimbladders of injected cod were cut open and transparent plastic was laid over the gas gland. Bubbles of gas could be seen collecting under the plastic.

It will be shown that at least in eels, the process of gas secretion into the swimbladder involves secretion of lactic acid into the blood passing through the gas gland. This lactic acid tends to release gases. Because it is an electrolyte it must reduce the solubility of all gases in blood, just as the salts in seawater make gases less soluble in it than in fresh water. This effect is small, but it is the only known basis for secretion of nitrogen and other

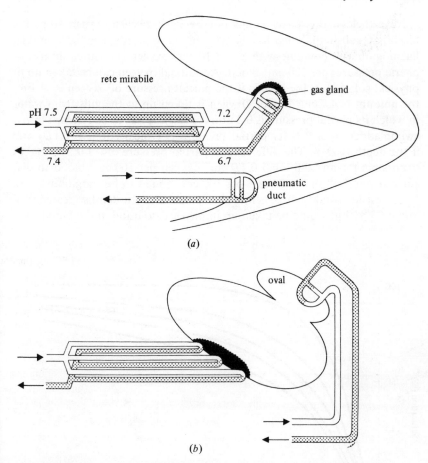

Fig. 5-7. Diagrams of the swimbladder and its blood supply in (*a*) the eel *Anguilla*, and (*b*) a typical advanced teleost with no swimbladder duct. The actual numbers of capillaries are, of course, very much larger than the numbers shown.

inert gases. Because it is an acid, lactic acid tends to release carbon dioxide from bicarbonates in the blood. Also because it is an acid, lactic acid tends to release oxygen from haemoglobin; Fig. 5-8*a* shows the effect of a fall in pH on the amount of oxygen which teleost blood will hold. This sort of effect is not peculiar to teleosts, but also operates in other vertebrates. It facilitates release of oxygen in capillaries where accumulation of carbon dioxide reduces the pH. It is particularly marked in most teleosts but not (for reasons explained on p. 215) in ones which breathe air instead of relying solely on gills. The gas secreted into most swimbladders is mainly oxygen, and by far the most important effect of the lactic acid secreted by the gas gland is its effect on haemoglobin.

Nevertheless, this effect could not by itself release oxygen from the blood at substantial depths. Teleost blood typically contains enough haemoglobin to combine with about 10 cm³ oxygen (measured at atmospheric pressure) per 100 cm³ blood. Additional oxygen can be taken up in physical solution in the blood; if the partial pressure of oxygen is X atm, the amount is $4X$ cm³ blood. Oxygen is taken up at the gills from water in which its partial pressure is 0.2 atm (or less). It has to be released into a swimbladder in which its partial pressure is typically $(0.2 + 0.1D)$ atm if the depth is D m. The difference in partial pressure of $0.1D$ atm makes it possible for an additional $0.4D$ cm³ oxygen per 100 cm³ blood to dissolve. Even if all of the 10 cm³ per 100 cm³ bound to haemoglobin could be released it would not pass into the swimbladder at depths greater than about 25 m, but would be taken up in physical solution in the blood.

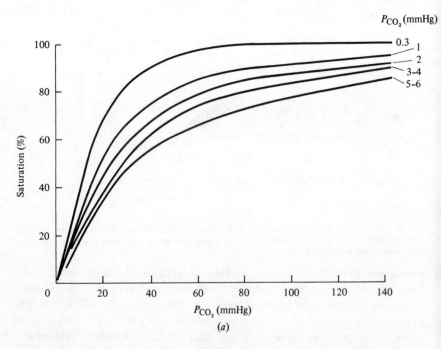

Fig. 5-8. Graphs of oxygen content against partial pressure of oxygen for the blood of (*a*) Rainbow trout (*Salmo gairdneri*) at 15°C, and (*b*) the African lungfish (*Protopterus aethiopicus*) at 25°C. In each case graphs are given for several different partial pressures of carbon dioxide; note that much lower partial pressures of carbon dioxide were used in the experiments on trout than in the experiments on lungfish. Partial pressures are given in millimetres of mercury (760 mmHg = 1 atm). Trout blood equilibrated with air ('100% saturation'), contained 90 cm³ oxygen l^{-1} (9 vol. oxygen %). (*a*) is from F. B. Eddy (1971). *J. exp. Biol.* **55**, 695–712, and (*b*) from C. Lenfant & K. Johansen (1968). *J. exp. Biol.* **49**, 437–52.

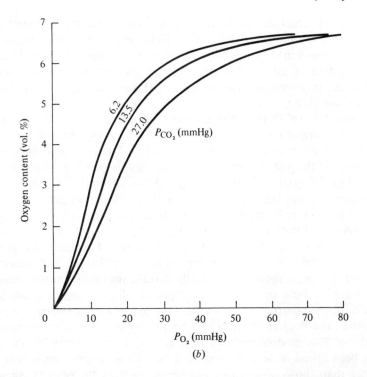

(*b*)

Secretion of oxygen at substantial depths depends on the arrangement of the blood vessels to the gas gland (Fig. 5-7). The artery breaks up into long parallel capillaries which mingle with venous capillaries leading from the gland. The whole group of capillaries is known as a rete mirabile (plural: retia mirabilia). Eels have a slightly different arrangement from other species (compare *a* with *b* in Fig. 5-7); the capillaries join up into a small number of arteries and veins between the rete and the gas gland. This trivial peculiarity has made possible experiments with eels which would otherwise have been a great deal more difficult. Dr J. B. Steen fixed eels (*Anguilla*) in a special holder and opened them to expose these blood vessels. He injected the drug yohimbine to stimulate gas secretion and when this was in progress withdrew for analysis small (40 mm^3) samples of blood from the four positions indicated by numbers in Fig. 5-7*a*. These numbers give the mean pH of samples from each position. Oxygen, carbon dioxide and lactic acid content were also determined, and their values were also in accordance with the theory of gas secretion which will be outlined. The theory was formulated by Drs W. and H. J. Kuhn, but the best evidence for it comes from experiments by Steen and his collaborators. It has been modified a little in the light of the experiments. It is apparent from the pH values shown in Fig. 5-7a that acid is secreted into the blood passing through the gas gland. This is lactic acid; the blood entering the gland contains 8 mM lactic acid, and the blood leaving it 13 mM. There is also diffusion of ions in the rete, between the venous and the arterial capillaries: this accounts for the fall in pH along the arterial capillaries and the rise along the venous ones.

As blood passes through the gas gland and lactic acid is added to it the partial pressure of the oxygen in it is increased, mainly through the effect of the acid on haemoglobin. When this blood passes into the venous capillaries of the rete it contains oxygen at a higher partial pressure than in the parallel arterial capillaries. Oxygen diffuses from it into the arterial capillaries, raising the oxygen concentration and partial pressure of the blood there before it arrives at the gas gland. The process is repeated with this blood, and the concentration and partial pressure of oxygen build up higher and higher at the gas gland end of the rete. Eventually the partial pressure exceeds the partial pressure in the swimbladder, and oxygen is released. It can be shown theoretically that the maximum partial pressure which can be built up (and so the depth at which secretion is possible) should increase sharply as the length of the rete increases. In large *Anguilla* the retia (and their capillaries) may be 10 mm long, but teleosts living at depths of 2000 m or more tend to have longer retia. Retia 25 mm long have been found in several deep-sea fishes. These lengths are immensely greater than those of ordinary capillaries, such as the ones in *Anguilla* muscle which are generally 0.5 mm long or less.

The fall in the lactic acid concentration and the consequent rise in the pH of the blood as it travels along the venous capillaries of the rete might be expected to have an adverse effect, partly cancelling out the effect of the rise in concentration in the gas gland. In fact this adverse effect is probably negligible because of a rather surprising property of eel blood. When the pH is decreased, for instance by adding lactic acid, the partial pressure of oxygen is affected very rapidly indeed: the effect has gone half way to completion in 0.05 s. Reversal of the effect by an increase in pH is much slower, and about 15 s is needed for it to go half way. It can be calculated that each blood corpuscle takes a second or thereabouts to travel the length of the rete. This is plenty of time for a fall in pH to take effect, but not for recovery when the pH rises again.

Secretion and resorption are necessarily slow processes, as can be understood by considering the quantities of gas involved. Consider a 1 kg fish with a 50 cm^3 swimbladder. Suppose that it moves deeper in the sea, keeping the volume of the swimbladder constant. Every increase in depth of 10 m increases the pressure by 1 atm and requires secretion of a quantity of gas which would have a volume of 50 cm^3 at atmospheric pressure.

Experiments on Brook trout (*Salvelinus*) showed that 1 kg specimens in well aerated water at 20 °C use about 50 cm^3 oxygen h^{-1} when resting and up to 150 cm^3 h^{-1} when active. The latter is presumably the maximum rate at which the gills can take up oxygen and the blood can transport it to the muscles. Even if one-third of the blood pumped by the heart could be directed to the gas gland, it seems most unlikely that oxygen could be supplied fast enough to allow secretion of more than 50 cm^3 (measured at atmospheric pressure) h^{-1}. This rate would fill an empty swimbladder in an hour, or maintain the volume of the swimbladder in a descent at the very slow rate of 10 m h^{-1}. Rates of secretion which have been measured are lower even than this. The fastest seems to be that of 1 kg Bluefish (*Pomatomus*) which refilled their swimbladders after experimental emptying in 4 h, fast enough for a descent at 2.5 m h^{-1}. Smaller fish might be expected to be able to fill their swimbladders in less time, since their metabolic rates (per unit body weight) tend to be higher, but none are known to be able to do so.

Resorption of gas in shallow, well aerated water can be expected to be even slower than secretion, since analysis of blood samples indicates that the haemoglobin of arterial blood in teleosts is generally at least 80% saturated with oxygen. The arterial blood arriving at the oval is therefore not capable of taking up much additional oxygen. Perch (*Perca*) are freshwater fish with no pneumatic duct and swimbladders which occupy about 75 cm^3 (kg body weight)$^{-1}$. Specimens trapped in Lake Windermere at a depth of 3 m (pressure 1.3 atm) took an average of 9 h to adjust their density when brought to the surface and put in a shallow aquarium. They must have resorbed gas at a rate of $0.3/9 = 0.03$ swimbladder volumes per hour or 2 cm^3 (kg body weight)$^{-1}$ h^{-1}. Resorption could presumably proceed faster at greater depths, for the reasons that make diffusion losses faster there (p. 146).

Some teleosts make substantial daily changes of depth. Herring (*Clupea*) spend the night near the surface and the day at depths which vary with locality but may exceed 150 m. Since swimbladders reflect sound well, these vertical movements can be followed by echo-sounding. Even bigger vertical migrations are made by the small lantern fishes (Myctophidae) which are very plentiful in oceanic waters. They can be followed by echo-sounding and have even been observed directly from deep submersible vehicles. Lantern fishes which spend the night within 50 m of the surface may spend the day at 300 m or more. The changes of depth at dawn and dusk seem far too fast for compensation by secretion and resorption of swimbladder gases to be possible.

When herring and such species of lantern fishes as have gas-filled swimbladders are caught at night at the surface, they have enough gas in the swimbladders to give them about the same density as seawater. It is not known how much gas they have in the swimbladder at their daytime depths, but it seems unlikely that enough extra gas is secreted to compensate for the increased depth. More probably the quantity of gas is kept constant, so that the swimbladder is compressed during the day to a small fraction of its night-time volume. If so, it would be almost totally ineffective by day. The fish would be denser than seawater and would have to depend on the hydrofoil action of fins to avoid sinking further. In these circumstances the advantage of a swimbladder over materials like squalene might disappear. Some lantern fishes have gas-filled swimbladders only as juveniles. As they grow the swimbladder regresses and wax esters are deposited around it. These are quite different chemically from squalene, but have about the same density. Some lantern fishes with no gas in the swimbladder have enough wax ester to give them about the same density as seawater. Their density is virtually unaffected by changes of pressure.

A great many bottom-living teleosts have also lost the swimbladder. The angler fish *Lophius* and the flatfishes (Pleuronectiformes) are among them. A fish which has the same density as water may rest in contact with the bottom but it needs no vertical force to support it, so there are no frictional forces to hold it in place. It will move with every water movement, and must swim to keep stationary in a current. If it were denser than the water it would use more power when it swam but would not need to swim all the time and so might make a net saving of energy. This is particularly likely to be the case for fish which find their food near bottoms exposed to currents, in rivers and on tidal shores.

FINS AND SWIMMING

A selachian fin is a relatively thick and rigid structure. A teleost fin consists of widely spaced, mobile rays joined only by extremely thin webs of tissue. A selachian fin is extended permanently but a teleost one can be folded and unfolded like a fan. Part of a typical teleost fin is shown in Fig. 5-9. A dorsal or anal fin is represented; the other fins differ in some details of structure.

The pterygiophores consist of cartilage bone and are homologous with the pterygiophores of selachians. The rays are not simple collagenous rods like the ceratotrichia of selachians but are branched and consist of short lengths of bone joined by collagen fibres. Each consists of two half-rays one on each side of the fin. Except at the ends of the fin, each ray articulates with a separate group of pterygiophores. The most distal pterygiophores consists of two halves, tightly bound together by collagen fibres. The bases of the two halves of the ray lie on either side of it, attached by ligaments so as to form rather a loose hinge joint. This joint bends when the fin is

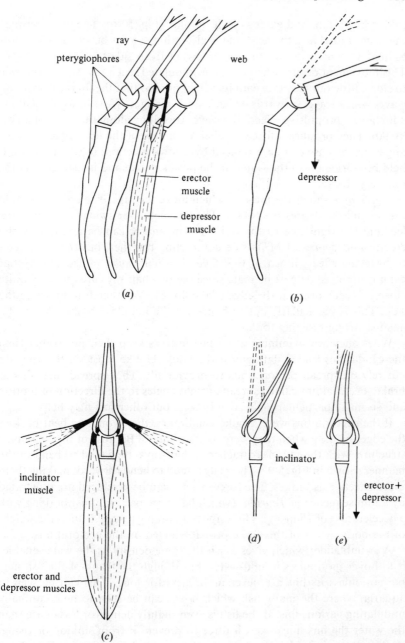

ray

pterygiophores

web

erector
muscle

depressor
muscle

depressor

(a)

(b)

inclinator
muscle

inclinator

erector and
depressor muscles

erector+
depressor

(c)

(d)

(e)

Fig. 5-9. Diagrams of the skeleton and muscles of a teleost fin, showing movements which are possible. (a) and (b) show rays in side view while (c), (d) and (e) represent transverse sections.

folded like a fan, and extends as it unfolds (Fig. 5-9*b*). The joint between the distal radial and the next one is also a hinge joint but its axis is longitudinal so that it allows the ray to swing from side to side (Fig. 5-9*d*). This movement occurs when the fin is undulated: each ray rocks from side to side a little out of phase with its neighbours so that the fin is thrown into waves which travel forwards or backwards along it. Waves travelling along a fin have a propulsive effect, like waves travelling along the body of an eel (p. 90). They produce a force parallel to be base of the fin. Forces at right angles to the base can be produced by a rowing action, in which the fin is held broadside on to the water in the power stroke and is feathered in the recovery stroke.

Fig. 5-9*a*, *c* shows the muscles which move the rays. The erectors unfold the fin and the depressors fold it. The left inclinator swings the ray to the left and the right one to the right. An inclinator can be assisted by the erector and depressor of its side contracting together, but they also tend to have the effect shown in Fig. 5-9*e*. The half-ray is pulled basally but cannot simply slide past its mate since the two half-rays are attached quite closely to each other at the edge of the fin. The ray therefore bends to the side. This is particularly apt to happen if the muscles are moving the fin against strong resisting forces.

When they are swimming at all fast teleosts keep their fins (other than the caudal fin) folded flat against the body. Unlike most sharks, they do not rely on spread pectoral fins to provide lift. They spread their fins to brake. Pectoral fins can be spread at right angles to the direction of motion and so make particularly effective brakes, but other fins may help.

If the pectoral fins were simple membranes, they would be bent back at the edges by drag when they were used as brakes. Because of their complex structure, with the muscles attached to half-rays, they tend to bend in the manner shown in Fig. 5-9*e*. Their edges tend to bend forward, making them more effective as brakes. The forces which can be produced are illustrated by a film sequence of *Taurulus* (which has large pectoral fins) braking with a deceleration of 15 m s^{-2}. This implies a braking force 1.5 times the weight of the body. Most of this force probably acted on the pectoral fins.

A swimbladder which gives a fish the same density as the water enables it to hover motionless in mid-water, but though the trunk of the fish may be motionless its fins are never motionless for long. This can be seen in aquaria, where the many fish which hover can be seen to be continually undulating various fins. If the fish is even slightly denser or less dense than the water the fins must exert a force to prevent it from sinking or rising, as the case may be. Such a force can be provided by waves passing down or up the pectoral and caudal fins (Fig. 5-10). Even if the fish is at the particular depth at which it has precisely the same density as the water, its equilibrium is unstable. A slight accidental rise in the water will reduce the pressure on the swimbladder, allowing it to expand and reducing the density of the fish so that it tends to go on rising. Similarly, slight sinking will make the fish denser and inclined to sink further. The fish cannot keep at a constant level unless it compensates for accidental vertical movements: it does this by means of its fins. Breathing by pumping water in through the mouth and out through the gills has a jet-propulsion effect, so a teleost

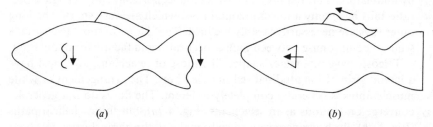

(a)   (b)

Fig. 5-10. Diagrams showing how a teleost can use undulatory movements of its fins (a) to overcome a downward force or to swim upwards, and (b) to overcome a forward force or to swim backwards.

hovering in mid-water will tend to move slowly forwards unless it exerts a compensating backward force. Waves travelling forward along the dorsal and anal fins or backward rowing movements of the pectoral fins can serve this purpose (Fig. 5-10b).

Though such fin movements are used by hovering fish to keep themselves stationary, they are also used for propulsion. They make possible a wide variety of manoeuvres that are beyond the capability of selachians. Movements like those of Fig. 5-10a can make the fish rise vertically, and opposite movements can make it sink. Waves travelling upwards in the pectoral fins and downwards in the caudal one, or vice versa, can tilt it. Movements like those of Fig. 5-10b can be used for backward propulsion. Some teleosts also use fin movements as their normal means of slow forward swimming. The South American knife-fishes (Gymnotoidei) swim by undulating their long anal fins, except in bursts of speed.

Experiments have been done to measure the vertical forces which perch (*Perca*) can produce with their fins. A perch was put into a large bottle full of water, where it hovered in mid-water. The pressure was reduced so that the swimbladder expanded (perch have no pneumatic duct, so no gas escaped). As the pressure fell the perch made more and more vigorous fin movements to avoid floating to the surface, but eventually fin movements alone were not enough and the fish had to swim with its tail, nose down. This happened when the pressure had been reduced enough to reduce the density of the fish to 990 kg m^{-3}. Thus the maximum downward force produced by fin movements was 1% of the body weight.

Teleosts do not have heterocercal tails like selachians. The vertebral column ends anterior to the caudal fin and the upper and lower halves of the fin are more or less symmetrical about a horizontal plane through it. This type of tail is called homocercal.

A caudal fin with a high aspect ratio is most efficient for forward propulsion (p. 90). Tunnies (*Thunnus*) and some other teleosts have such caudal fins but they are rigid and incapable of producing vertical forces by undulation. This requires a fin with mobile, horizontal rays. High aspect ratio tails have only a few horizontal rays, which are very short. The long outer rays are necessarily steeply inclined. The shapes of most teleost tails seem to compromise between high aspect ratio and facility for undulation.

Teleosts have myomeres shaped like those of selachians, and most have a lateral strip of longitudinal red muscle fibres. The arrangement of white muscle fibres is generally completely different. The fibres do not generally converge on tendons as in selachians (Fig. 4-7a), but follow helical paths (Fig. 4-7b). Both arrangements seem to make all the fibres shorten by about the same fraction of their length when the fish bends, but the fraction is less, for a given bend, with the helical pattern than with the selachian pattern. Typical swimming movements involve about 15% changes in fibre length in selachians but only about 5% in typical teleosts. Fibres arranged in the helical, teleost pattern only have to contract one-third as fast to move the tail at a given speed but they have to exert three times as much force to produce a given force on the tail. The difference between the two patterns resembles the difference between two levers of different proportions; a change which alters the ratio of the velocities at the ends of a lever also alters the ratio of forces.

The bodies of most teleosts taper to a slender caudal peduncle just anterior to the caudal fin. The last few myomeres are here, and their white fibres are arranged as in selachians, even though the rest of the myomeres have the typical teleost arrangement. Otherwise they probably could not transmit to the caudal fin the forces which the myomeres in the fatter parts of the body can produce. To produce a transverse force F on the caudal fin the muscles at a distance x anterior to the centre of the fin must be able to exert a moment Fx about the vertebral column. Fx is least near the tail so some tapering is appropriate but typical teleosts have caudal peduncles that would be far too slender to allow full use of the more anterior myomeres, if the fibres were arranged in the same way all along the body. In the extreme case of the tunnies the peduncle is so slender that even myomeres arranged in selachian fashion could not exert large enough moments, and forces are transmitted along the peduncle by tendons from the myomeres in the thicker, more anterior parts of the body. A tendon can transmit forces over 100 times as great as a bundle of parallel striated muscle fibres of the same cross-sectional area can exert. The advantage given by a slender caudal peduncle has already been considered (p. 87).

SILVERY TELEOSTS

Bottom-living teleosts often have colours which merge with the bottom, and patterns of stripes or blotches which help to make them inconspicuous by distracting attention from the outline of the fish. The teleosts that were cited as examples of bottom dwellers lacking swimbladders (p. 154) are also examples of this. In contrast, teleosts such as the herring and its close relatives (*Clupea* spp.) which habitually swim in mid-water are often brilliantly silvery. This silveriness is a very effective camouflage. Divers report that silvery fishes, swimming gently, are very difficult to see, except from below.

The problem of camouflage in mid-water is the problem of looking bright against a bright background and dark against a dark background. All the light comes ultimately from above. If a submerged animal in deep water looks upwards towards the sky its eye will receive far more light than if it looks down into the depths. To be inconspicuous from all directions, a fish must match the bright background of the surface water when seen from below, and the dark background of the depths when seen from above. It could of course achieve this by perfect transparency. The paragraphs which follow explain why silveriness is almost as good. Some teleosts are largely transparent but none is completely transparent. Some organs such as eyes and red muscle cannot be made transparent because pigments are essential to their functioning.

Close to the surface on a sunny day, the brightest light comes not from directly overhead but more nearly from the direction of the sun. At depths of 30 m or more in clear water (or less in turbid water) the direction of greatest brightness is much more nearly vertical, whatever the altitude of the sun. The distribution of light is then more or less symmetrical about the vertical (Fig. 5-11a). In clear oceanic water the vertical downward light is about 200 times the vertical upward light.

Fig. 5-11c shows that if the light distribution is symmetrical about the vertical, and if a fish has vertical sides which are perfect mirrors, these sides will match their background perfectly from whatever direction they are viewed. If they are viewed from below the reflected light is as bright as the background and if from above the reflected light is as dim as the background. The horizontal back of a fish can only be seen from above. Fig. 5-11b shows that if it is to match its background it must reflect only a very small fraction of the light which falls on it from above. The backs of herrings and other silvery fishes are appropriately dark. This figure also shows that the horizontal ventral surface of a fish cannot be camouflaged

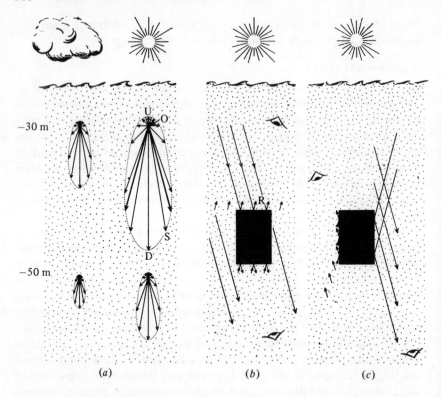

Fig. 5-11. (*a*) Diagrams of light distribution at two depths in the sea, on a cloudy and on a sunny day. The length of each arrow represents the brightness of light travelling in the direction of the arrow at the depth in question. The lengths are not, however, strictly proportional to brightness: if they were, OD would be about 200 times OU.

   (*b*), (*c*) Diagrams of a 'fish' of rectangular cross-section in the sea where it is illuminated from above by bright light (long arrows) and from below by much dimmer light (short arrows). (*b*) shows that it can be made to match its background when viewed from above if the light R reflected from its dorsal surface is a sufficiently small fraction of the light incident on it. However, it cannot be made to match its background when viewed from below, even if the ventral surface is a perfect mirror. (*c*) shows that if the sides of the fish are perfect mirrors they will match their background, whether the fish is viewed from below or from above. From E. J. Denton (1970). *Phil. Trans. Roy. Soc.* B, **258**, 285–313.

by reflecting surfaces, but must be silhouetted against the bright background when the fish is seen from below. Herring and many other silvery teleosts have very narrow bodies so the ventral strip which cannot be camouflaged is narrow.

   The silveriness of teleosts is due to reflecting platelets attached to the inner surfaces of the scales, or in the skin under the scales. If these are scratched off a scale they break up into very thin crystals of guanine (the purine which is also found in swimbladder walls, p. 145) but it has been shown by electron microscopy that undamaged platelets consist of about five of these crystals, with intervening layers of cytoplasm (Fig. 5-12a). The effectiveness of this arrangement as a reflector depends on the thickness and spacing of the crystals and on their refractive index. More precisely, it depends on the thickness of each crystal and of each space being about one quarter of the wavelength of light. Consider, for instance, red light of wavelength 720 nm. This is the wavelength in a vacuum. In water or cytoplasm of refractive index 1.33 its wavelength would be 720/1.33 = 540 nm,

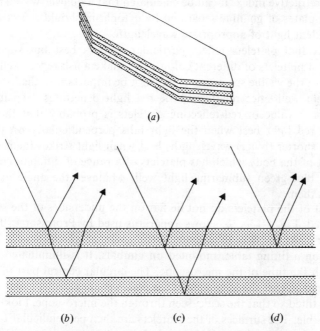

Fig. 5-12. (*a*) A diagram (not drawn to scale) of a reflecting platelet from a teleost. (*b*), (*c*), (*d*) Diagrammatic section through the platelet showing possible pathways for light. Further explanation is given in the text.

while in guanine crystals of refractive index 1.8 it would be 720/1.8 = 400 nm. Alternate layers of cytoplasm 540/4 = 135 nm thick and of guanine 100 nm thick would make the best possible platelet for reflecting this light, provided the light was falling perpendicularly on it.

When light reaches an interface between two materials of different refractive index, a proportion of it is reflected. A platelet includes a large number of interfaces, each of which reflects only a small proportion of the light that reaches it, but quarter-wavelength spacing ensures that interference effects do not dim the reflected beam. The reflected light from all the interfaces is in phase, as Fig. 5-12b–d shows. Fig. 5-12b shows rays reflected from the upper surfaces of two successive guanine crystals. The path of one is the longer by four quarter-wavelengths (i.e. by one complete wavelength) so the two rays are in phase. Fig. 5-12c shows that the same is true for rays reflected from the lower surfaces of successive crystals. Fig. 5-12d shows rays reflected from opposite surfaces of the same crystal. The path difference is only half a wavelength but the reflected rays are nevertheless in phase, because the phase of light is reversed when it is reflected from a material of higher refractive index, but not from one of lower refractive index. It can be calculated that a quarter-wavelength stack of five plates of guanine separated by cytoplasm should reflect over 80% of incident light of appropriate wavelength.

Individual platelets reflect particular colours best but superimposed layers of platelets of different colours can make a fish reflect well, over the whole of the visible spectrum. It may not be important to the fish to reflect red light well since relatively little red light penetrates deep in the sea. The main value of red-reflecting platelets is probably that though they reflect red light best when the light falls perpendicularly on them they reflect shorter (bluer) wavelengths best when light strikes them obliquely. A part of the body which has platelets of a range of different colours will reflect blue-green submarine light well, whatever the angle at which it strikes them.

Most of the platelets do not lie flat on the undersides of the scales, but are tilted. Fig. 5-13a, b, shows apparatus used by Professor E. J. Denton and Dr J. A. C. Nicol to investigate the angles at which they lie. The scale is put on a tilting table, mounted on gimbals. It is illuminated vertically through the tube of the microscope. The circular central part of the table is rotated so that the long axes of the platelets are parallel to bb. The table is then tilted so that the scale, seen through the microscope, looks as bright as possible. The surfaces of the platelets are then perpendicular to the tube of the microscope (Fig. 5-13b) and the angles they make with the scale about the axes aa and bb can be read from the dials. If the position the scale occupied on the body is known, the orientation of the platelets on the body can be worked out (Fig. 5-13c). The orientation of the platelets which lie in the skin, not attached to scales, has not been investigated by the tilting table technique but by examining microscope sections of skin.

Fig. 5-13c shows a typical arrangement of platelets. They are set at varying angles to the body surface in such a way that all are more-or-less vertical. They will thus reflect light in the manner indicated in Fig. 5-11c, so that the fish will match its background when viewed from the side, whether horizontally or obliquely from above or below. They are edge-on to an observer looking at the fish from directly above so they do not reflect light from the sky back up into his eyes. There is dark pigment deeper in

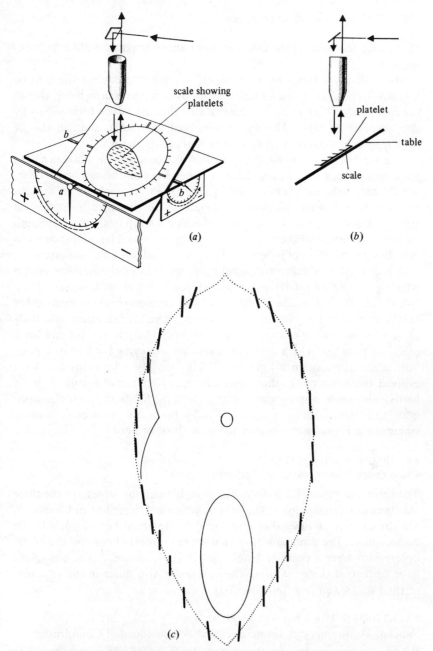

scale showing platelets

platelet

table

scale

(a)

(b)

O

(c)

Fig. 5-13. (*a*), (*b*) Apparatus used to investigate the orientation of reflecting platelets under the scales of fishes. From E. J. Denton (1970). *Phil. Trans. Roy. Soc.* B, **258**, 285–313. (*c*) Transverse section of a herring (*Clupea harengus*) showing (by thick black lines) the planes in which the platelets under the scales lie. This diagram is based on measurements made by the method shown in (*a*), (*b*). From E. J. Denton & J. A. C. Nicol (1965). *J. mar. biol. Assoc. U.K.* **45**, 711–38.

the skin of the back so the fish seen from above looks dark, like its background.

Many silvery teleosts form shoals. It is advantageous for them to be invisible to predators, but shoaling seems to depend on their being able to see each other. It seems to be a general rule that fish which form shoals by day, disperse by night. This can be observed in aquaria, and can also be shown by echo-sounding to be true of herrings in the sea.

The pupil of the eye must always remain as a conspicuous (though small) spot on a silvery fish. Many shoaling fishes have additional black spots, which may help members of the species to locate each other without making them too conspicuous to predators. For instance, *Pristella* is a partly transparent, partly silvery teleost which lives in fresh water in South America. It has a conspicuous black spot on the dorsal fin, and shoals less readily with members of its species which have had this fin amputated.

It is not clear whether the shoaling habit reduces the number of fish which get eaten by predators, or not. It does seem clear, however, that a fish at the centre of a shoal is relatively safe because there are many other fish between it and any predator which approaches. A fish which habitually pushes its way to the centre of a shoal is more likely to survive and leave offspring than one which does not. Constant jockeying for a central position tends to maintain a tight shoal. This may be why many fish have evolved the habit of shoaling, and why many terrestrial mammals form herds. The open waters where many shoaling fishes live and the open grasslands where many mammals live in herds are both environments where there is no cover – except within a shoal or herd.

ACTINOPTERYGIAN FISHES
Class Osteichthyes, subclass Actinopterygii

The actinopterygians form only one (though much the largest) of the three subclasses of Osteichthyes. The others, which are described in Chapter 7, are the Dipnoi or lungfishes and the Crossopterygii (which include the coelacanths). The dipnoans have characteristic toothplates and the crossopterygians have a peculiar hinge joint in the cranium. Actinopterygians have neither of these features. The subclasses also differ in the structure of their scales and fins (pp. 206, 208).

PALAEONISCOIDS AND STURGEONS
Subclass Actinopterygii, infraclasses Palaeoniscoidei and Chondrostei

Palaeoniscoids are among the earliest known fossil Osteichthyes, from the early Devonian. By the early Carboniferous they were the most plentiful fishes in freshwater. They remained plentiful until the Trias but the only modern fish included in the group are *Polypterus* (Fig. 5-14*b*) and

Calamoichthys, both African freshwater fish. These are so far from being typical palaeoniscoids that many zoologists do not even include them among the actinopterygians, but give them a subclass to themselves.

Some palaeoniscoids grew to a length of 1 m but most were considerably smaller. They vary considerably in shape but the example shown in Fig. 5-14a is typical. Most have thick rhomboidal (diamond-shaped) ganoid scales (Fig. 3-17) with a thick outer layer of enameloid substance as in acanthodians. Also like acanthodians, most had long jaws and eyes set well forward leaving little room for olfactory organs. The gills were covered by an operculum borne on the hyoid arch, very much as in teleosts. There were no subsidiary opercula on other gill arches, like those of acanthodians. There seems to have been a spiracle. The maxilla was sutured to the dermal bones of the cheek and cannot have moved as it does in teleosts, when the mouth was opened (Fig. 5-15a).

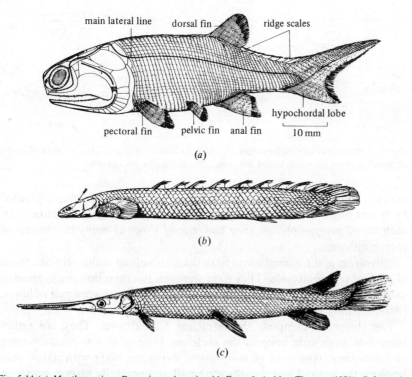

(a)

(b)

(c)

Fig. 5-14 (a) *Moythomasia*, a Devonian palaeoniscoid. From J. A. Moy-Thomas (1971). *Palaeozoic fishes*, 2nd edit., revised by R. S. Miles. Chapman & Hall, London.
(b) *Polypterus senegalus*. The arrow points to the spiracle. From T. W. Bridge (1904). Fishes. In S. F. Harmer & A. E. Shipley (eds.), *The Cambridge Natural History*, Macmillan, London.
(c) *Lepisosteus osseus*. From C. Arambourg & L. Bertin (1958). In P.-P. Grassé (ed.), *Traité de Zoologie*, vol. 13, pp. 2172–203. Masson, Paris.

In several features palaeoniscoids resembled selachians more than teleosts. The spiracle is one. Other were stiff fins, apparently incapable of folding, and a heterocercal tail. These are probably all primitive characters shared by the (unknown) common ancestor of Selachii and Osteichthyes. The fins rays consisted of bone and had essentially the same structure as in

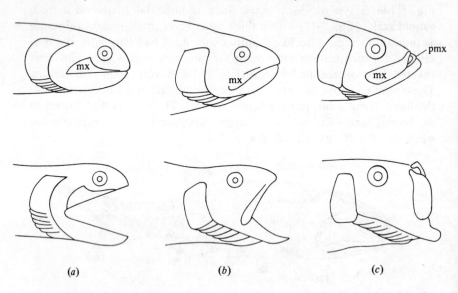

(a)   (b)   (c)

Fig. 5-15. Sketches of (*a*) a palaeoniscoid, (*b*) a typical holostean or primitive teleost, and (*c*) a herring (*Clupea*), showing the mouth closed and open. pmx, premaxilla; mx, maxilla.

teleosts, but they were packed closely together instead of being separated by broad webs of thin tissue. Instead of having one ray articulating with each set of pterygiophores, they had several times as many rays as sets of pterygiophores.

Polypterus and *Calamoichthys* have thick rhomboid scales very like those of ancient palaeoniscoids. They have spiracles but their tails are symmetrical and their pectoral and dorsal fins are peculiar. They have a pair of lungs. (Air-breathing by fishes is discussed in Chapter 7.)

The sturgeons comprise the infraclass Chondrostei. They are rather large fish, with little bone in the skeleton. Instead of a normal covering of scales they have rows of bony plates along the body with naked skin between the rows. Their mouths are small and ventrally placed, posterior to an elongated snout. They find a lot of their food by rooting about with their snouts in sand or mud. Though peculiar in so many features, sturgeons have heterocercal tails and closely-packed bony fin rays just like those of ancient palaeoniscoids. They cannot fold their fins and they depend on the heterocercal tail and the pectoral fins (or possibly the shape of the snout) for lift. They have a swimbladder (which does not serve as a lung) but it is too small to reduce the density of the fish to that of the water. They have spiracles.

Some primitive features shared by palaeoniscoids and selachians have already been noted. The soft parts of ancient palaeoniscoids have not been preserved but were probably broadly similar to those of *Polypterus* and sturgeons, which show some additional primitive features. The intestine has a spiral valve. The white fibres of the myomeres are arranged in selachian fashion. There is a conus arteriosus in the heart.

Polypterus has lungs and sturgeons have swimbladders. Lungfishes have lungs and the ancient crossopterygians from which the amphibians evolved presumably had lungs too. It seems likely that the structure which is a swimbladder in sturgeons and teleosts evolved first as a lung. Of course an air-filled lung reduces the density of a fish, just as a swimbladder does.

HOLOSTEANS AND TELEOSTS
Subclass Actinopterygii, infraclasses Holostei and Teleostei

The evolution of teleosts from palaeoniscoids was a gradual process. Similar features were often evolved independently by several related groups of actinopterygians. The infraclasses Palaeoniscoidei, Holostei and Teleostei represent three rather arbitrarily defined stages in evolution, and the distinctions between them are not sharp. The holosteans replaced the palaeoniscoids as the most plentiful fish in the Trias, and were in turn replaced by the teleosts in the Cretaceous. Only two modern genera are included in the holosteans. Both live in fresh water, though the ancient holosteans were also plentiful in the sea. They are the garpike, *Lepisosteus* of North and Central America (Fig. 5-14c) and the bowfin, *Amia*, of North America.

Teleosts generally have thin roundish scales, consisting simply of a thin layer of bone without dentine or enameloid substance. *Lepisosteus* and many extinct holosteans have thick scales much more like those of palaeoniscoids. *Amia* has thin scales like those of teleosts. Teleosts have homocercal tails but *Lepisosteus* and most holosteans have heterocercal ones though not as markedly heterocercal as in typical palaeoniscoids. Holosteans and teleosts have highly mobile fins that can be undulated and folded, with rays well spaced in the fin web.

Though *Lepisosteus* has a heterocercal tail it keeps its density very close to that of water, as can be seen by watching it hovering in aquaria. Few teleosts hover with so little fin movement. There is plainly no need for a heterocercal tail to provide lift. Presumably the upper and lower lobes of the tail are matched in stiffness so that the vertical components of the forces they produce in swimming cancel out. This example makes it seem unwise to assume that because ancient palaeoniscoids had heterocercal tails they must have been denser than water.

Fish with thick scales need big swimbladders if they are to have the same density as the water. Scales have a specific gravity of about 2, so 2 g of scale occupy 1 cm³ and call for an additional 1 cm³ of swimbladder gas. The thick scales of small *Polypterus* weigh about 100 g (kg body weight)$^{-1}$, but those of typical teleosts of similar size only 40 g kg^{-1} or less. *Polypterus* are generally denser than water but sometimes take enough air into their lungs to float almost motionless in an aquarium. Experiments with the apparatus shown in Fig. 5-5 have shown that the volume of air required in the lungs is 130 cm³ kg^{-1}. If the scales were as thin as in typical teleosts, 100 cm³ kg^{-1} or less would suffice. The disadvantages of a large swimbladder are that it increases the bulk of the fish and so the energy needed for swimming, and that it increases the amount by which the density of the fish is changed by a given change of depth. This may be why their scales became generally thinner as the actinopterygians evolved.

Lepisosteus and *Amia* both have lungs, but very few teleosts do. *Lepisosteus* and *Amia* retain a vestige of the spiral valve but this is lost in the teleosts. No known holostean or teleost has a spiracle.

Lepisosteus and *Amia* have a conus arteriosus in the heart but, apart from vestiges in a few very primitive familes, teleosts do not. Instead of having a conus with several rows of valves between the ventricle and the aorta they have a structure known as the bulbus arteriosus. There are valves between the ventricle and the bulbus, but none in the bulbus itself. There is smooth muscle in the wall of the bulbus but no cardiac muscle, and it does not make active contractions in time with the heart beat. However, there is a great deal of elastin in its wall, and it is highly distensible. It helps to smooth the flow of blood around the body, in the same way as the elastic ventral aortas of sharks are believed to do (p. 115).

In palaeoniscoids the premaxilla and maxilla were firmly attached to the cranium and the dermal bones of the cheek (Fig. 5-15a). In typical holosteans and in the more primitive teleosts such as trout (*Salmo*) and African knife-fish (*Papyrocranus*) the premaxilla is still fixed to the cranium but the maxilla is free from the cheek bones (Figs. 5-2, 5-3). The maxilla is hinged at its anterior end to the cranium, and attached by the lip to the lower jaw at a point anterior to the jaw articulation. When the mouth opens it swings forward, pulled by the lip, as shown in Fig. 5-15b. Some other teleosts such as the herring (*Clupea*) have both the maxilla and the premaxilla hinged to the cranium, and both swing forward as the mouth opens (Fig. 5-15c). In many of them the premaxilla is long and excludes the maxilla from the edge to the mouth. In such cases the maxilla is toothless. The premaxillae have become even more mobile in many of the advanced teleosts described in Chapter 6, in which they have become protrusible. The movements of protrusible premaxillae are described on p. 187.

Fig. 5-15 shows how movements of the maxilla and premaxilla affect the shape of the open mouth. In the palaeoniscoids the mouth opens in a sort of grin, but in typical holosteans and teleosts the opening is more nearly circular. This is particularly so in teleosts such as the herring. The grinning type of mouth may be best for grabbing prey. The round type is probably better for sucking food in, since water can only enter from in front and not through the corners of the mouth.

FURTHER READING

General

Greenwood, P. H., D. E. Rosen, S. H. Weitzman & G. S. Myers (1966). Phyletic studies of teleostean fishes, with a provisional classification of living forms. *Bull. Am. Mus. nat. Hist.* **131**, 339–456.
Marshall, N. B. (1971). *Explorations in the life of fishes.* Harvard University Press, Cambridge, Mass.
See also the general list for fishes at the end of Chapter 3.

Feeding and respiration

Alexander, R. McN. (1969) Mechanics of the feeding action of a cyprinid fish. *J. Zool., Lond.* **159**, 1–15.
Alexander, R. McN. (1970). Mechanics of the feeding action of various teleost fishes. *J. Zool., Lond.* **162**, 145–56.
Ballintijn, C. M. & G. M. Hughes (1965). The muscular basis of the respiratory pumps in the trout. *J. exp. Biol.* **43**, 349–62.
Muir, B. S. (1969). Gill dimensions as a function of fish size. *J. Fish. Res. Bd. Canada,* **26**, 165–70.
Osse, J. W. M. (1969). Functional morphology of the head of the perch (*Perca fluviatilis* L.): an electromyographic study. *Neth. J. Zool.* **19**, 289–392.

Buoyancy

Alexander, R. McN. (1966). Physical aspects of swimbladder function. *Biol. Rev.* **41**, 141–76.
Alexander, R. McN. (1972). The energetics of vertical migration by fishes. *Symp. Soc. exp. Biol.* **26**, 273–94.
Berg, T. & J. B. Steen (1968). The mechanism of oxygen concentration in the swimbladder of the eel. *J. Physiol.* **195**, 631–8.
Denton, E. J., J. D. Liddicoat & D. W. Taylor (1972). The permeability to gases of the swimbladder of the conger eel (*Conger conger*). *J. mar. biol. Assoc. U.K.* **52**, 727–46.
Kuhn, W., A. Ramel & E. Marti (1963). The filling mechanism of the swimbladder. *Experientia,* **19**, 497–511.

Fins and Swimming

Arita, G. S. (1971). A re-examination of the functional morphology of the soft-rays in teleosts. *Copeia,* 1971, 691–7.

Harris, J. E. (1936). The mechanical significance of the position and movements of the paired fins in the Teleostei. *Pap. Tortugas Lab.* **31**, 171–89.

McCutchen, C. W. (1970). The trout tail fin: a self-cambering hydrofoil. *J. Biomechanics,* **3**, 271–81.

Silvery teleosts

Denton, E. J. (1970). On the organization of reflecting surfaces in some marine animals. *Phil. Trans. Roy. Soc.* **73**, 258–313.

Hamilton, W. D. (1971). Geometry for the selfish herd. *J. theor. Biol.* **31**, 295–311.

Keenleyside, M. H. A. (1955). Some aspects of the schooling behaviour of fish. *Behaviour,* **8**, 183–248.

6

Carps and perches

This chapter is about two superorders which between them include the majority of species of fish. One is the Ostariophysi, which includes some 5000 of the 20000–30000 teleost species. Its members include the carps and minnows, the catfishes and other related groups comprising the great majority of freshwater teleost species, but very few of them live in the sea. The other superorder is the Acanthopterygii which has fewer freshwater species, but more species overall. The perch is a typical member of the group.

FISH WITH WEBERIAN OSSICLES
Infraclass Teleostei, superorder Ostariophysi

The distinctive feature of the Ostariophysi is a group of small bones known as the Weberian ossicles, which connect the swimbladder to the ear. The system aids hearing, as can be demonstrated by hearing tests.

These test are not easy to perform. The fish must be trained to perform some clear-cut movement in response to sound. For instance, it may be kept in an aquarium partly divided by a barrier and trained to cross the barrier whenever a loudspeaker sounds. If it does not cross, it is given a mild electric shock. When it has been trained in this way to respond to clearly audible sounds the intensity and frequency of the sound is varied. In this way the experimenter discovers how faint a sound will still elicit a response, and how this is affected by the pitch of the sound. The results are only meaningful if the tests are made in carefully controlled acoustic conditions. Interference of sound with its reflections from the walls of an aquarium can make sound intensity very different at points quite close together. This difficulty does not arise if the aquarium is small compared to the wavelength of the sound, but another difficulty does. This is that the vibrations of a submerged loudspeaker force the surrounding water to vibrate in phase with it (the near-field effect, see p. 53), as well as starting propagated sound waves. The amplitude of vibration corresponding to a given sound pressure is very much greater close to the loudspeaker than it is further away. This difficulty can be overcome by placing the loudspeaker in the air over the aquarium, instead of submerging it.

The results of careful hearing tests on four species of teleost, are shown in Fig. 6-1. *Carassius* (the goldfish) and *Ictalurus* (a N. American catfish) are Ostariophysi. They have lower sound thresholds (i.e. they can hear fainter sounds) than the other two species which are not Ostariophysi. In other experiments the Weberian apparatus of *Ictalurus* was put out of action by deflating the swimbladder. This made the fish much less sensitive to sound, over a wide range of frequencies.

The possibility of using the swimbladder to aid hearing depends on

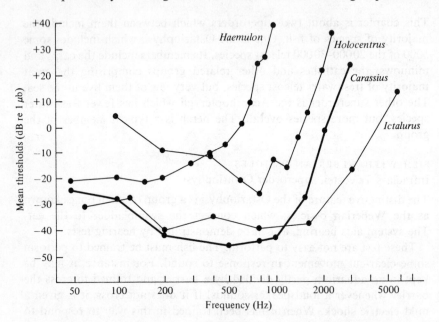

Fig. 6-1. The results of hearing tests on various teleost fishes. The auditory threshold (i.e. the minimum intensity of sound to which the fish responded) is plotted against the frequency of the sound. The intensity is expressed in decibels; every increase of 20 dB represents an increase in the amplitude of the sound vibrations by a factor of 10. Frequency is expressed in Hertz (cycles s⁻¹). From D. W. Jacobs & W. N. Tavolga (1967). *Anim. Behav.* **15**, 324–35.

the facts that air is far more compressible than water, and that air is elastic; it is compressed when the pressure is increased but expands again when the original pressure is restored. Where mass is combined with elastic stiffness, there is the possibility of resonance. A mass hanging from a spring has a natural frequency at which it will vibrate up and down, if set in motion by some disturbance. Similarly the water around a bubble has mass and must move radially in and out if the bubble contracts or expands. The bubble has a natural frequency at which it will pulsate, expanding and contracting, if disturbed. The same applies to swimbladders. The frequency depends on the volume of the bubble or swimbladder, but hardly at all on the shape. At small depths where the pressure is little more than atmospheric the natural frequency is about 530 Hz for a volume of 1 cm³ (for instance, for the swimbladder of a typical 20 g fish) and about 110 Hz for a volume of 100 cm³ (for instance, for the swimbladder of a 2 kg fish).

When sound travels through water or any other fluid it makes the fluid vibrate, and also causes fluctuations of pressure. If a submerged bubble or swimbladder is exposed to sound, the pressure fluctuations make it pulsate. If the frequency of the sound is reasonably near the natural frequency of the bubble, the amplitude of the pulsations may be very much greater than that of the vibrations which would occur in the water, if the swimbladder were not there. These pulsations involve radial movements of the surrounding water, so the vibrations due to the sound are amplified in the neighbourhood of the bubble. Fish hear by means of otolith organs, which are sensitive to vibrations rather than pressure fluctuations. The presence of a swimbladder in the body can therefore enable a fish to hear better. The effect falls off with distance from the centre of the swimbladder, in accordance with an inverse square law. The swimbladders of most fish are some distance posterior to the ears but in herring and their relatives (Clupeomorpha) and in some other teleosts the swimbladder has extensions which reach forward to the ear. In Clupeomorpha they actually penetrate the skull. They presumably make the fish more sensitive to sound but there is little experimental evidence. *Holocentrus* (Fig. 6-1) has forward extensions of the swimbladder and is more sensitive to sound of some frequencies than *Haemulon* (which has not), but it is less sensitive at all frequencies than *Equetus* (which also has no extension).

Fig. 6-2 shows the basic structure of the Weberian apparatus. Most Ostariophysi have a two-chambered swimbladder as illustrated, though most catfishes and a few others have only one chamber. The posterior chamber has a single wall. The anterior one has a double wall composed of an inner tunica interna, which is complete, and a tunica externa, which has a slit in it. These layers are connected only by a very delicate network of oily connective tissue. The swimbladder is filled with gas at a pressure slightly above that of the surrounding water so the tunica interna is kept taut and smooth. The tunica externa can therefore slide freely over it and does not have to stretch to the same extent as the tunica interna if the swimbladder is allowed to expand by reduction of the pressure around it; expansion is accommodated largely by widening of the slit.

The Weberian ossicles are attached by bone or cartilage to the vertebrae, but their attachments are so slender that they pivot freely. The posterior ossicles are attached to the tunica externa, at the edges of the slit, and successive ossicles are connected by short ligaments. When the swimbladder expands and the slit widens, all the ossicles pivot forwards. A ligament of elastin (elastic ligament, Fig. 6-2) probably helps to pull them and the edges of the slit back again, when the swimbladder contracts. When the swimbladder pulsates in a sound field, the ossicles must vibrate. The most anterior ossicles form the side walls of a fluid-filled chamber from which their vibrations are transmitted hydraulically to one of the otolith organs of each ear.

The amplifying effect of a bubble in a sound field has a sharp peak at

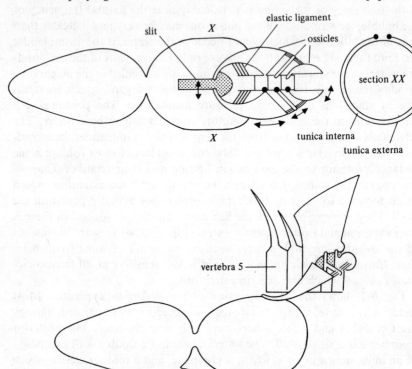

Fig. 6-2. Diagrammatic dorsal and lateral views of the swimbladder and Weberian ossicles of a typical characin. The ossicles are shown disproportionately large.

the resonant frequency (close to the natural frequency of free vibration). At the resonant frequency, the effect may be very large indeed (Fig. 6-3). This is only possible because there is little damping. The pulsations of a swimbladder are much more heavily damped, by the swimbladder wall and by surrounding tissues. This damping has been measured by investigating the effect of the presence of a fish with a swimbladder on an underwater sound field. The swimbladder graph in Fig. 6-3 has been calculated from the results of such experiments with goldfish. The amplitude which has been calculated is the amplitude of pulsation of the swimbladder wall, but the amplitude of vibration of the fluid in the ear is probably very similar.

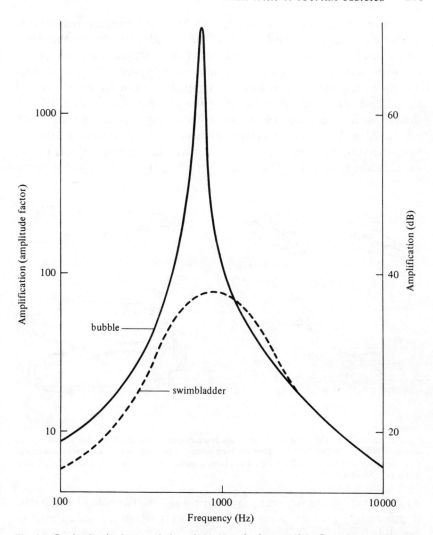

Fig. 6-3. Graphs showing how much the swimbladder of a 5 cm goldfish (*Carassius auratus*), and a bubble of the same volume, can be expected to amplify sound of various frequencies. The ordinate on the left shows the factor by which the amplitude of vibration in the sound field would be increased at the surface of the bladder or bubble. The ordinate on the right shows the same information converted to the decibel scale, for convenience of comparison with Fig. 6-1.

The calculation is based on goldfish 5 cm long, which is about the size of the goldfish used for hearing tests of which the results are shown in Fig. 6-1. The calculated amplification factors are of the right order of magnitude to account for differences in auditory thresholds between Ostariophysi and other teleosts.

In many Ostariophysi including catfishes (Siluriformes) and loaches (Cobitidae) there are what seem to be refinements of the Weberian apparatus. There is a gap in the muscles of the body wall on either side of the swimbladder, so that only skin intervenes between the swimbladder and the water. This must reduce the damping effect of the body wall and so increase sensitivity to sound near the resonant frequency (Fig. 6-3 shows that it is only near the resonant frequency that damping has much effect). Also these fish have smooth sheets of bone extending laterally from the vertebral column, over the dorsal and anterior surfaces of the swimbladder (Fig. 6-4a). The bone covers the parts of the tunica externa,

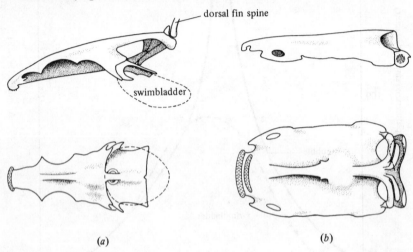

Fig. 6-4. Lateral and ventral views of the skulls and anterior vertebrae of two catfish, (a) *Bagrus docmac* which has its swimbladder partly covered by sheets of bone, and (b) *Clarias mossambicus* which has a reduced swimbladder enclosed in a bony capsule. The broken lines indicate the position of the swimbladder.

on either side of the slit, that move most when the swimbladder pulsates. It presumably provides a smooth surface over which the tunica externa can slide freely.

Many bottom-living Ostariophysi have very small swimbladders, but none seem to have lost the swimbladder altogether, like so many bottom-living members of other groups (p. 154). There is no known fish in which the Weberian apparatus appears to be a functionless vestige, nor any which lack a Weberian apparatus but have other features indicating relationship to Ostariophysi. Various catfishes and loaches have very small hourglass-shaped swimbladders, with one globe of the hourglass on either side of the body (Fig. 6-4b). Primitive members of both groups have larger swimbladders with overlying sheets of bone as in Fig. 6-4a. Reduction of the swimbladder has not been accompanied by proportionate reduction of the sheets, which have come to enclose it in a bony capsule open at the ends where it is covered only by skin.

The Weberian apparatus makes Ostariophysi very sensitive to sound but apparently gives no indication of the direction the sound comes from. It cannot indicate direction since, even when the swimbladder is divided into left and right halves with only a narrow duct between, the vibrations of the two halves are not transmitted separately to the ears. Nevertheless, the huge numbers of Ostariophysi and the fact that none seem to have lost the Weberian apparatus, suggest that the apparatus confers a substantial selective advantage. Perhaps faint sounds made by predators or potential prey alert the fish which then use other senses to locate their source.

The Weberian apparatus is a remarkably complex system in which every part has its function. How can it have evolved? The homologies of the parts can be established satisfactorily enough by comparing the anterior vertebrae with more posterior, normal, vertebrae, in embryos and in adults. The first two ossicles are the modified neural arches of the first two vertebrae, and the third is the much shortened rib of the third vertebra together with a process of the vertebra. An early stage in the evolution of the apparatus seems to be retained in the Gonorhynchiformes, a small group of tropical fish only recently recognized as related to the Ostariophysi.

As well as the Weberian apparatus, there is another characteristic of Ostariophysi which distinguishes them from all other fish. Nearly all of them have it, including the primitive Gonorhynchiformes. No other fish are known to have it but it has been evolved (obviously independently) by toad tadpoles (*Bufo*). It is the fright reaction.

If an injured minnow (*Phoxinus*) is added to a shoal which has settled down in an aquarium, the shoal shows signs of alarm. Its members swim away to a hiding place, or crowd together and sink to the bottom. The same effect is obtained by adding to the tank some water which has been shaken up with chopped minnow skin (but not with minnow muscle, or other tissues). This fright reaction is apparently brought about by an unidentified substance (the alarm substance) which is released by damaged skin. If a minnow is attacked by a sharp-toothed predator such as a pike (*Esox*), the alarm substance is released and nearby minnows are put on their guard. The substance is apparently detected by the sense of smell. Removal of the olfactory nerves from minnows destroys the fright reaction. This could of course be due to post-operative shock, but this possibility has been more or less eliminated by sham operations in which all the surgery needed to get at the nerves was performed without actually damaging them, and the wound was then closed in the same way as in the real operations. The sham operations had no apparent effect on the fright reaction.

The fright reaction does not take the same form in all Ostariophysi. Some bottom-living species become motionless. Other species which inhabit still water over muddy bottoms swim nose down, stirring up the mud. Yet other species flee to the surface rather than the bottom. Alarm substances from other species will elicit the reaction but the substance from the same species seems always to be the most effective. Minnows do not respond to tadpole alarm substance, and vice versa.

Though the groups included in the Ostariophysi are very diverse in other respects, the Weberian apparatus and fright reaction are strong evidence of common ancestry. The Weberian apparatus is particularly convincing, because it is so complex. Though chains of bones connecting swimbladder and ear might well evolve independently in several groups of fish, it seems unlikely that they would evolve in every case from precisely the same parts of precisely the same vertebrae.

CHARACINS

Superorder Ostariophysi, order Cypriniformes, suborder Characinoidei

Apart from the few Gonorhynchiformes, the characins are the most primitive of modern Ostariophysi. They are found in the fresh waters of Central and South America and tropical Africa. An African example is shown in Fig. 6-5*a*. They are particularly plentiful in tropical America, and include some 45% of the freshwater fish species of Guyana. The small characins known as tetras are very popular in tropical aquaria.

Tetras and other typical characins have remarkably complicated teeth (Fig. 6-6). Instead of being simple cones like most teleost teeth, the larger teeth each bear a row of five to seven cusps. The lines of cusps form serrated cutting edges. The lower teeth close inside the upper ones and alternate with them. The outer faces of the lower teeth and the inner faces of the main upper ones are vertical, so the cutting edges pass close to each other as the mouth closes. The arrangement seems admirable for cutting and it is used for biting pieces from leaves of water plants. The small upper teeth, anterior to the cutting ones, close against a thick lower lip and are presumably used for gripping rather than cutting.

The more primitive characins such as *Alestes* (Fig. 6-5*a*) and *Creatochanes* feed on insects and water plants, in various proportions, and sometimes on fruits which fall into the water from overhanging trees. Others are adapted to a remarkable range of ways of life. The South American piranhas (*Serrasalmus*) are notorious predators that do not always swallow prey whole, like most fish, but bite pieces from large animals including, occasionally, people. The cutting teeth which are typical of characins are most perfect in them, triangular blades with knife-like edges. I have used a piranha jaw successfully for sharpening pencils. Some other characins are more conventional predators with long jaws and tall single-pointed teeth which are used for seizing prey, which is swallowed whole. There are others with weak jaws and no lower teeth which feed on detritus. There are characins living in swamps where there is very little oxygen dissolved in the water that have evolved the ability to breathe air; the swimbladder, which probably originated as a lung, (p. 167), has evolved back into one. There is even a group of characins which, when alarmed, can jump out of the water and fly a few yards, beating wing-like pectoral fins with extraordinarily large fin muscles. The characins present one of the best examples of adaptive radiation: that is, of adaptation to varied ways of life of closely related organisms.

Though their ways of life vary so much in other respects most characins are active only by day. At night they lie inactive, hiding motionless under a river bank or among vegetation. Some are so inactive that an accepted method of fishing in Guyana is to patrol the banks of river pools with a torch and a cutlass, slashing at inactive fish.

AMERICAN KNIFE-FISHES
Superorder Ostariophysi, order Cypriniformes, suborder Gymnotoidei

The gymnotoids are slender fish with enormously long anal fins, which suggest the blade of a knife of which the head is the handle. They often swim slowly with the body straight, by undulating this fin. They are found only in Central and South America. (There is another group of similarly shaped fish living in Africa and Asia which have been given the same English name but belong to the superorder Osteoglossomorpha.) Unlike the characins, they hide by day under river banks or among roots, or even bury themselves in sand, emerging only at night. Many characins are silvery and so are inconspicuous by day as they swim in mid-water in rivers and lakes which are often turbid. Knife-fishes are generally near rocks and vegetation when it is light, and are dark in colour or have patterns of vertical stripes which tend to disrupt their distinctive outline. Most characins have big eyes and apparently depend largely on them for finding food, and finding their way about. Knife-fishes have small eyes, and make use of a remarkable electric sense. They can detect objects which differ in electrical conductivity from the water they are swimming in, by detecting distortions of an electric field set up by special electric organs.

These organs consist of modified muscle fibres in some knife-fishes, and modified nerve fibres in others. In either case, the cells involved are known as electrocytes, and in both cases the electric discharge is produced by action potentials. Only the muscle-fibre type of electric organ will be described.

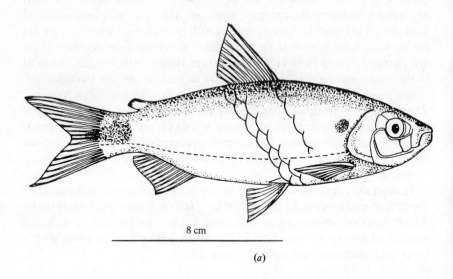

8 cm

(a)

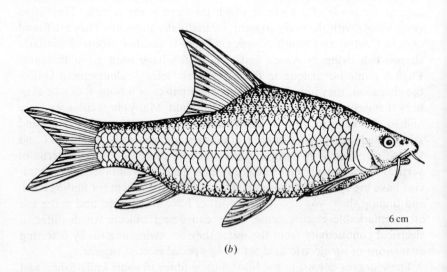

6 cm

(b)

Fig. 6-5. A selection of teleosts from Uganda. (a) A characin, *Alestes jacksonii*; (b) a cyprinoid, *Barbus bynni*; (c) a catfish, *Synodontis afrofisheri*; and (d) an acanthopterygian, *Tilapia nilotica*. Drawings by Barbara Williams from P. H. Greenwood (1966). *The fishes of Uganda*. Uganda Society, Kampala.

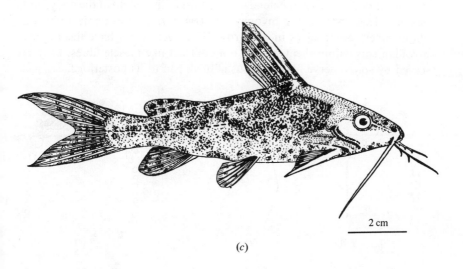

2 cm

(c)

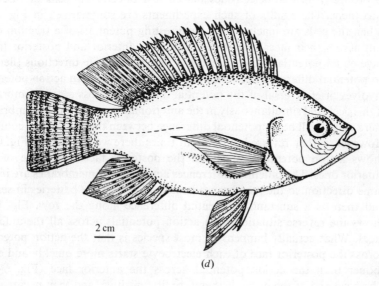

2 cm

(d)

These consist of rows of drum-shaped electrocytes, often four rows on each side of the body. Each row extends along most of the length of the body, and has one electrocyte in each myomere. In *Gymnotus*, for example, there are about ninety electrocytes in each row. The electrocytes have disoriented myofilaments within them, but do not contract like muscle fibres. Each is served by spinal nerve fibres that usually all run to its posterior face.

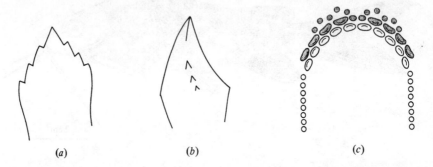

(a)   (b)   (c)

Fig. 6-6. Teeth of a typical characin, *Creatochanes*. (*a*) and (*b*) are two views of one of the larger teeth of the lower jaw. (*c*) is a plan showing how the teeth of the upper jaw (stippled) and those of the lower jaw fit together in the closed mouth. Redrawn from R. McN. Alexander (1964). *J. Linn. Soc.* (*Zool.*), 45, 169–90.

The electric properties of the electrocytes have been investigated by recording from microelectrodes placed just outside the cells or inserted into them. The results of such experiments are summarized in Fig. 6-7. When the cells are inactive they have resting potentials of a fraction of a volt across their faces (Fig. 6-7*a*). Since the anterior and posterior faces have equal potentials across them but face in opposite directions there is no potential difference between the ends of the organ. An action potential involves brief reversal of the potential difference across a cell membrane. If this occurred simultaneously in the anterior and posterior cell membranes there would still be no potential difference between the ends of the column. However, if they reversed at different times there would be one. Fig. 6-7*b* shows action potentials across all the posterior faces but none of the anterior ones. The potential differences across all the membranes are in the same direction so the membranes behave like so many batteries in series, and there is a substantial potential difference along the row. Fig. 6-7*c* shows the reverse situation with action potentials across all the anterior faces. What actually happens in most species is that the action potential across the posterior face of each electrocyte starts more quickly and ends sooner than the action potential across the anterior face (Fig. 6-7*d*). The head end of the organ is therefore first positive, and then negative, to

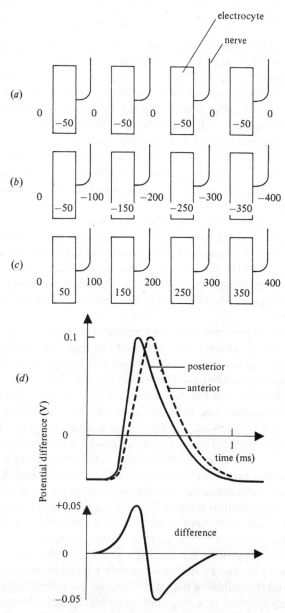

Fig. 6-7. (a) A diagram of part of an electric organ in the resting state, showing electrical potentials in and between the electrocytes. The potential difference across all the electrocyte membranes is 50 mV (inside negative). (b) The same organ, with action potentials in the posterior faces only of the electrocytes. The potential difference across each posterior face is 50 mV (inside *positive*), but the potential differences across the anterior faces are unaltered. (c) The same, with action potentials in the anterior faces only. (d) Schematic graphs showing the sequence of events in a typical electrocyte during discharge. Potential differences across each face and the difference between them (which is the potential difference across the whole cell) are shown.

the tail end. In at least two species there is no action potential in the anterior face, and so no head-negative phase in the discharge.

A typical action potential involves a change in potential difference of 100–150 mV, but because the action potentials across the opposite faces of an electrocyte overlap in time the net potential difference across the cell may never be more than about 50 mV (Fig. 6-7d). Even so, a row of about ninety electrocytes, as in *Gymnotus*, might be expected to produce about 4.5 V. The potential differences which can be measured between the head and tail of *Gymnotus* are considerably smaller than this, because of the electrical resistance of the electric organ and because of short-circuiting through surrounding tissues.

The electric organs of knife-fishes discharge regularly nearly all the time. Frequencies range from a few per minute to about 600 Hz for the muscle-fibre type of organ, and from about 700 Hz to the extraordinarily high value of 1700 Hz for the nerve-fibre type. Species which have low frequencies at rest generally increase the frequency when they are swimming or feeding, and when they are stimulated for instance by a change of light intensity.

The electric organs are used in conjunction with electric receptors, which are scattered all over the surface of the body but most densely on the head. There are two types. There are ampullary organs which resemble the ampullae of Lorenzini of selachians (p. 123). They have very short canals, more like those of the freshwater stingray *Potamotrygon* than those of marine selachians. The skin has a rather high electrical resistance and the ampullary organs are sensitive to d.c. and low-frequency a.c. potential differences across it. They are insensitive to high-frequency a.c. and also to the brief, biphasic electric discharges which most electric organs produce. They are ill suited for use in conjunction with the electric organ, but probably have essentially the same function as ampullae of Lorenzini.

There are also tuberous receptors, which are much more numerous than the ampullary ones. They have no open canal to the surface of the skin, but neither are they covered by the layer of flattened epithelial cells which is believed to be responsible for the high electrical resistance of the skin. Each is covered instead by a plug of loosely-packed epithelial cells. They do not respond throughout a long d.c. stimulus, but merely to its beginning and end. They respond to a.c. stimuli up to much higher frequencies than the ampullary receptors, and they respond to the discharges of the electric organ. In species with a reasonably low frequency of discharge, each discharge evokes a separate burst of action potentials in the nerve from the receptor. A nearby electrical conductor or insulator can alter the number of action potentials per burst, by distorting the electric field and so altering the current flowing through the receptor. Fig. 6-8 shows the

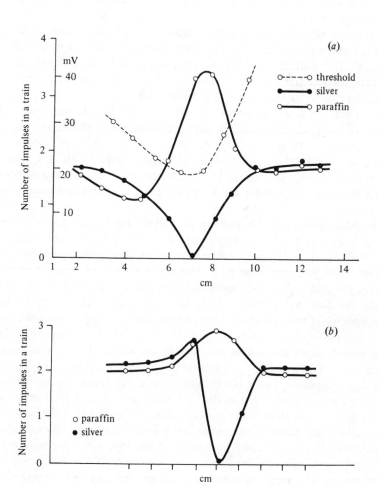

Fig. 6-8. Graphs of the mean number of nerve impulses elicited by each discharge of the electric organ of *Steatogenys*, against the position of a silver plate or paraffin block in the water alongside the body. Graphs for two electric receptor axons are shown. The broken line in (*a*) shows thresholds for response to stimuli from an electrode in the water: it shows that the sense organ served by the axon was at the 7 cm position. From S. Hagiwara & H. Morita (1963). *J. Neurophysiol.* **26**, 551–67.

results of an experiment which demonstrated this. A *Steatogenys* was fixed in a tilted trough of water so that its body, except for the head, was submerged. It was kept alive by passing aerated water into the mouth and over the gills. The nerve to the electric receptors was exposed close to the head and a microelectrode was inserted into it, to record from a single fibre. A silver plate or a block of paraffin wax, held in the water beside the fish, altered the number of action potentials per burst: each might increase or decrease the number, depending on its position.

The electric eel, *Electrophorus*, grows to more than 1.5 m, and is the largest of the gymnotoids. It has huge electric organs, which may occupy 40% of the volume of the body. An adult may have 6000 electrocytes in each row, and produce pulses of 500 V. These large pulses are used in defence, or in attacking prey. There is a small subsidiary electric organ which produces the regular (10 V) pulses used to locate conductors and insulators.

The Mormyriformes are a group of African freshwater teleosts which do not belong to the Ostariophysi but have electric organs and use them, like knife-fishes, in electric location. One of them, *Gymnarchus*, was used by Drs H. W. Lissmann and K. E. Machin in a series of tests. They put a porous pot and a piece of food simultaneously into the aquarium. They taught the fish that it could safely take the food if the pot contained aquarium water but that it would be chased away if it approached the food when the pot contained a material of lower electrical conductivity. They were able to train the fish in this way to distinguish between aquarium water and a mixture of 75% aquarium water, 25% tapwater. This distinction was presumably made electrically, since a fish trained to refuse food presented with a porous pot of distilled water also refused food presented with a pot of aquarium water also containing a glass rod. (The electrical conductivity of glass is of course low.) Great care was taken in these experiments to eliminate the possibility of unconscious signalling by the experimenter to the fish. Apparatus was constructed which enabled the experimenter to present the pot without being seen, and the response of the fish was recorded automatically. In some experiments the experimenter was given a pot without being told its contents: he tested the fish's response to it and was only then told what it contained.

CARPS AND MINNOWS

Superorder Ostariophysi, order Cypriniformes, suborder Cyprinoidei

The cyprinoids are the most numerous freshwater fishes of Europe, Asia, Africa and North America, but do not occur in South America where the characins are so plentiful. An example is shown in Fig. 6-5*b*.

Nearly all characins have jaws that work like those of primitive teleosts (Fig. 5-15*b*), but cyprinoids have protrusible premaxillae. Instead of merely swinging forward with the maxillae as in herring (Fig. 5-15*c*) they move bodily forward, away from the maxillae to which they are connected only by extensible skin (Fig. 6-9). Fig. 6-10 shows a cyprinoid feeding. In (*a*) it is approaching a piece of food with its mouth still closed. In (*b*) its mouth is

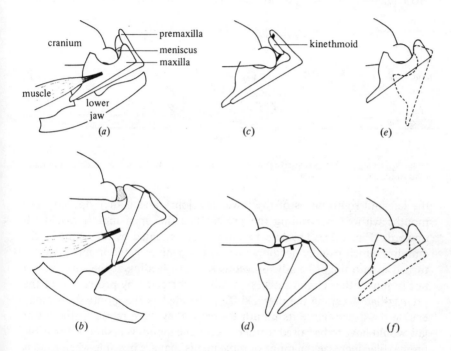

Fig. 6-9. Diagrams based on the orfe (*Idus*) showing how cyprinoid fish protrude their premaxillae. (*a*) and (*b*) show jaw bones and a muscle in side view, (*a*) with the mouth closed and premaxillae retracted and (*b*) with the mouth open and premaxillae protruded. (*c*) and (*d*) are similar but with the right premaxilla and maxilla removed to show the kinethmoid. (*e*) and (*f*) are explained in the text.

open, the premaxillae are fully protruded and the food is being sucked in. Notice that the mouth opening is at the end of a tube projecting forwards from the head and edged by the protruded premaxillae. In (*c*) the mouth is closed again with the food inside, but the premaxillae are still protruded. Protrusible premaxillae have also been evolved by the acanthopterygians (described later in this chapter) and by certain other teleosts, and in each case the sequence of movements is the same: the premaxillae are protruded as the mouth opens and kept protruded as it closes. Indeed, the majority of living teleosts have protrusible premaxillae. Selachians also have protrusible jaws, but of a very different sort: the whole of the palatoquadrate and Meckel's cartilages move forward (p. 110).

Though protrusible premaxillae are so widespread it is not at all clear what selective advantage they give. Shooting the premaxillae forward at the last moment may sometimes help a fish to effect a capture, but the distance they move seems generally too small to give much advantage. Most teleosts have the jaw articulation ventral to the symphysis (where

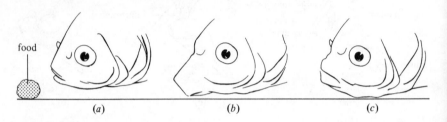

Fig. 6-10. Tracings of three successive frames of a film taken at 60 frames s⁻¹ of a goldfish (*Carassius auratus*) feeding.

the left and right halves of the lower jaw join), and if they opened their mouths without protruding the premaxillae the open mouth would tilt upwards (Fig. 5-2d). This effect would be accentuated by the tilting of the cranium which usually accompanies wide opening of the mouth. An upturned mouth might be advantageous to a fish feeding at the surface, but not in most other circumstances. It can be corrected by protrusion of the premaxillae. Keeping the premaxillae protruded as the mouth closes may enable the fish to close its mouth faster; if they were retracted the lower jaw would have to be raised further, before the mouth was shut. Protrusible premaxillae increase the range of movements that are possible when food is being manipulated between the lips.

The mechanism of protrusion is shown in Fig. 6-9. The head of each maxilla rests on a pad of cartilage (the meniscus) which in turn rests on a rounded projection of the cranium. The other end of the maxilla is attached by ligament to the lower jaw. When the mouth opens, the lower jaw pulls the maxilla forward and down. Two types of movement are involved: the maxilla rotates about its head (Fig. 6-9e) and the whole maxilla including the head moves ventrally by sliding of the meniscus on the cranium (Fig. 6-9f). The latter type of movement can also be produced by contraction of the muscle shown in Fig. 6-9a. (This is a branch of the adductor mandibulae that has shifted its insertion, in the course of evolution, from the lower jaw to the maxilla.)

It is this type of movement of the maxillae that causes protrusion of the premaxillae. A small median bone, the kinethmoid, is connected by ligaments both to the cranium and to the premaxillae (Fig. 6-9c). There are also ligaments between the kinethmoid and the heads of the maxillae. When the latter move ventrally the kinethmoid is made to rock, pushing the premaxillae forward (Fig. 6-9d).

A fresh dead cyprinoid can be made to protrude its premaxillae by opening its mouth widely or by pulling on the tendon of the muscle shown in Fig. 6-9a. It has been shown by electromyography that this muscle contracts as a carp (*Cyprinus*) protrudes its premaxillae. Retraction of the premaxillae is apparently brought about by elasticity of the tissues rather than by any muscle.

Most of the typical cyprinoids with jaws like the goldfish seem to feed largely on bottom-living invertebrates (including insect larvae), but many also eat insects taken at the surface, and various plants. There are other cyprinoids with more downwardly-directed mouths, which scrape algae off rocks with their lips or go over the bottom like a vacuum cleaner, sucking up loose detritus.

Cyprinoids have no teeth in their jaws but have large and peculiar lower pharyngeal teeth (Fig. 6-11). There are no upper pharyngeal teeth but their place is generally taken by a horny pad, mounted on a projection from the back of the skull. The left and right teeth interdigitate. They generally have hooked ends and sharp upper edges. They are not attached to separate toothplates as in most other teleosts, but directly to the bones of the lower part of the last gill arch. Many muscles insert on these bones, and only the two largest are illustrated. The movements they cause can be demonstrated by stimulating them electrically in a fish which has just been killed. Muscles *A* pull the teeth dorsally and laterally. If both are used at once they must press the food against the horny pad and tear it by opposite movements of the hooked left and right teeth. Muscles *B* pull the teeth dorsally and posteriorly. They are presumably used in swallowing but may also be used to rasp food, by drawing the upper edges of the teeth along the horny pad. A cyprinoid that has eaten a relatively large piece of food can be seen to make throat movements for a few minutes thereafter. It is presumably chewing with its pharyngeal teeth. Pieces of earthworm fed to orfe (*Idus*), and recovered a few minutes later by killing and dissecting the fish, were found to be severely lacerated, with the gut showing through tears in the body wall. These lacerations were presumably made by the action on the pharyngeal teeth of muscles *A*. They could be expected to speed digestion by admitting enzymes to the interior of the food.

The Chinese grass carp, *Ctenopharyngodon*, is a large cyprinoid which feeds on aquatic plants and (during floods) on land vegetation. It has

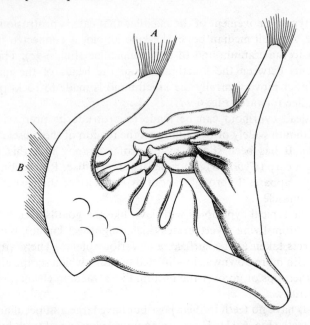

Fig. 6-11. Pharyngeal teeth of a typical cyprinoid (the orfe, *Idus*) with their supporting bones and principal muscles.

peculiar pharyngeal teeth, which it uses to break up the leaves into fragments, and to rasp their surfaces. In this way many, but not all, of the cells are broken open. Since the fish has no enzyme which will digest cellulose, the cell walls pass undigested through the gut and even the contents of intact cells are protected. Only the contents of broken cells get digested. Nevertheless, grass carp fed a grass of which 4% of the dry weight was silica, passed faeces of which 9% of the dry weight was silica. This indicates that more than half of the organic content of the food was digested. The pharyngeal teeth must be remarkably effective.

Many cyprinoids have barbels projecting from the corners of their mouths, with numerous sense organs in their skin which resemble the taste buds of other vertebrates and are believed to be organs of taste (Fig. 6-5b). The bases of the barbels are attached to the maxillae so that they lie neatly against the side of the head when the premaxillae are retracted. They swing forward with the maxillae when the premaxillae are protruded and project on either side of the mouth, where they can be used to search for food and to taste things before they are eaten. Some cyprinoids such as the Stone loach, *Nemacheilus*, have taste buds in the skin all over the body.

CATFISHES

Superorder Ostariophysi, order Siluriformes

The catfishes are widely distributed over the world but are commonest in the tropics. Most live in fresh water but a few live in the sea. The marine fish *Anarhichas* is known in English as the catfish, but is not a member of the Siluriformes; it simply happens to have been given the same common name. The name 'catfish' is used in this book exclusively for the Siluriformes. An example is shown in Fig. 6-5*c*.

Most catfishes seem to be active mainly at night. For instance, the North American catfish *Ictalurus* is caught in nets more often by night than by day, and comparison of the amounts of food in the stomachs of specimens caught at different times indicates that it feeds mainly at night. In South American rivers many catfishes can be found by day in cavities in dead tree trunks which have fallen into the water. They emerge at night with the gymnotoids. The catfishes most commonly kept in tropical aquaria belong to the genus *Corydoras* and are exceptions to the general rule that catfishes are most active at night; indeed, their activity by day helps to make them popular.

Like the gymnotoids, catfishes seem to rely less on their eyes than on other senses. Most have rather small eyes. They do not have the special electric location system of gymnotoids but have barbels that are well supplied with taste buds. There are up to four pairs of barbels, two on the upper and two on the lower jaw. The longest are often as long as the head, and sometimes very much longer. Their small eyes and whisker-like barbels give some catfish the cat-like appearance which is probably responsible for their name. The resemblance is helped by their being mostly dark in colour, or blotched in a camouflage pattern, rather than silvery.

Most catfishes feed on bottom-living invertebrates, and on other fish. They use their barbels in searching for food. The ones on the lower jaw are particularly well placed for searching the bottom since they can be trailed along, below and to either side of the mouth. Catfish also seem to use their barbels as organs of touch, for feeling their way around; they often swim with their barbels extended in front of them.

Each barbel has a core of calcified cartilage, which makes it reasonably stiff. All can be moved a little but the most mobile are the longest (maxillary) barbels. The cartilage cores of these are continuous at the base with a slender bone, which is in fact the maxilla. In the rare *Diplomystes*, which is in various respects the most primitive catfish known, this bone bears teeth and is very much like the maxillae of characins and primitive teleosts. In almost all the others it is toothless and in most it is a simple rod, no longer involved in the jaw mechanism. In teleosts in general the bases of the maxillae rest against the palatine bones at the anterior ends of the suspensoria, and are attached to them by ligaments. Catfish retain this connection, but the palatine has become detached from the rest of the suspensorium. It can be slid forward and back by muscles which are arranged in different ways in different groups of catfishes, but have evolved from parts of the adductor arcus palatini and the adductor mandibulae. The maxilla passes through the tough connective tissue of the upper lip, which prevents it from moving bodily forward and back with the palatine. Instead it pivots about the point where it passes through the lip, so that backward movement of the palatine swings the barbel forwards, and forward movement swings it back.

In most catfish the anterior rays of the pectoral and dorsal fins have become spines. They are much stouter than ordinary rays, and they have lost their flexibility by fusion of the small pieces of bone of which ordinary rays consist into a single whole. They are often formidably strong and sharp, and seem likely to protect catfish against being eaten. Evidence that they do give protection was obtained by observers in the USA, who watched herons (*Ardea*, etc.) feeding and were able, using binoculars, to identify the fish they caught and see which escaped without being eaten. Out of a total of 3023 fish caught, 47 (1.6%) escaped and the rest were swallowed. However, of the 239 catfish (*Ictalurus*) which were caught, 20 (8%) escaped. Presumably their spines made them difficult to swallow. Eels (*Anguilla*) which coiled themselves around the herons' beaks, were also more successful than average in escaping.

If spines are to give really effective protection they must be held firmly erect and not be folded out of the way by pressure of the predators' jaws. Catfish erect their spines when alarmed, and though they use muscles for this they do not depend on muscular strength to keep them erect. The joints at the bases of the spines incorporate friction-locking devices. The pectoral spines articulate with (and lock in position on) an exceptionally strong pectoral girdle which is attached firmly to the skull. The dorsal spine is close behind the head. It articulates with (and locks in position on) a strong pterygiophore, which is firmly attached to the vertebral column and often also to the back of the skull (Fig. 6-4*a*).

The principle of friction locks is the principle of knots, which are pulled tighter by tension so that tension increases the frictional forces between the ropes. If the coefficient of friction between the ropes is high enough the knot will never slip, even though the ropes are pulled until one breaks. The pectoral and dorsal friction locks work differently from each other, and only the former will be described. The base of the spine bears a flange, tilted at an angle to the length of the spine, which fits a slot in the pectoral girdle (Fig. 6-12). Both flange and slot have rough surfaces. When the spine is

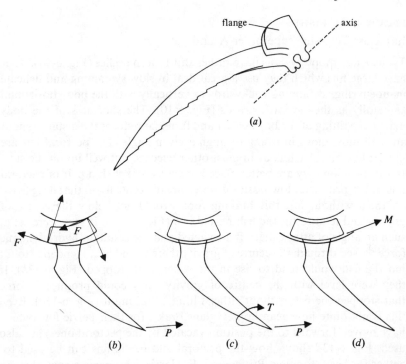

Fig. 6-12. Diagrams to show the mechanism of locking of the pectoral spines of catfishes. (*a*) represents a spine, while (*b*), (*c*) and (*d*) represent the spine in position in the slot in the pectoral girdle, with various forces acting on it.

erected at right angles to the body, or folded back along the body, the flange rotates in the slot, about the axis indicated in Fig. 6-12*a*. Fig. 6-12*b* shows that a force *P* on the end of the spine tends to twist the flange out of alignment with the slot. Frictional forces *F,F* act on the flange and if the coefficient of friction is large enough the spine jams and cannot be moved. To erect or depress it by forces acting on its tip, a torque *T* must be applied as well as the force *P* (Fig. 6-12*c*). This can be demonstrated by manipulating the spines of dead catfish. The spine can also be moved by a single force *M* acting at an appropriate point close to the flange (Fig. 6-12*d*). It is in such positions that the muscles that erect and depress the spine insert.

The rough surfaces on which the pectoral friction lock depends are rubbed together by many catfishes to make rasping sounds. No function is known with certainty for these sounds but they may serve as a warning to predators, which may learn that fish making such sounds have unpleasant spines and are best left alone.

PERCH-LIKE FISHES

Infraclass Teleostei, superorder Acanthopterygii

Typical acanthopterygians have rather short, deep bodies (Fig. 6-5*d*). They have large fins which they use a great deal in slow swimming and delicate manoeuvring, swimming backwards or vertically with the body horizontal as readily as they swim forwards (Fig. 5-10). The shortness of the body may make tilting of the body easier and faster: it reduces the distance snout and tail must move in tilting through a given angle. The pectoral fins are not set low on the trunk as in most other teleosts, but well up on the side. In this position they are better placed to act as brakes. If a pair of pectoral fins in the primitive, low position were spread vertically so that drag acted on them without lift, this braking force would act below the centre of gravity and tend to tilt the fish nose down (Fig. 6-13*a*). If they were set at such an angle of attack that lift also acted and the resultant hydrodynamic force passed through the centre of gravity there would be no tendency to tilt but the fish would tend to rise in the water as it stopped (Fig. 6-13*b*). If they were level with the centre of gravity, they could produce a force that stopped the fish without either tilting or raising it (Fig. 6-13*c*). Even this is not quite how acanthopterygians brake, for their pelvic fins (which have moved forward to the position vacated by the pectoral ones) are also used. Fig. 6-13*d* shows how the pectoral and pelvic fins can be used together for level stopping. Professor J. E. Harris, who made a special study of the functions of fins, found that if the pelvic fins of the acanthopterygian *Lepomis* were amputated it rose in the water whenever it braked. This is what one would expect from Fig. 6-13*d*, which shows an upward lift

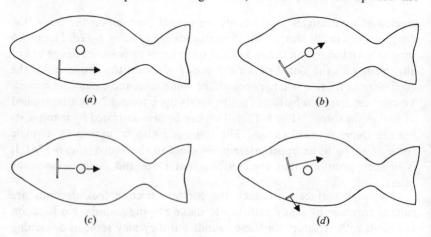

(*a*)   (*b*)

(*c*)   (*d*)

Fig. 6-13. Diagrams of fish using their pectoral and pelvic fins as brakes. The hydrodynamic forces acting on the fins are indicated. ○ represents the position of the centre of gravity of the fish.

component of the force on the pectoral fins, balanced by a downward lift component on the pelvics.

Most acanthopterygians have spines in their dorsal and pelvic fins. These are modified fin rays like those of catfish but are generally much more slender and lack locking devices. They are, however, much more numerous: several anterior rays in the dorsal and anal fins and the first ray of each pelvic fin are generally spines. The changed position of the fins makes the pelvics, rather than the pectorals, the more appropriate to bear protective spines. Spines pointing up from the dorsal surface and down from the ventral surface seem likely to make a deep-bodied fish particularly awkward to eat.

It is assumed that the spines have some protective value, but there is actually very little evidence for it. The spiny part of the dorsal fin has another function, in assisting turning by making the fish hydrodynamically unstable. This is perhaps best explained by referring to an arrow, with flights at the rear end. If the arrow is deflected so that it is no longer in line with its path through the air, lift acts on the flights and tend to pull it back into line. The flights give it aerodynamic stability. Flights at the other end of the arrow would have the opposite effect, tending to accentuate any deflection and so making the arrow unstable.

The spiny part of the dorsal fin lies largely anterior to the centre of gravity. Acanthopterygians often turn by raising it and one pectoral fin. The pectoral acts as a brake on its side of the body and so starts the turn. The spiny part of the dorsal fin speeds the turn by making the fish hydrodynamically unstable. The spines make it more effective because they are stiffer than ordinary rays, and because they are hinged to the pterygiophores and can only be raised and lowered: they cannot be deflected to the side like ordinary rays.

Most acanthopterygians have protrusible premaxillae which they use in the same way as cyprinoids do (Fig. 6-10). They are protruded as the mouth opens, and kept protruded as it closes. The mechanism, however, is quite different. The premaxillae are not attached to a tilting kinethmoid bone but to an ethmoid cartilage that can slide forward and back on the top of the cranium. Each maxilla is attached at point A to the suspensorium and at point B to the lower jaw, through the lip (Fig. 6-14). As the mouth opens B is pulled forward and the maxilla rotates about a transverse axis through A. If a pull is exerted part way along the maxilla, as indicated by the arrow in the diagram, the maxilla also rotates about the line AB. This makes a process on its head press on a process of the premaxilla, causing protrusion (see the dorsal views in Fig. 6-14). The requisite pull is exerted by tension in the skin and ligaments, if the mouth is opened widely enough. It can also be exerted by the superficial part (A_1) of the

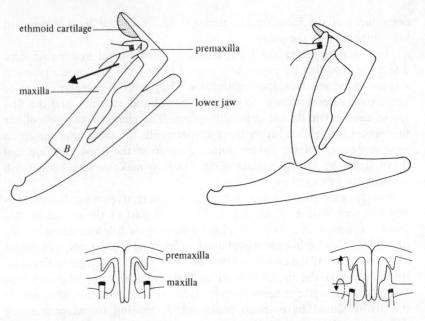

Fig. 6-14. Diagrammatic lateral and dorsal views of the skeleton of the jaws of a typical acanthopterygian showing (left) the mouth closed with the premaxillae retracted and (right) the mouth open with the premaxillae protruded.

adductor mandibulae, which has a tendon inserting on the maxilla. The electromyographic experiments of Osse showed that as a perch opens its mouth to feed and protrudes its premaxillae, this muscle contracts (p. 137).

Though most acanthopterygians have jaws which work in the manner described there are considerable variations in features such as mouth size and degree of protrusibility. The wide range is illustrated by the acanthopterygians of coral reefs. There are predatory groupers (*Epinephalus*) with big mouths which they use for eating other fish. There are butterfly-fishes (Chaetodontidae) which have tiny mouths on the ends of long snouts, with protruding teeth which are used to bite off coral polyps. There is a wrasse (*Epibulus*) with a small but extraordinarily protrusible mouth which it uses to pick shrimps and crabs out of the interstices between pieces of coral. There are also parrot-fishes (Scaridae) with the jaws modified as a beak that is not protrusible at all, and is used to scrape polyps and algae off the coral surface.

Acanthopterygians do not fall into a small number of large, sharply distinct groups, as Ostariophysi do. The following sections are not a survey of the superorder but brief accounts of two groups selected as being particularly interesting.

MACKEREL AND TUNA

Superorder Acanthopterygii, order Perciformes, family Scombridae

The mackerels and tunas seem more highly adapted for sustained swimming than any other fish. Species which have been kept in aquaria never stop swimming, day or night. They keep water flowing over their gills by holding their mouths open as they swim and seem unable to pump water fast enough over the gills when kept stationary. If mackerel (*Scomber scombrus*) are kept in well aerated water but prevented from swimming by being enclosed in a cage, the oxygen content of their arterial blood falls very low.

Power for sustained swimming is provided by the red muscle of the myomeres. This has been demonstrated by electromyographic experiments with skipjack tuna (*Katsuwonus*), like Bone's experiments with dogfish (p. 94). Mackerel and tunas have larger proportions of red fibres in their myomeres than most other teleosts. They have correspondingly large gill areas, which are presumably large enough to obtain oxygen as fast as the red muscles can use it. Tunas (the larger scombrids) have gills that are peculiar in structure as well as large in area.

Freshly caught tuna often have the deeper parts of their myomeres substantially warmer than the water in which they were swimming. Other teleosts have their muscle very close indeed to the temperature of the water. This has been investigated by probing fish, which have just been caught and killed, with long needles with thermistors at their ends. Fig. 6-15 shows temperatures measured in a typical Bluefin tuna. The water and the surface of the body were at 19 °C but much of the red muscle was above 29 °C (notice that the red muscle is not simply a superficial strip as in most fishes, but extends in to the vertebral column). Within limits, an increase in temperature increases the power that muscle can exert. An excised frog leg muscle, stimulated electrically to lift a given weight, lifts it 3.3 times as fast and so exerts 3.3 times as much power, at 20 °C as at 0 °C.

How are the high temperatures maintained? Blood must circulate constantly between the muscles and the gills. If it is warmed in the muscles it must (in a fish with a normal blood circulation) be cooled in the gills, and it will only be possible to maintain a very small temperature difference between the muscles and the water. A quick calculation will demonstrate this.

Let the mean difference in partial pressure of oxygen between the water and the blood at the gills be ΔP. Let the area of the gill surface be A, the distance for diffusion between water and blood d and the permeability constant for diffusion of oxygen D. Then by equation 3.1 (p. 48) the rate

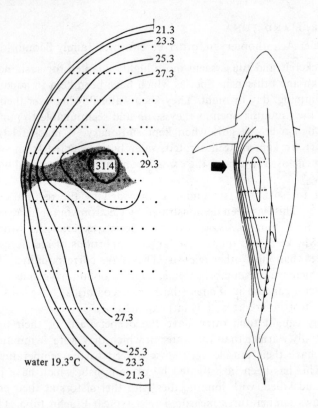

Fig. 6-15. The distribution of temperatures (in °C) in a Bluefin tuna (*Thunnus thynnus*). Dots represent the points at which measurements were made. The red muscle is stippled in the transverse section. From F. G. Carey & J. M. Teal (1969). *Comp. Biochem. Physiol.* **28**, 205–13.

M at which oxygen diffuses from water to blood is given by

$$M = DA \cdot \Delta P/d. \tag{6.1}$$

Similarly, if the mean temperature difference between the water and the blood is ΔT and the thermal conductivity of water and tissue is C, the rate H of loss of heat at the gills is given by

$$H = CA \cdot \Delta T/d. \tag{6.2}$$

From (6. 1) and (6. 2)

$$\Delta T = (HD/MC) \, \Delta P. \tag{6.3}$$

A similar equation can be obtained relating the differences in partial pressure of oxygen ($\Delta P'$) and temperature ($\Delta T'$) between the blood and the muscle. Hence the total temperature difference between the water and the muscle is given by

$$(\Delta T + \Delta T') = (HD/MC) (\Delta P + \Delta P'). \tag{6.4}$$

Metabolism using 1 cm³ oxygen produces 20 J heat, so if the system is in equilibrium $H/M = 20$ J cm^{-3}. The permeability constant for oxygen diffusing through water is 2×10^{-3} cm^2 atm^{-1} h^{-1} and this will be taken as D, although the diffusion is partly through tissue in which the permeability constant is lower. The thermal conductivity of water is about 20 J cm^{-1} h^{-1} K^{-1}, and this will be taken as C. $(\Delta P + \Delta P')$ cannot be greater than 0.2 atm. Hence

$$(\Delta T + \Delta T') \leqslant 20 \times 2 \times 10^{-3} \times 0.2/20$$
$$\leqslant 4 \times 10^{-4} \text{ K.}$$

Taken at its face value, this indicates that the muscles could not be more than a tiny fraction of a degree warmer than the water. However, the main trunk muscles of *Tilapia* (a teleost with quite ordinary blood circulation) have been found to be generally 0.4 K warmer than the water. The discrepancy between theory and observation probably arises because the arteries to the trunk muscles run quite near the veins leaving them. It is by means of a quite different and much more highly developed arrangement of arteries running alongside veins that the high muscle temperatures of tunas are maintained.

This peculiar arrangement of blood vessels is shown in Fig. 6-16. The myomeres do not get their blood from the aorta, but from two arteries

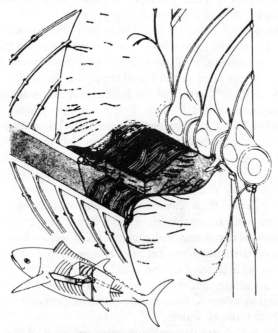

Fig. 6-16. A diagram showing the blood supply of the swimming muscles of a Bigeye tuna (*Thunnus obesus*). From F. G. Carey & J. M. Teal (1966). *Proc. nat. Acad. U.S.A.* **56**, 1461–9.

running close under the skin on either side of the body. They are not drained by the usual veins but by ones running alongside these arteries. Fine arteries and veins run inwards from these vessels to the red muscle, parallel to each other and in close contact. The blood supply to the white muscle is more sparse but also consists of parallel arteries and veins in contact with each other, extending inwards from larger vessels under the skin. The arrangement is that of a countercurrent heat exchanger (p. 46). The superficial arteries and veins are cool, close to the temperature of the water. The deeper muscle is warm. Blood in the fine arteries going to the muscle is warmed by blood leaving in the parallel veins. As it leaves again in the veins it is cooled by blood arriving in the arteries. Thus the muscle is kept much warmer than the water but the blood is close to water temperature by the time it returns to the gills.

There might seem to be a snag in the system. If heat can be conducted between the arteries and the veins, oxygen can also diffuse. The countercurrent exchanger which keeps the muscle much warmer than the gills must also be expected to make the partial pressure of oxygen in the muscle fall below that in the gills. The likely extent of this can be worked out in the same way as equation 6. 4 was derived. It emerges that a 10 °C temperature difference could in principle be obtained for the penalty of only a 0.001 atm drop in partial pressure of oxygen.

The tunas are not quite the only fish to keep their muscles warmer than the water. The sharks *Lamna* and *Isurus* do the same, but the temperature differences are generally smaller. They also have counter-current heat exchangers incorporated in the blood supply to the muscles. The streamlined shape of these sharks and the high aspect ratios of their caudal fins have already been noted (p. 87). Tunas are very similar in shape. The sharks cannot fold their fins against the body but the tunas can: their pectoral fins fit into shallow depressions so that they are flush with the sides of the body, and the anterior dorsal fin folds down into a longitudinal slot.

Curiously, many mackerels and tunas have no swimbladder and are considerably denser than seawater. This must increase the energy needed for swimming but presumably gives some other, compensating advantage. One possibility is that it may make them less conspicuous to toothed whales which feed on them (dolphins eat mackerel and killer whales, *Orcinus*, eat tuna). These whales seem to use echolocation in finding food (p. 435). Swimbladders scatter underwater sound strongly and so are conspicuous to echolocation systems, because the acoustic properties of air are very different from those of water.

Scombrids which are denser than water get most of the lift they need by swimming with their pectoral fins extended as hydrofoils.

FLATFISHES

Superorder Acanthopterygii, order Pleuronectiformes

Though both groups belong to the Acanthopterygii, the flatfishes are strikingly different from the scombrids. They start their life as very normal-looking, symmetrical larvae but then undergo a remarkable transformation. One eye with its optic nerve moves across the top of the head, through the still unossified tissues that will form the bones of the skull roof. Thus both eyes come to lie on one side of the body. The skin of this side becomes pigmented, while that of the other side does not, but remains white. The adult fish lies on the bottom with its pigmented, eyed side uppermost.

Like many other teleosts that spend a lot of time on the bottom, adult flatfishes have no swimbladder. (Some species have a swimbladder in the symmetrical larval stage.) Their shape as they lie on the bottom is flattened, like that of rays, but while rays are flattened dorsoventrally and lie on their bellies flatfish are flattened transversely and lie on their sides. The fins generally have no spines. The dorsal and anal fins are very long and together run most of the way around the body. Slow swimming over the bottom is achieved by undulating these fins, but the whole body is undulated in more vigorous swimming. The fish also use their dorsal and anal fins in the process of burying themselves in sand.

Flatfishes often bury themselves with only the mouth, eyes and upper operculum uncovered. This conceals them from predators and from potential prey. *Bothus* lies in ambush in this way, swimming up suddenly to chase passing fish. While flatfish are buried, they can pump water out directly from the uncovered upper opercular opening, but any leaving the lower opercular opening would have to be pumped out through the sand. In practice, only the upper opening seems to be used when the fish is buried or even simply resting on the bottom. If a drop of ink is placed in front of a flatfish resting quietly on the transparent bottom of an aquarium, it is drawn into the mouth and can be seen to emerge from the upper opening only. However, both opercula keep moving, apparently pumping water over their gills. It seems that water pumped over the lower gills travels through a passage connecting the two opercular cavities, to emerge from the upper opercular opening. Other teleosts do not have this passage.

The sole (*Solea*) and its immediate relatives are the most advanced of all flatfishes (Fig. 6-17). The dorsal fin extends forward over the head. An overhanging, hook-like snout effectively divides the mouth into left and right halves. The jaw symphysis is loose enough to allow the oddly shaped lower jaw of the left (lower) side to rotate about the axis indicated by the broken line in Fig. 6-17b, opening the lower side only of the mouth.

Similarly the right (upper) side can be opened separately. Sole resting on the bottom open only the upper side of the mouth for breathing, but the lower side for feeding. The upper side is obviously better placed for taking in respiratory water without sand grains but the lower half is better placed for taking food (mainly polychaete worms) from the bottom.

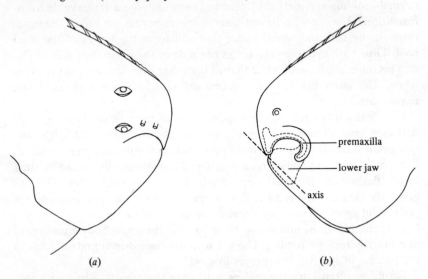

(a)   (b)

Fig. 6-17. (*a*) The right (upper) side and (*b*) the left (lower) side of the head of a sole (*Solea solea*). Outlines of the left premaxilla and lower jaw are shown in (*b*). Further explanation is given in the text.

OTHER ADVANCED TELEOSTS

Infraclass Teleostei, superorders Paracanthopterygii and Atherinomorpha

There are two more superorders of teleosts which are generally considered advanced. The Paracanthopterygii resemble the Acanthopterygii in many respects but have various distinguishing features including a peculiar additional jaw muscle. The best known ones are the cod and its close relatives (family Gadidae). The Atherinomorpha are mostly small surface-living fishes and include the toothcarps (Cyprinodontoidei) which are often kept in tropical freshwater aquaria. The guppy (*Lebistes*) and the swordtail (*Xiphophorus*) are particularly popular. Gadids have slightly protrusible premaxillae which work just like those of acanthopterygians. (Many accounts of them say that their jaws are not protrusible, but an unpublished cinematograph film of *Pollachius* feeding, taken by Miss S. Khrinam, shows protrusion.) Toothcarps also have protrusible premaxillae, but they work quite differently.

FURTHER READING

General

See the general lists for fishes (Chapter 3) and for teleosts (Chapter 5).

Fish with Weberian ossicles

Alexander, R. McN. (1966). Physical aspects of swimbladder function. *Biol. Rev.* **41**, 141–76.

Pfeiffer, W. (1963). Alarm substances. *Experientia,* **19**, 113–23.

Rosen, D. E. & P. H. Greenwood (1970). Origin of the Weberian apparatus and the relationships of the ostariophysan and gonorhynchiform fishes. *Amer. Mus. Novit.* **2428**, 1–25.

Characins

Alexander, R. McN. (1964). Adaptation in the skulls and cranial muscles of South American Characinoid Fish. *J. Linn. Soc. (Zool.),* **45**, 169–90.

American knife-fishes

Bennet, M. V. L. (1971). Electric organs and Electroreception. In Hoar, W. S. & Randall, D. J. (eds.), *Fish physiology,* vol. 5, pp. 347–574. Academic Press, New York.

Lissmann, H. W. (1963). Electric location by fishes. *Scient. Am.* **208**, 50–9.

Carps and minnows

Ballintijn, C. M., A. van den Burg & B. P. Egberink (1972). An electromyographic study of the adductor mandibulae complex of a free-swimming carp (*Cyprinus carpio* L.) during feeding. *J. exp. Biol.* **57**, 261–83.

Hickling, C. F. (1966). On the feeding process in the White amur, *Ctenopharyngodon idella. J. Zool., Lond.* **148**, 408–19.

Catfishes

Alexander, R. McN. (1966). Structure and function in the catfish. *J. Zool., Lond.* **148**, 88–152.

Recher, H. F. & J. A. Recher (1968). Comments on the escape of prey from avian predators. *Ecology,* **49**, 560–2.

Perch-like fishes

Alexander, R. McN. (1967). The functions and mechanisms of the protrusible upper jaws of some acanthopterygian fish. *J. Zool., Lond.* **151**, 43–64.

Hiatt, R. W. & D. W. Strasburg (1960). Ecological relationships of the fish fauna on coral reefs of the Marshall Islands. *Ecol. Monogr.* **30**, 65–127.

Mackerel and tuna

Carey, F. G., J. M. Teal, J. W. Kanwisher, K. D. Lawson & J. S. Beckett (1971). Warm-bodied fish. *Am. Zoologist,* **11**, 137–45.

Walters, V. (1962). Body form and swimming performance in the scombroid fishes. *Am. Zoologist,* **2**, 143–9.

7

Lobe-finned fishes

This chapter is about two groups of fishes, the lungfishes and the crossopterygians, which have between them only four modern genera. Both groups were far more numerous in the Palaeozoic, but they would not deserve a chapter to themselves were it not for the light they throw on the origin of the tetrapods – that is, of the amphibians and their descendants the reptiles, birds and mammals. There is clear evidence, which will be considered later, that the amphibians evolved from the crossopterygians. The lungfishes resemble the crossopterygians in various ways (for instance, both have the 'lobe fins' which give this chapter its title) but there is disagreement among zoologists as to whether the groups are closely related. Whether they are or not the lungfishes breathe air and resemble the amphibians in many other ways. Study of the lungfishes helps us to understand how amphibians may have evolved from fishes.

LUNGFISHES
Class Osteichthyes, subclass Dipnoi

The lungfishes appeared at the beginning of the Devonian period as a group of fish not very different in appearance from the palaeoniscoids, which have already been described, or from the early crossopterygians, which will be described shortly (Fig. 7-1a; compare Figs. 7-1d, e and

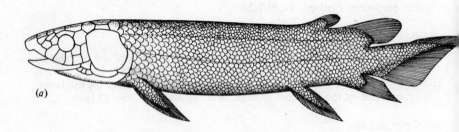

(a)

Fig. 7-1. A selection of lobe-finned fishes. (a) Dipterus, (b) Neoceratodus and (c) Protopterus (all Dipnoi); (d) Holoptychius and (e) Eusthenopteron (Rhipidistia); and (f) Latimeria (Coelacanthini). From E. Jarvik (1960). Théories de l'Évolution des vertébrés. Masson, Paris and J. R. Norman & P. H. Greenwood (1963). A history of fishes, 2nd edit. Benn, London.

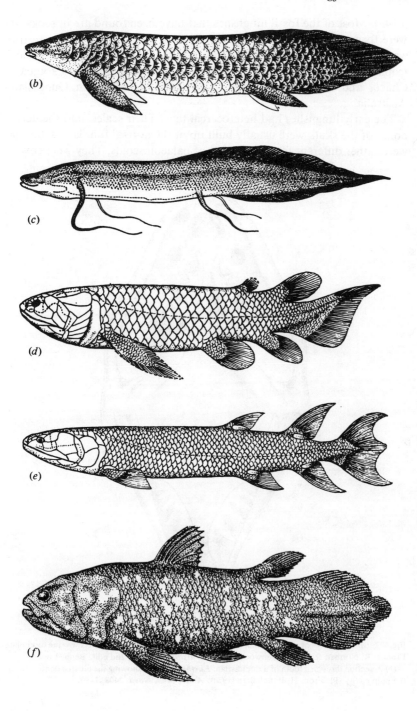

(b)

(c)

(d)

(e)

(f)

5-14*a*). Most of the fossil lungfishes that have been found are in rocks that were formed in fresh water, and the three modern genera all live in fresh water. These are *Protopterus* (Fig. 7-1*c*) of which there are several species in tropical Africa, *Lepidosiren* from the Amazon basin and the Paraguayan Chaco, and *Neoceratodus* (Fig. 7-1*b*) from two rivers in Queensland, Australia.

The early lungfishes had heterocercal tails. Their scales, and the dermal bones of the skull, were usually built up of the normal four layers, but they were rather different from the scales of palaeoniscoids. They were covered

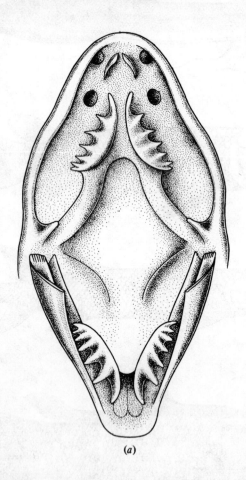

(*a*)

Fig. 7-2. (*a*) The mouth of an Australian lungfish, *Neoceratodus*, forced open to show the toothplates. From J. R. Norman & P. H. Greenwood (1963). *A history of fishes*, 2nd edit. Benn, London.
(*b*) A section through a tooth of a rhipidistian, *Eusthenopteron*, showing the pleated dentine (O) and the pulp cavity (P). From H.-P. Schultze (1970). *Amer. Mus. Novit.* **2408**, 1–10.

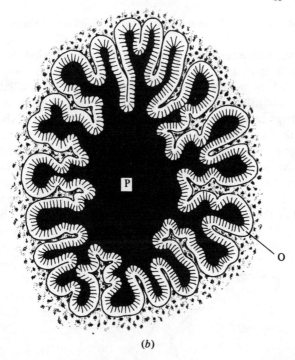

(*b*)

by a single layer of enameloid substance like the scales of crossopterygians
(Fig. 3-17*d*). Palaeoniscoid scales had a thick covering of enameloid substance laid down as a series of superimposed layers (Fig. 3-17*g*). The primitive lungfish (thin enameloid) type of scale is known as cosmoid and the
palaeoniscoid (thick enameloid) type as ganoid. Modern lungfishes, like
teleosts, have lost all the layers of the primitive scale except the laminated
bone. *Neoceratodus* has large scales but the others have small ones.

The most characteristic features of the lungfishes are their toothplates
(Fig. 7-2*a*). They consist of a very hard material which, like enameloid substance, contains a high proportion of calcium phosphate and very little
collagen, but which does not grow in the same way. It is very like the
material of the toothplates of Holocephali (p. 128). As in Holocephali, the
upper toothplates are borne by palatoquadrates which are fixed rigidly to
the cranium. The lower plates, on the lower jaw, have sharp ridges that
interdigitate when the mouth closes with corresponding ridges on the
upper plates. Most early lungfishes had toothplates like modern ones, but
the earliest lungfish known, *Uranolophus*, had only a narrow border of
toothplate material on palatoquadrates and lower jaw. Modern lungfishes
have large jaw muscles. They chew food between the toothplates to break
it up before it is swallowed. *Protopterus* feeds mainly on molluscs, while

Neoceratodus and *Lepidosiren* seem to eat various invertebrates and plant material.

Lungfishes, with a few exceptions in the Devonian period, have no teeth in the mouth apart from the toothplates. There are no teeth along the outer edges of the mouth, such as are borne by the premaxillae, maxillae and dentaries of other bony fishes. The premaxillae and maxillae are absent except perhaps in a few fossils that seem to have vestiges of them.

The nasal sacs of most teleosts each have two openings on the top of the head. Water enters at the anterior nostril and leaves by the posterior one, driven by the movement of the fish through the water, by cilia or by the indirect action of respiratory movements. In lungfishes the anterior nostril is on the upper lip and the posterior one has moved right inside the mouth. Reduction of the maxilla meant that there was no bone to obstruct this shift. When a lungfish breathes water it opens its mouth only slightly and some water must be drawn in through the nasal sacs. Thus breathing is

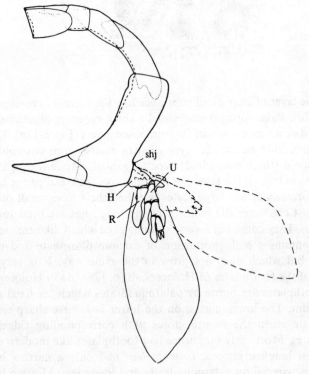

Fig. 7-3. Pectoral girdle and fin of the rhipidistian *Eusthenopteron*. From S. M. Andrews & T. S. Westoll (1970). *Trans. Roy. Soc. Edinb.* **68**, 207–329. H, humerus; R, radius; U, ulna; shj, shoulder joint.

made to help smelling, though by a mechanism quite different from those of lampreys (p. 50) and some teleosts. When lungfish come to the surface to breathe air they open their mouths widely and little or no air must pass through the nasal sacs.

Most lungfishes and crossopterygians have pectoral and pelvic fins quite unlike any fins mentioned so far (Figs. 7-1, 7-3), though some early selachians had rather similar fins. These fins have a relatively thick central axis or lobe containing pterygiophores and muscles, edged by a thin blade stiffened only by fin rays. *Protopterus* and *Lepidosiren* have vestigial pectoral and pelvic fins consisting of slender filaments without blades, but *Neoceratodus* has pectoral and pelvic fins of the primitive type. Early lungfish probably used their fins in the same way as it does. It uses its pectoral fins to row itself slowly along and to 'walk' along the bottom, but the fins are only strong enough to support the body when it is submerged in water: *Neoceratodus* cannot walk on land. The pectoral fins can be moved up and down or forward and back from the shoulder, they can be rotated about their long axis (i.e. moved as an oar is moved when it is feathered) and they can be bent. The pelvic fins are less mobile. Lungfish pectoral and pelvic fins have probably evolved from fins much more like those of placoderms and early selachians. We have seen how shark fins became more mobile in the course of evolution by narrowing of their bases (Figs. 4-11c, 4-21c).

BREATHING AIR

The best-known fact about the lungfishes is that they have lungs and use them to breathe air. They are by no means the only fish which breathe air. We have already seen that the swimbladders of teleosts probably evolved from lungs (p. 167). Some of the characins in the swamps of tropical America have modified swimbladders which they use for breathing air: the swimbladder has, in effect, evolved back into a lung. Some catfishes living in the same swamps use their intestines as lungs. There are many other examples. The ability to breathe air enables *Protopterus* and *Lepidosiren* to survive for many months in burrows when the swamps in which they live dry up in the dry season. However, most air-breathing fish including *Neoceratodus* seem to spend their whole lives in water. Air breathing probably evolved in most cases as an adaptation to life in water containing little dissolved oxygen, rather than as an adaptation to life out of water.

Fresh water that is saturated with air contains about 8 cm³ dissolved oxygen per litre, depending on the temperature, but there are many environments where water is far from being saturated with air. If this water is far below the surface (as in the oxygen minimum layer of the oceans, or in the depths of lakes in summer) the ability to breathe air can be no help to fish that live in it. If, however, it is near enough the surface for it to be feasible to go up for breaths of air, air-breathing fish can survive in it however low the oxygen concentration may fall. Particularly low oxygen concentrations occur in tropical swamps overhung by dense vegetation. During the day these swamps are shaded from the sun. At night, they do not radiate heat to a cold sky but to this same vegetation, which is never much cooler than the swamp and radiates back almost as much heat as it receives from below. Daily temperature fluctuations in the water are therefore small and there is relatively little mixing by convection currents at night, which would distribute oxygen through the whole body of water as the surface water cooled and sank. The water is stagnant, so there is no turbulence to mix it, and the vegetation shields it from disturbance by wind. The vegetation also makes the light so dim that little oxygen can be produced by photosynthesis in the water. Concentrations below 1 cm^3 dissolved oxygen per litre are usual except within a few millimetres of the surface, where a higher concentration is maintained by diffusion from the air.

If there is too little dissolved oxygen in the water, it can be unprofitable to breathe it. The lower the concentration of dissolved oxygen, the faster water must be pumped over the gills to supply the needs of the fish. Pumping faster means pumping under greater pressure, and more work must be done to pump each litre of water over the gills (the work done by a pump is the volume pumped multiplied by the pressure change in the pump). The fish does more work on each litre of water and gets less oxygen from it. The water, travelling faster over the gills, has less time to equilibrate with the blood, and this exaggerates the effect. Eventually, as oxygen concentration falls and pumping gets faster there must come a time when most or all of the oxygen obtained is needed to drive the respiratory pump. Fig. 7-4 shows evidence that at low concentrations of dissolved oxygen the respiratory muscles may use a substantial proportion of the oxygen uptake; the best explanation for the initial rise in oxygen uptake at rest as oxygen concentration falls seems to be that the extra oxygen is needed to drive the respiratory pump. At very low oxygen concentrations uptake inevitably falls and the fish eventually asphyxiates. Fig. 7-4 suggests that air breathing would be advantageous at concentrations of 4 cm^3 l^{-1} (partial pressure 0.1 atm) or less and essential for life at substantially lower concentrations. There are of course differences between species but most fish which are not air breathers asphyxiate between about 0.3 and 1.5 cm^3 oxygen l^{-1}.

Neoceratodus has a single lung like *Amia*, while *Protopterus* and *Lepidosiren* each have a pair of lungs that merge at the anterior end into a single chamber. If oxygen is to diffuse rapidly into the blood the area available for diffusion must be large and the distance from air to blood short. Complicated partitions in the lungs greatly increase this area (Fig. 7-5). I

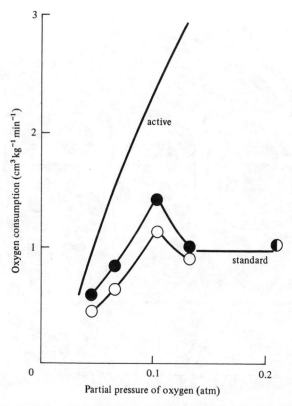

Fig. 7-4. Graphs of oxygen consumption of a brook trout, *Salvelinus*, against partial pressure of dissolved oxygen in the water. Both consumption during swimming at the maximum sustainable rate ('active') and at rest ('standard') are shown.
●, Fish acclimated to air-saturated water; ○, fish acclimated to the oxygen concentration in question. From F. W. H. Beamish (1964). Reproduced by permission of the National Research Council of Canada from *Can. J. Zool.* **42**, 355–66.

know no estimates of this area in lungfishes, but the lung areas of many amphibians and reptiles are similar to the gill areas of fishes of the same weight, around 4 cm² (g body weight)$^{-1}$. The lung wall is richly supplied with blood capillaries and the blood is separated from the air only by the walls of these capillaries and the lung epithelium, a distance in *Protopterus* of about 0.5 μm. The corresponding distance in fish gills is commonly over 2 μm (p. 45), but the lungs are in a more protected position and can be more delicate. Also, there is more advantage in having the blood very close to the respiratory surface in lungs than in gills. This is because oxygen diffuses immensely faster through air than through water: its permeability constant in air is about 3×10^5 times as high as in water. In gills an important part of the resistance to diffusion is due to most of the

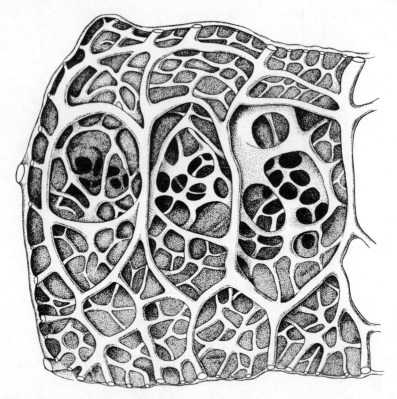

Fig. 7-5. Lung of *Protopterus* opened to show the internal partitions and alveoli. From M. Poll (1962). *Ann. Reeks Zool.* Wetenschap. **108**, 131–72.

oxygen having to diffuse through 5 μm or more of water before reaching the gill surface (p. 45). Shortening the distance from this surface to the blood beyond a certain point has only a marginal effect. In a lungfish lung oxygen might well have to diffuse through 3 mm of air before reaching the respiratory surface, but because of the difference in permeability constants this is equivalent to only $3/(3 \times 10^5)$mm $= 0.01$ μm water. The layer of tissue between the air and the blood offers the major part of the resistance to diffusion, and there is a great advantage in having it as thin as possible.

Partitions which increase the surface area of a lung could substantially increase the pressure needed to inflate it, owing to the effects of surface tension. The partitions divide the wall of the lung into pockets known as alveoli, of which the smallest in *Protopterus* seem to have radii around 50 μm. If the fluid coating an alveolus of this size had the same surface tension as water, 73 dyn cm^{-1}, the pressure (2 × surface tension/radius) needed to overcome this surface tension as the alveolus was inflated would be $(2 \times 73)/(5 \times 10^{-3}) = 3 \times 10^4$ dyn cm^{-2} = 30 cmH$_2$O. The actual pressure which has been recorded from the mouth cavity is much less, about 15 cmH$_2$O. The reason is the presence in the lung of a surface active material (apparently a lipoprotein) which forms a film coating the inner surface of the lung and greatly reduces the surface tension. Similar material is present in the lungs of tetrapods. It is deficient in some human babies, which may die because of the difficulty they have in inflating their lungs. There is, of course, a duct leading from the lungs to the mouth. The glottis, its opening into the floor of the pharynx, has muscles to open and close it.

Neoceratodus makes far less use of air breathing than the other modern lungfishes. It has a full set of gills and it relies almost entirely on them for respiration when it is kept in well aerated water. It takes a breath of air occasionally, but these breaths are often more than an hour apart. It can survive indefinitely when prevented from reaching the surface. If it is put into water of low dissolved oxygen content it pumps water faster over its gills, just like normal fish that breathe only water, but it also takes breaths of air more often. The ability to breathe air does not seem to be of much importance to it in its natural habitat. The rivers where it lives seem never to dry up, and never to contain less than about 5 cm^3 oxygen l^{-1}.

Protopterus and *Lepidosiren* commonly live in swamps which are much less well aerated. They have reduced gills, and indeed *Protopterus* has no gill filaments on two of its gill bars. Even in well aerated water they visit the surface to breathe air every 3–10 min, and they asphyxiate if prevented from reaching the surface. It has been shown that even in well aerated water *Protopterus* only gets 10% of its oxygen from the water: this was done by keeping the fish in a closed aquarium and following the reduction of partial pressure of oxygen in the water and in the air above.

It might be thought remarkable that a gulp of air every 3–10 min should provide 90% of the oxygen a lungfish needs. Teleosts breathing water make many cycles of breathing movements every minute. For instance, a large trout at 10 °C took 82 breaths min^{-1} to obtain 0.9 cm^3 oxygen kg^{-1} min^{-1}. *Protopterus* at 24 °C used 1.0 cm^3 oxygen kg^{-1} min^{-1} (getting 0.9 cm^3 of this from the lungs) but only took a breath of air about once in 5 min. The trout, obtaining oxygen through the gills at the same rate as the lungfish obtained it from the lungs, took about 400 breaths of water for every breath of air taken by the lungfish. However, the water used by the trout contained only 8 cm^3 oxygen l^{-1} while the air used by the lungfish contained 200 cm^3 l^{-1} or 25 times as much. The trout took only 1.8 cm^3 water (kg body weight)$^{-1}$ in each cycle of respiratory movements while the lungfish took very much larger volumes of air. *Protopterus* normally holds about 35 cm^3 air (kg body weight)$^{-1}$ in its lungs, and most of this is changed at each breath. The air taken in by *Protopterus* at each breath may well have contained 400 times as much oxygen as the water taken in by the trout.

Protopterus can remove a very high proportion of the oxygen from the air in its lungs. This has been demonstrated by collecting the air it breathes out in an inverted water-filled funnel. Analysis of this air shows that if the interval since the preceding breath has been a long one, almost all of the oxygen has been removed. About 60% of the oxygen seems to be removed between breaths at average intervals.

Suppose a lungfish takes in at each breath 10 cm^3 air containing 8 cm^3 nitrogen, 2 cm^3 oxygen and a negligible amount of carbon dioxide. Suppose that 1 cm^3 of the oxygen is removed in the lungs. Metabolism which uses 1 cm^3 oxygen generally produces about 1 cm^3 carbon dioxide, so the fish must get rid of about 1 cm^3 carbon dioxide at each breath. If all this leaves through the lungs the air breathed out must contain 8 cm^3 nitrogen, 1 cm^3 oxygen and 1 cm^3 carbon dioxide: that means that the partial pressure of carbon dioxide in it is 0.1 atm. Carbon dioxide could only diffuse into the lungs to this partial pressure if its partial pressure in the blood was even higher. However, the experiments that showed that 90% of the oxygen is taken up in the lungs also showed that 70% of the carbon dioxide is eliminated through the gills. A lot of carbon dioxide can be dissolved in only a little water passing over the gills without giving it a high partial pressure, because carbon dioxide is highly soluble in water. The partial pressure of carbon dioxide in the blood of *Protopterus* in well aerated water has been found to be about 0.03 atm.

The partial pressure of carbon dioxide may rise very high in the waters of tropical swamps, to values around 0.1 atm. In such conditions irrigating the gills can do nothing to keep down the partial pressure of carbon dioxide in the blood: rather, it may tend to raise the partial pressure by allowing carbon dioxide from the water to diffuse into the blood. Air-breathing fish tend to stop irrigating their gills when the partial pressure of carbon dioxide in the water is raised. *Protopterus* does this when air containing 0.05 atm carbon dioxide is bubbled through the water. When this is done, or when the fish is removed from water, it takes more frequent breaths and so prevents the partial pressure of carbon dioxide in the lungs and blood from rising too high. The partial pressure in the blood of a *Protopterus* out of water has been found to be 0.06 atm.

Contrast *Protopterus* with an ordinary water-breathing fish in well aerated water. This fish would probably remove 30–80% of the dissolved oxygen from the water passing over its gills. Suppose it removed 50%, reducing the partial pressure of oxygen in the water by 0.1 atm. Carbon dioxide is 25 times as soluble in water as oxygen, so if carbon dioxide was released into the water as fast as oxygen was removed it would raise the partial pressure of carbon dioxide in the water by only $0.1/25 = 0.004$ atm. The partial pressure in the blood can be correspondingly low, and is generally about 0.005 atm in the venous blood of water-breathing fishes. This is far lower than the values found in *Protopterus*, and a lungfish out of water or in a carbon dioxide-rich swamp would have to take exceedingly frequent breaths to get the partial pressure of carbon dioxide in its blood down to so low a value.

The high partial pressures of carbon dioxide mean that lungfishes and other swamp-living fishes are best served by haemoglobin rather different from that of most other fishes. The effect of carbon dioxide on the ability of haemoglobin to take up oxygen has already been described (p. 149). In most teleosts the effect is very marked, and a partial pressure of carbon dioxide of 0.03 atm (such as is usual in the blood of *Protopterus*) may prevent full oxygenation occurring even in the presence of 0.2 atm oxygen. The air in the lung of *Protopterus* will never contain more than 0.2 atm oxygen and will usually contain less. The haemoglobin of *Protopterus* and other air breathers is much less sensitive to carbon dioxide than that of most teleosts (Fig. 5-8).

Dr Kjell Johansen and his colleagues have carried out a series of experiments on lungfish in which they have been able to sample blood from various blood vessels, and record the pressures in the same vessels, as the fish swam freely in an aquarium. The information on carbon dioxide contents already referred to was obtained from these experiments, and more information from the same experiments will be referred to soon. The fish was anaesthetized and opened, cannulae were fixed in the chosen blood vessels and the surgical wounds were sewn up. When the fish recovered from the anaesthetic it was able to swim in its aquarium with long tubes connecting the cannulae to pressure transducers. There were also arrangements for withdrawing blood samples through the same tubes. Fig. 7-6 shows some of the results of these experiments and of similar experiments on trout. It also shows the arrangement of the blood vessels concerned. Note that blood travels to the lungs of lungfish along pulmonary arteries which branch off the last efferent branchial arteries. It is returned direct to the atrium by pulmonary veins, by-passing the sinus venosus.

The experiments show that in trout and in *Neoceratodus* there is a substantial difference in pressure between the ventral aorta and the dorsal aorta. (It seems safe to assume that the pressure in the dorsal aorta of *Neoceratodus* is about the same as the measured pressure in the pulmonary artery.) This pressure difference is required to drive the blood through the capillary spaces of the gill lamellae. In *Protopterus* two of the gill bars lack

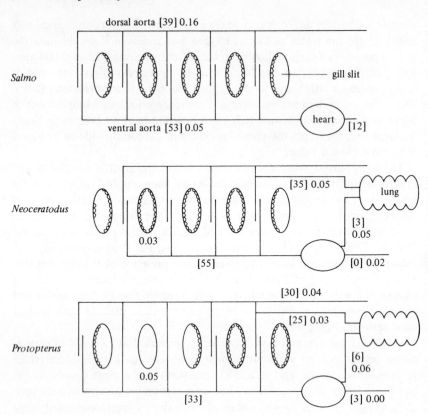

Fig. 7-6. Diagrams showing the arrangement of the branchial and pulmonary arteries in trout (*Salmo*) and the lungfishes *Neoceratodus* and *Protopterus*. Also shown are the results of experiments with cannulated, free-swimming fish in well-aerated water: the peak (systolic) pressure in cmH$_2$O in various vessels is shown in brackets [ ] and the partial pressure in atm of oxygen in the vessels is shown without brackets. Data from E. D. Stevens & D. J. Randall (1967). *J. exp. Biol.* **46**, 307–15; G. F. Holeton & D. J. Randall (1967). *J. exp. Biol.* **46**, 317–29; and K. Johansen, C. Lenfant & D. Hanson (1968). *Z. vergl. Physiol.* **59**, 157–86.

gills and the arteries run directly through them, connecting the ventral aorta to the dorsal aorta without intervening capillaries. The pressure difference between the aortas is consequently small and the heart does not have to pump the blood at as high a pressure as in teleosts, so energy is saved. Presumably only a small proportion of the blood flows through the gills that remain. A small gill area is enough for getting rid of carbon dioxide because carbon dioxide diffuses through water very much faster than oxygen does.

We have already noted that *Neoceratodus* in well aerated water gets nearly all its oxygen from its complete set of gills. Since it takes breaths of air very seldom, the partial pressure of oxygen in the lungs must tend to fall until it equals the partial pressure in the blood arriving at the lungs from the gills. This seems to happen, for the partial pressures in the pulmonary artery and pulmonary vein are found to be approximately equal (Fig. 7-6). However, in experiments with *Neoceratodus* in water of low oxygen content, taking much more frequent breaths, the partial pressure of oxygen was found to be only 0.03 atm in the pulmonary artery but 0.13 atm in the pulmonary vein.

In these experiments, in which *Neoceratodus* was depending largely on its lungs, and in the experiments on *Protopterus*, the blood in the anterior branchial arteries was found to be richer in oxygen than the blood arriving at the lungs. Since blood in these arteries is less likely to go to the lungs than blood in the last branchial arteries, this indicates that blood that is oxygenated by the lungs in one circuit from the heart tends to be directed away from them on the next. We have here the beginnings of a double circulation like the familiar arrangement in mammals. In mammals the heart is partitioned so that no mixing is possible between the oxygen-rich blood from the lungs and the oxygen-poor blood from the rest of the body. All the blood which goes to the lungs on one circuit from the heart avoids them on the next, and vice versa. In lungfish a good deal of mixing does occur (otherwise the partial pressure of oxygen would be as high in the dorsal aorta as in the pulmonary vein and as low in the pulmonary artery as in the sinus venosus). Complete mixing is prevented by incomplete partitions in the atrium, ventricle and conus arteriosus (Fig. 7-7). The atrium and ventricle are partially divided into left and right halves. The pulmonary veins bring blood from the lungs directly to the atrium on the left side of the partition. The rest of the blood enters the right side through the sinus venosus. The partition in the conus twists so as to become horizontal at its anterior end, with oxygen-rich blood from the left side of the heart ventral to it and oxygen-poor blood from the right side dorsal to it (Fig. 7-8). The first three pairs of branchial arteries start from below the partition and the last two pairs start from above it. Since the pulmonary arteries are branches of the last efferent branchial arteries they receive mainly oxygen-poor blood.

The partition in the conus is known as the spiral valve (it must not, of course, be confused with the spiral valve of the intestine, which lung-fishes also possess). It consists in *Protopterus* and *Lepidosiren* of two ridges rising from opposite walls of the conus and almost meeting in the middle (Fig. 7-7a). In *Neoceratodus* only one of the ridges runs the length of the conus and it is obviously formed from a series of valves just like the valves in the conus of selachians (Fig. 7-7b).

The advantage of separating the two streams of blood seems plain. The

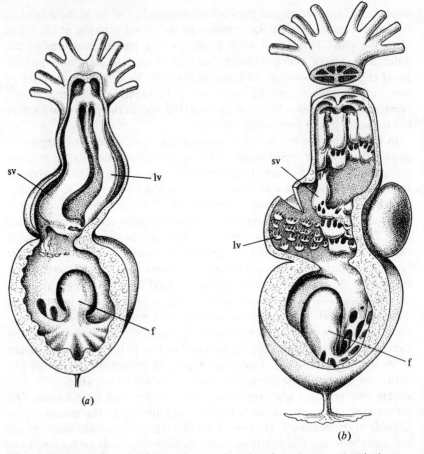

Fig. 7-7. The hearts of two lungfishes, opened to show the spiral valves (s.v., l.v.) in the conus arteriosus and the partial division (f) between the two halves of the ventricle. (*a*) is *Protopterus* and (*b*) *Neoceratodus*. From E. S. Goodrich (1909). In E. R. Lankester (ed.), *Treatise on Zoology*, part 9. Macmillan, London.

lower the partial pressure of oxygen in the blood entering the lungs, the faster will oxygen diffuse into it. The higher the partial pressure in the blood arriving at other tissues, the faster can oxygen diffuse into the tissues. If mixing is prevented the lungs will receive blood with a low partial pressure of oxygen and the rest of the body will receive blood with a high partial pressure. If complete mixing occurs they will receive blood with the same, intermediate partial pressure and diffusion of oxygen in both will be slower. If complete mixing occurred lungfish would need a bigger lung area, and more capillaries in their tissues, to maintain the same rate of consumption of oxygen. Also, a given volume of blood arriving at the tissues would have less oxygen to give up and a given volume arriving at the lungs would have less capacity for taking up oxygen, so the heart would have to pump blood faster.

In normal water-breathing fishes blood goes first to the gills and then to the rest of the body before returning to the heart, so there is no mixing of

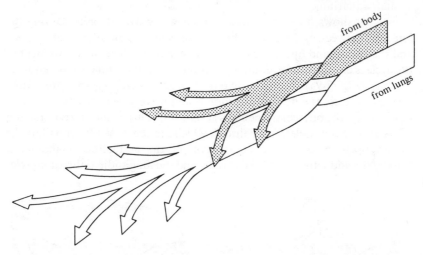

Fig. 7-8. A diagram representing the paths of the two streams of blood through the conus arteriosus and ventral aorta of the lungfish *Protopterus*.

oxygen-rich and oxygen-poor blood. It would presumably have been possible for the lungs of lungfish to have been fitted into the circulation in the same way as gills, so that the blood went from them to the rest of the body before returning to the heart. The disadvantage of such an arrangement would have been that the blood would have to be pumped out of the heart at sufficient pressure to drive it successively through two sets of capillaries, in the lungs and in the rest of the body. There would be a bigger pressure difference between the blood in the lung capillaries and the air in the lungs, so the lung capillaries would have to be stronger to prevent bursting. They would need thicker walls so oxygen would diffuse more slowly into them. (It was explained on p. 211 why the length of the diffusion path through tissue has much more effect on the rate of uptake of oxygen in lungs than in gills.) A bigger lung area would be needed and more blood would have to be pumped through the lungs.

It remains to describe how lungfish pump water over their gills and air into their lungs. Since the palatoquadrates are fixed rigidly to the cranium they cannot be involved in pumping movements. The main bones involved are the hyoid bars and pectoral girdle, and their movements have been demonstrated by X-ray cinematography. The fish had to be placed in a very narrow, thin-walled aquarium as thick walls or an unnecessary thickness of water would have spoilt the contrast of the X-ray image and made it impossible to see anything. The muscles responsible for the movements have been identified by electromyography of *Protopterus* moving freely in a small aquarium.

Fig. 7-9 shows how *Protopterus* pumps water over its gills. During the intervals between cycles of breathing movements the mouth is slightly open and the buccal and opercular cavities are expanded. The first movement in the cycle is closure of the mouth by contraction of the adductor mandibulae muscles (Fig. 7-9a). Next the hyoid bars and pectoral girdle are swung forward to raise the floor of the mouth and drive water over the gills and out through the opercular openings (Fig. 7-9b). This is done by contraction of longitudinal muscles in the throat, which are prevented from pulling the mouth open by continued contraction of the adductores mandibulae. Finally the adductors mandibulae relax and the longitudinal throat muscles

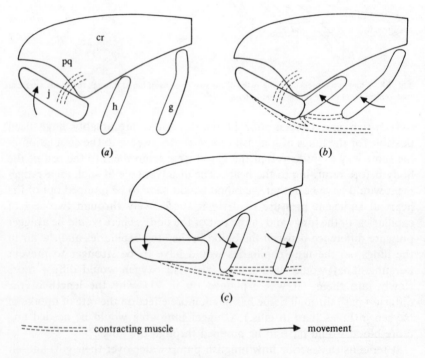

Fig. 7-9. Diagrams showing how the lungfish *Protopterus* pumps water over its gills. Muscles which have been shown by electromyography to be active at each stage in the cycle of movements are indicated. Based on the findings of McMahon (reference on p. 231), cr, cranium; g, pectoral girdle; h, hyoid bar; j, lower jaw; pq, palatoquadrate.

open the mouth (Fig. 7-9c). Muscles posterior to the pectoral girdle and others from the girdle to the hyoid bars swing girdle and bars back to their original position. This lowers the floor of the mouth, drawing water in.

Air breathing has also been investigated by X-ray cinematography. Air is much more transparent to X-rays than water so air in mouth or lungs can be distinguished in the films. The sequence of events is shown in Fig. 7-10. The fish, still submerged, closes its mouth and drives water from its mouth and opercular cavities out through the opercular openings (Fig. 7-10a). The opercula are then closed by contraction of a muscle which runs transversely across the throat, and they are held closed from now until the air in the lungs has been changed. Next the mouth is thrust above the surface of the water and opened widely, and the mouth cavity is enlarged and

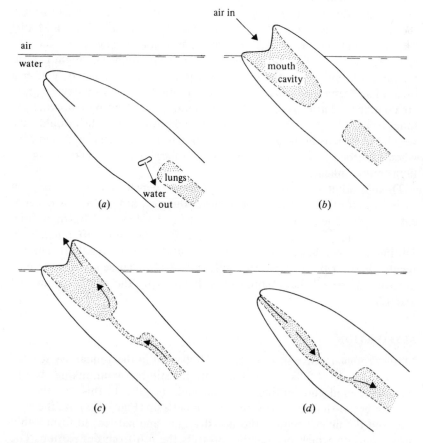

Fig. 7-10. Diagrams showing how the lungfish *Protopterus* changes the air in its lungs, based on the findings of McMahon (reference on p. 231).

filled with air (Fig. 7-10*b*). The lungs are then nearly emptied, becoming almost indistinguishable in the X-ray pictures (Fig. 7-10*c*). The mouth is closed and the floor of the mouth is raised. The air inside cannot be driven out through the opercula since these are being held closed, so it is forced into the lungs (Fig. 7-10*d*). This sometimes completes filling the lungs but a second mouthful of air is sometimes needed.

The pressure inside the lungs has been recorded through a cannula leading to a pressure transducer. It was found that between breaths this pressure was always about 10–15 cmH$_2$O greater than the pressure in the water around the fish. The lung walls must be taut and the glottis must be closed. The elasticity of the lung walls must help to empty the lung when the glottis is opened. There is smooth muscle in the lung walls which may be involved as well, as may the muscles of the body wall.

It seems odd at first sight that the air from the lungs is expelled through the mouthful of air that is to be forced into the lungs (Fig. 7-10*c*). Mixing seems bound to occur, so that some of the air from the lungs goes back into the lungs. It seems that rather less mixing occurs than one might expect, for air withdrawn from the lungs through a cannula immediately after a breath has been taken generally contains more than 0.15 atm oxygen (pure air contains 0.2 atm oxygen). Mixing could be avoided by emptying the lungs before filling the mouth, but this would mean that the fish would have very little air in its body, and be denser than usual, at precisely the moment when it needed to get its mouth to the surface. It might make filling the lungs more difficult.

Though adult *Protopterus* and *Lepidosiren* depend almost entirely on their lungs for their oxygen supply, their young are at first entirely dependent on gills. The rather large eggs are laid in submerged burrows (Fig. 7-11*a*) or pits, and the young live there for some time after hatching. During this time they do not breathe air, but their internal gills are supplemented by four pairs of external gills (Fig. 7-11*b*). These are pinnate outgrowths of the gill bars, supplied with blood by the efferent branchial arteries.

AESTIVATION

Many *Protopterus* live in swamps which dry out in the annual dry season. As the water level falls they burrow into the still-soft mud, taking mouthfuls of the mud and spitting it out through the gills. In this way the fish forms a bottle-shaped burrow in which it curls up (Fig. 7-12*a*). As the mud dries out the mucus coating the skin dries out and hardens to form a thin brown cocoon which presumably protects the fish from desiccation. The cocoon does not cover the mouth but turns in, forming a tube through which

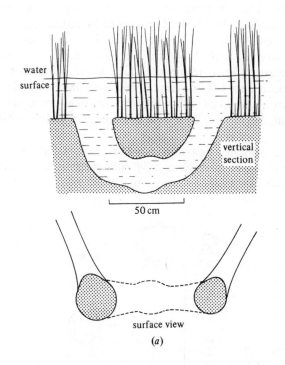

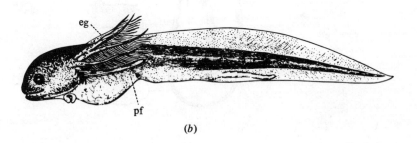

Fig. 7-11. (a) Section and plan of a nest burrow of *Protopterus*. From A. G. Johnels & G. S. O. Svensson (1954). *Ark. Zool.* **7**, 131–64.

(b) Larva of *Protopterus*. eg, external gills; pf, pectoral fin. From T. W. Bridge (1904). In S. F. Harmer & A. E. Shipley (eds.), *The Cambridge Natural History*, vol. 7. Macmillan, London, after Budgett.

the fish breathes. The fish in the burrow has a low metabolic rate and if it is dug up it is found to be torpid. It is said to be aestivating. Aestivation normally lasts 4–6 months (the length of the dry season) but aestivating lungfish have survived in the laboratory for more than a year. *Protopterus* do not live exclusively in environments which dry up: those that live in the large East African lakes have no need to aestivate, and do not do so. Some *Lepidosiren* live in swamps which dry up seasonally and burrow like *Protopterus* but they do not form mucus cocoons and do not become so torpid.

Neoceratodus does not aestivate, but the ability to aestivate was apparently evolved by at least some lungfishes early in the history of the group; Devonian and Carboniferous lungfish burrows have been found.

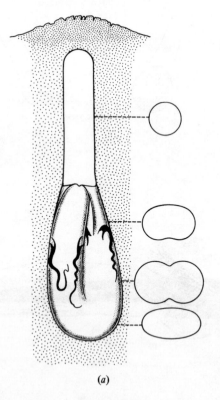

(*a*)

Fig. 7-12. Sections of aestivation burrows of *Protopterus*.

(*a*) A natural burrow. The thin lid of mud is cracked and so allows gas exchange. From A. G. Johnels & G. S. O. Svensson (1954). *Ark. Zool.* **7**, 131–64.

(*b*) A burrow in a container supplied in a laboratory. From P. A. Janssens (1964), *Comp. Biochem. Physiol.* **11**, 105–17.

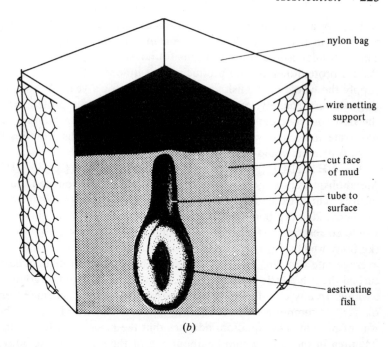

nylon bag

wire netting
support

cut face
of mud

tube to
surface

aestivating
fish

(*b*)

Figure 7-12*b* shows apparatus in which *Protopterus* has been made to aestivate in the laboratory, for physiological investigation. The container of mud, with the fish on top, was placed in a sink with water covering the mud. The water level was lowered gradually over a period of about two weeks. The fish burrowed into the mud, but came to the mouth of the burrow to breathe so long as the water was above the mud surface. Thereafter it remained invisible. The mud was allowed to dry out and fish kept in this way for seven months revived when returned to water.

Protopterus aestivating in the laboratory lose weight much more slowly than ones kept starving in an aquarium for the same period. Starving lungfish use about 20 cm^3 oxygen (kg body weight)$^{-1}$ hr^{-1}, and the proportion of glycogen in the body falls. Aestivating ones removed from the mud with as little disturbance as possible have been found to use on average only 8 cm^3 oxygen kg^{-1} hr^{-1}, and neither the glycogen nor the fat in the body seems to diminish. Not only is the metabolic rate much reduced but metabolism is apparently changed so that virtually nothing but protein is consumed.

Protein metabolism seems strangely unsuitable in aestivation, since it produces nitrogenous waste which cannot be eliminated from the body. *Protopterus* in water excrete most of their waste nitrogen as ammonia, which would be toxic if it were allowed to accumulate during aestivation.

It does not accumulate but urea does, reaching about 2% of the weight of the body in twelve months aestivation in experiments by H. W. Smith. This is similar to the concentrations found in marine selachians (p. 115). Would protein metabolism producing this amount of waste be enough to supply the whole of the fish's metabolic needs for a year? We shall work this out very roughly, assuming for simplicity that the protein was synthesized entirely from alanine ($CH_3.CHNH_2.COOH$), an amino acid of moderate molecular weight. The equation for metabolism of this hypothetical protein would be

$$2\,(-NH.CH(CH_3)CO-) + 6\,O_2 = CO(NH_2)_2 + 5\,CO_2 + 3\,H_2O.$$

Metabolism using 6 mol (6×22.4 l) of oxygen would produce 1 mol (60 g) of urea. The urea actually produced was about 2% of the final weight of the body, 20 g (kg body weight)$^{-1}$ or 1/3 mol (kg body weight)$^{-1}$, which would correspond to the consumption in the year of $6 \times 22.4/3 = 45$ l O_2 (kg body weight)$^{-1}$. This is 5 $cm^3 O_2$ kg^{-1} h^{-1}. This is rather less than the average rate of 8 cm^3 kg^{-1} h^{-1} measured by Smith, but he may have disturbed the fish a little in removing them from the mud and so got too high a value. In any case the correspondence between urea accumulation and oxygen consumption seems to confirm that at least most of the metabolism was of protein. The equation indicates that the protein needed to produce 2% urea in the body would be about 5% of the weight of the body. The aestivating fish actually lost on average 27% of their weight: presumably most of this loss was water.

In more recent experiments than Smith's, aestivating lungfish lost weight more slowly and accumulated urea more slowly. This may be due to differences in the conditions in which the fish were kept.

CROSSOPTERYGIANS

Class Osteichthyes, subclass Crossopterygii, orders Rhipidistia and Coelacanthini

The earliest crossopterygians belong to the order Rhipidistia. They appeared in the Devonian period and they were in some ways very like the contemporary lungfishes: they had similar pectoral and pelvic fins, similar (cosmoid) scales and usually similar heterocercal tails (Fig. 7-1*d*, *e*). However, the structure of their heads was very different. Though they had pharyngeal teeth and teeth on the palate they had nothing like the toothplates of lungfish. There were large teeth along the edges of the jaws with a characteristic structure: the dentine was folded into the pulp cavity in a series of pleats (Fig. 7-2*b*). The palatoquadrates were hinged to the cranium, not fixed rigidly to it as in lungfish. They could presumably be swung laterally to enlarge the mouth cavity.

The most peculiar feature of these fish was the intracranial joint. The cranium was divided into two halves, a posterior one penetrated by the notochord and an anterior one, anterior to the notochord. The halves were connected by a hinge joint (the intracranial joint) which apparently allowed the anterior half to be tilted upwards (Fig. 7-13*b*). The palato-quadrate was hinged to the anterior half of the cranium and the hyo-mandibular to the posterior half. There seems to have been a spiracle between them; evidence for this is provided by small toothed plates of bone on the part of the cranium indicated by hatching in Fig. 7.13*b*, which must have lain in the wall of the spiracle.

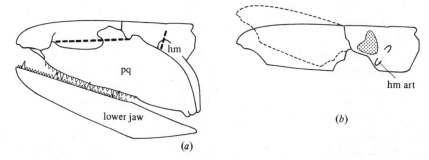

Fig. 7-13. Diagrams showing the mechanism of the intracranial joint of rhipidistians. (*a*) shows the cranium with the palatoquadrate (pq), hyomandibular (hm) and lower jaw in place. The axes of the hinge joints between the palatoquadrate and hyomandibular and the cranium are indicated by broken lines. (*b*) shows flexure of the intracranial joint. hm.art., facet for articulation of hyomandibular. The hatching is explained in the text.

Notice that the axes of the hinges between the hyomandibulars and the cranium were almost vertical (Fig. 7-13*a*). When the mouth cavity was enlarged the hyomandibulars swinging outwards about these hinges pre-sumably pushed the jaws forward and so bent the intracranial joint, raising the snout. This tilting up of the anterior part of the cranium would pre-sumably give the same advantage in feeding as the tilting of the whole cranium which occurs in teleosts (Fig. 5-15).

Rhipidistians had internal nostrils. Some had only one external nostril on each side but some had two, which suggests that the internal nostrils were new openings and not simply originally external nostrils that had shifted into the mouth as in lungfish. The internal nostrils presumably arose as an adaptation to aid smelling by making respiratory movements draw water through the nasal sacs.

The rhipidistians mostly lived in fresh water. The coelacanths are a mainly marine group, probably descended from them, that appeared in the Carboniferous. It was believed that the coelacanths had been extinct for as long as the dinosaurs until one was caught off South Africa in 1938. It was named *Latimeria*, after a Miss Latimer who played an important part in its discovery. This specimen had probably strayed from its normal habitat, for the few dozen which have been caught since have all been caught off the Comoro Islands, between Madagascar and Mozambique. They are large fish, weighing up to 80 kg (Fig. 7-1*f*). They have all been caught close to the rocky bottom, on lines up to 400 m long.

Coelacanths are easily recognized by the characteristic shape of the tail. They have the same type of pectoral and pelvic fins as most rhipidistians and lungfishes, and their scales, though thin, show signs of having evolved from the cosmoid type. There is an intracranial hinge, just like the intra-cranial hinge of rhipidistians. In some other respects coelacanths are very different from rhipidistians: disappointingly different to zoologists who would like to find in *Latimeria* hints about the origin of the tetrapods. There are two external nostrils on each side and no internal ones. There are no maxillae, and the mechanism of the jaws and of the intracranial hinge seems to be quite different from the rhipidistian mechanisms.

Latimeria has not so far been kept alive for more than a few hours after capture, and there has been no opportunity to study the jaw movements of living specimens. However, it has been possible to show by manipulating the heads of dead ones that widening the head by swinging the palato-quadrates laterally does not flex the intracranial joint in the way it is believed to have done in rhipidistians. This proves to be because the forward-pushing effect of movement at the joint between the hyomandibu-lar and the cranium is cancelled out by movement at a hinge joint further down the hyoid arch (Fig. 7-14*a*). However, opening the mouth of *Latimeria* automatically flexes the intracranial joint (Fig. 7-14*b*). This is because the articulation of the lower jaw with the hyoid arch (*C* in Fig. 7-14*a*) is posterior to its articulation with the palatoquadrate (*B*), so the jaw cannot be lowered without bending the hyoid arch at the joint *D* (which is not present in rhipidistians). This pushes the jaws forward and bends the intra-cranial joint *A*. The system works like a system of rods hinged at *A*, *B*, *C* and *D* (Fig. 7-14*c*): that is, it is the very simple type of mechanism known to engineers as a four-bar crank chain.

Ability to breathe air would be no use to a fish which lives as far below the surface as *Latimeria*. Indeed, there are extremely few air breathers even among surface-living marine fish, for all water within easy reach of the surface of the sea is generally well aerated. *Latimeria* has no lung, but it has a peculiar fatty organ that seems to be a degenerate lung or swim-bladder. The internal cavity is only a narrow tube joining the oesophagus but the walls are immensely thick and loaded with lipids. These are not triglycerides like ordinary fats but wax esters such as are found in some lantern fishes (p. 154). They are also plentiful in the muscle of *Latimeria* and must make it a good deal more buoyant than it would otherwise be.

There is about as much urea in the blood as in marine selachians, enough to make the osmotic pressure of the blood about the same as that of seawater.

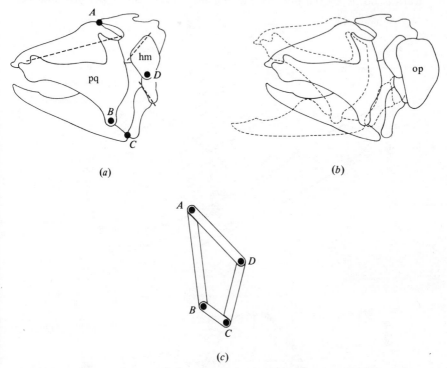

(a)

(b)

(c)

Fig. 7-14. (*a*). The skull of *Latimeria* with the operculum removed showing (broken lines) the axes of the hinge joints which bend when the palatoquadrates are swung laterally. *A, B, C, D,* axes of joints involved in mouth opening; hm, hyomandibular; pq, palatoquadrate.

(*b*) The same with the opercular bone (op) in place, showing the movements which occur when the mouth is opened.

(*c*) A four-bar crank chain.

DERMAL BONES OF THE SKULL

The skulls of bony fishes have an outer covering of dermal bones arranged in complicated patterns which generally have no obvious functional significance. Though these patterns may fascinate specialists in fish osteology they need not be described in a book like this, except for the pattern that is found in rhipidistian fishes. This pattern was inherited in modified form by the early tetrapods, and it is necessary to be familiar with the pattern to appreciate the changes, often of great functional interest, that have occurred in the course of evolution of tetrapod skulls. This unfortunately involves learning the names of rather a lot of bones, which are shown in Fig. 7-15.

The dermal bones in and around the mouth bear teeth. An outer row of teeth is borne by the premaxillae and maxillae of the upper jaw, and the

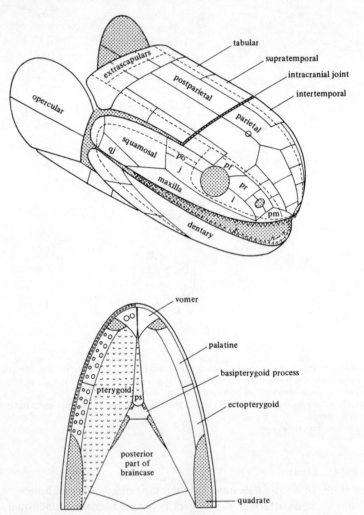

Fig. 7-15. Diagrams showing the arrangement of dermal bones in the skull of the rhipidistian *Eusthenopteron*. The lower diagram is of the palate seen from below. j, jugal; l, lacrimal; pf, postfrontal; pm, premaxilla; po, postorbital; pr, prefrontal, ps, parasphenoid; qj, quadratojugal. Broken lines represent lateral line canals.

dentaries of the lower. There is an inner row of big teeth on the vomers, palatines and ectopterygoids. These bones are dermal ones formed in the lining of the mouth rather than on the outer surface of the head (see Fig. 5-1). Other such bones, bearing smaller teeth, are the pterygoids and parasphenoid. The vomers and paraspheroid are borne on the underside of the cranium while the palatines, ectopterygoids and pterygoids cover the inner faces of the palatoquadrates.

Many of the dermal bones of the outer surface of the skull have lateral line canals running through them. The neuromasts of the lateral line system probably played an important part in the development of these bones, just as they seem to influence the development of many dermal bones in modern fishes. For instance, many bones in the skull of the trout first appear as a short length of bony tube in the wall of a lateral line canal around a neuromast, and a little plate of bone beneath. If the epithelium bearing the developing neuromasts is removed surgically at an early stage the bone fails to appear, and additional tubes of bone can be produced by grafting neuromast tissue.

FURTHER READING
General
Thompson, K. S. (1969). The biology of the lobe-finned fishes. *Biol. Rev.*, **44**, 91–158.
See also the general list for fishes at the end of Chapter 3.

Lungfishes
Johnels, A. G. & G. S. O. Svensson (1954). On the biology of *Propterus annectens* (Owen). *Ark. Zool.* **7**, 131–64.
White, E. I. (1966). A little on lungfishes. *Proc. Linn. Soc. Lond.* **177**, 1–10.

Breathing air
Bishop, I. R. & G. E. H. Foxon (1968). The mechanism of breathing in the South American lungfish, *Lepidosiren paradoxa*; a radiological study. *J. Zool., Lond.* **154**, 263–71.
Johansen, K. (1970). Air breathing in fishes. In Hoar, W. S. & D. J. Randall (eds.), *Fish physiology*, vol. 4, pp. 361–411. Academic Press, New York.
Johansen, K., C. Lenfant & D. Hanson (1968). Cardiovascular dynamics in the lungfishes. *Z. vergl. Physiol.* **59**, 157–86.
McMahon, B. E. (1969). A functional analysis of the aquatic and aerial respiratory movements of an African lungfish, *Protopterus aethiopicus. J. exp. Biol.* **51**, 407–30.
McMahon, B. R. (1970). Relative efficiency of gaseous exchange across the lungs and gills of an African lungfish, *Protopterus aethiopicus. J. exp. Biol.* **52**, 1–16.

Aestivation
Janssens, P. A. (1964). The metabolism of the aestivating African lungfish. *Comp. Biochem. Physiol.* **11**, 105–17.

Crossopterygians

Alexander, R. McN. (1973). Jaw mechanisms of the coelacanth *Latimeria*. *Copeia*, 1973, 156–8.

Andrews, S. M. & T. S. Westoll (1970). The postcranial skeleton of *Eusthenopteron foordi* Whiteaves. *Trans. Roy. Soc. Edinb.* **68**, 207–329.

Lutz, P. L. & J. D. Robertson (1971). Osmotic constituents of the coelacanth *Latimeria chalumnae* Smith. *Biol. Bull.* **141**, 553–60.

Millot, J. & J. Anthony (1958–65). *Anatomie de* Latimeria chalumnae. Centre de la Recherche Scientifique, Paris.

Nevenzel, J. C., W. Rodegker, J. F. Mead & M. S. Gordon (1966). Lipids of the living coelacanth, *Latimeria chalumnae*. *Science*, **152**, 1753–4.

Robineau, D. & J. Anthony (1973). Bioméchanique du crâne de *Latimeria chalumnae* (poisson crossopterygien coelacanthidé). *C.R. Acad. Sci., Paris*, **276**, 1305–8.

Dermal bones of the skull

Corsin, J. (1968). Rôle de la competition osseuse dans la forme des os du toit cranien des Urodeles. *J. Embryol. exp. Morph.* **19**, 103–8.

Devillers, C. (1946). Étude expérimentale du rôle morphogénétique des fossettes sensorielles (pit-organs) de *Salmo fario*. *C.R. Acad. Sci., Paris*, **223**, 1180–1.

Ørvig, T. (1972). The latero-sensory component of the dermal skeleton in lower vertebrates and its phyletic significance. *Zoologica Scripta*, **1**, 139–55.

8

Amphibians

This chapter is about the ancient fossil Amphibia which evolved from fish and gave rise to reptiles, and also the modern Amphibia (mainly frogs and newts) which are in many ways strikingly different from them.

EARLY AMPHIBIANS
Class Amphibia, subclasses Labyrinthodontia and Lepospondyli

The earliest known amphibians are three closely related genera of which one, *Ichthyostega*, is shown in Fig. 8-1. They were found in Greenland, in rocks that are either from the very end of the Devonian or the very beginning of the Carboniferous. They must have looked like large newts but they had fish-like features not found in modern amphibians: a fin round the tail and a rudimentary operculum. The fin is not just a fold of skin such as many newts have but a genuine fin with pterygiophores and rays. The operculum is a single small bone and it is not known whether anything remained of the gills. *Ichthyostega* also had small, fish-like scales; the only scales possessed by modern amphibians are the small scales of some of the worm-like Apoda. *Ichthyostega* had lateral line canals in the dermal bones of its skull which suggest that it spent much of its time in water. Lateral line sense organs serve to detect water movements and have no known function out of water. Modern amphibians which live out of

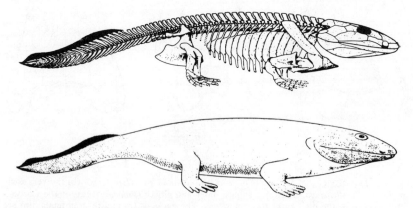

Fig. 8-1. Reconstructions of the skeleton and body of the very early amphibian *Ichthyostega*. From E. Jarvik (1955). *Science Monthly*, **80**, 141–54. *Ichthyostega* was about 1 m long.

water lack lateral line organs, though their aquatic larvae and aquatic adults (such as those of *Xenopus*, the clawed toad) possess them.

As well as generally fish-like features, *Ichthyostega* and its close relatives have features which link them with the rhipidistians. The dentine of their teeth is pleated in just the same way. There is no intracranial joint but there is a suture in the cranium in the corresponding position. There is a tunnel for the notochord in the posterior part of the cranium. The dermal bones of the skull are arranged in a modified version of the pattern found in rhipidistians, which is quite distinct from the patterns found in other bony fishes.

The skulls of *Ichthyostega* and a rhipidistian are compared in Fig. 8-2. It is quite easy to find the homologue in one skull of almost any dermal bone in the other. The parietal bones on either side of the aperture for the pineal organ make a convenient starting point. The remaining bones can be matched with their homologues by studying their positions relative to each other and to the lateral line canals which run through them. The main difference between the two skulls is one of proportions. In the fish, the parts posterior to the eye (including the operculum) are long so that there is room for large gills. In the amphibian they are short, leaving little or no room for gills. However, the snout, anterior to the eye, is relatively longer in the amphibian. There are fewer bones on the roof of the snout in the amphibian: note how a number of small bones (shown in Fig. 7-15) have been replaced by a pair of large frontals and a pair of large nasals. The

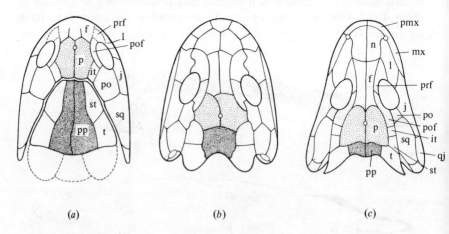

(a)   (b)   (c)

Fig. 8-2. Dorsal views of the skulls of (*a*) a rhipidistian fish (*Osteolepis*), (*b*) the very early amphibian *Ichthyostega*, and (*c*) a slightly later amphibian (*Palaeogyrinus*, of the Carboniferous) to show the differences in proportions. The parietals (p) and post-parietals (pp) are stippled. Names of other bones are given in Fig. 7-15 except f. frontal and n. nasal. From A. S. Romer (1941). *J. Morph.* **69**, 141–60.

palatoquadrates have ceased to be mainly vertical side walls to the mouth cavity and form, in the amphibian, a mainly horizontal palate. This change has involved movement of the basipterygoid processes of the cranium, with which the pterygoid bones articulate, from half-way up the sides of the cranium to its ventral surface. It is not clear why the pterygoids and the basipterygoid processes simply articulate and are not sutured together as in some other amphibians, for the rest of the skull seems too rigidly constructed to allow the hinge movement between palatoquadrates and cranium, which was apparently possible in rhipidistians.

The labyrinthodonts survived until the Trias but they quickly lost some of the fish-like features of *Ichthyostega*. From quite early in the Carboniferous they had no fins, no trace of an operculum and no suture half-way along the cranium. Some were as small as modern newts and some as large as alligators.

HEARING ON LAND

The hyomandibular bone had two functions in rhipidistians. It played a part in the mechanism of the intracranial joint and it supported the operculum. It is not required for either of these functions in amphibians, which have no intracranial joint and no operculum (apart from the rudiment in *Ichthyostega*). It has acquired a new function in connection with the ear. Early amphibians seem to have had eardrums, as modern frogs do, immediately posterior to the skull. Each eardrum occupied a notch between the tabular and squamosal bones, corresponding in position to the gap between the dermal bones of the skull roof and of the side of the head in rhipidistians. An ear ossicle, the stapes, has been found in position in some early amphibians, though not in *Icthyostega*. In most it is simply a rod running from the cranium to the position of the eardrum. In *Hesperoherpeton* of the Carboniferous, however, it is so like a rhipidistian hyomandibular that it is quite plainly a hyomandibular. The attachment to the eardrum corresponds to the attachment to the operculum and the lower parts of the bone, running to the jaw articulation, are rudimentary or absent in other amphibians. The homology of hyomandibular and stapes is also indicated by embryological evidence.

The stapes in tetrapods is surrounded by an air-filled cavity, the middle ear, which is connected to the mouth by the Eustachian tube (Fig. 8-3*a*). The middle ear and Eustachian tube develop in embryos from the spiracular gill cleft, which enlarges to surround the hyomandibular (stapes) instead of lying wholly anterior to it as in fishes.

The inner ear is enclosed in the cranium, as in fishes, but the perilymph cavity reaches the surface of the cranium at two 'windows' where it is bounded only by flexible membranes (Fig. 8-3*a*). These are the oval window, where the stapes ends on the lateral face of the cranium, and the round window on its posterior face. The endolymph cavity contains two neuromast organs not found in fishes, the basilar papilla and the amphibian papilla. Both have been shown by recording action potentials to be sensitive to sound. Both are papillae projecting from the endolymph cavity and

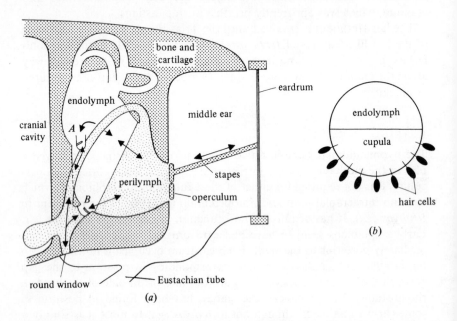

Fig. 8-3. (*a*) A diagram showing the structure of a frog ear. *A*, amphibian papilla; *B*, basilar papilla. (*b*) A diagrammatic section through the basilar papilla.

touching the perilymph cavity. The cupula of the basilar papilla is gelatinous, like the cupulae of semicircular canals and the lateral line system, but it is not conical. It has the form of a semicircular membrane reaching half-way across the papilla (Fig. 8-3*b*). That of the amphibian papilla (which is present only in amphibia) has a more complicated shape.

Vibrations of the eardrum are transmitted to the oval window by the stapes. When the oval window is driven in the round window must bulge out, and vice versa. There is a path entirely in the perilymph cavity from oval window to round window, but it is slender (Fig. 8-3*a*) and the membranes separating the perilymph from the endolymph are flexible, so the endolymph must vibrate as indicated in the Figure, through the amphibian and basilar papillae. Since the papillae are slender the amplitude of vibration will be higher in them than elsewhere. The cupulae will be made to vibrate by the endolymph, and stimulate the hair cells.

This radically new mechanism of hearing apparently evolved because the first mechanism, which depends on the whole body vibrating (p. 57), is very poor in air. An animal in air reflects nearly all the sound which strikes it and hardly vibrates at all. A satisfactory hearing organ must not reflect all the sound energy that strikes it, but must absorb a reasonable proportion. The same requirement applies to microphones, and we need some understanding of the physics of microphones if we are to understand the role of the eardrum and stapes.

When sound is passing through a material there are fluctuations of velocity in the vibrating material, and fluctuations of pressure. The ratio of the amplitude of the pressure changes to the amplitude of the velocity changes is known as the characteristic acoustic impedance of the material. It is very much higher in dense incompressible materials like water than it is in air. Sound is transmitted well from one material to another if they have similar characteristic acoustic impedances, but hardly at all if they have very different ones. When sound travelling in air strikes water, or vice versa, at least 99.9% of its energy is reflected. When sound travelling in air strikes a microphone some of its energy is used in driving the microphone and some is reflected. The proportions depend on the impedance of the microphone, which is the ratio of the amplitude of the force driving it to the amplitude of the velocity changes of the diaphragm. A substantial proportion of the sound energy is used in driving the microphone only if the impedance per unit area of the microphone diaphragm is close to the characteristic acoustic impedance of air.

The same applies to ears. Forces must act on the oval window to make the fluid of the inner ear vibrate so the inner ear, like a microphone, has an impedance. The impedance of frog ears has not, I think, been measured. That of the human inner ear has been measured by applying fluctuating pressures to the oval window and observing the bulging in and out of the round window. It varies with frequency, but its value per unit area of oval window at 500 Hz is about 250 times the characteristic acoustic impedance of air. If sound fell directly on the oval window, nearly all would be reflected, and the same would be true of frogs.

In the frog ear the sound actually falls on the eardrum. Eardrum and stapes have to be made to vibrate as well as the fluid of the inner ear, so they increase the impedance of the ear. However, the area of the eardrum is enormously larger than that of the oval window, so the somewhat increased impedance is spread out over a very much larger area, and the impedance per unit area is lower and closer to the characteristic acoustic impedance of air. A larger proportion of the sound energy will be used to drive the ear. Also, since the area of the eardrum is larger than that of the oval window, sound energy is collected from a larger area. The eardrum and stapes increase the sensitivity of the ear to sound travelling in air.

RESPIRATION

This section is about the respiration of adult modern amphibians, and about how *Ichthyostega* and other early amphibians probably breathed.

Among modern amphibians, frogs have been most thoroughly studied, but newts seem to breathe in essentially the same way. Frog breathing movements are in some ways very like the air-breathing movements of lungfishes, but frogs breathe through their nostrils instead of through the mouth. There are muscles which close the nostrils at the appropriate times.

Lungfishes pump air into their lungs by movements of their gill skeleton, contracting the mouth cavity to fill the lungs. Tadpoles have a cartilage gill skeleton, but it becomes greatly changed when the tadpole metamorphoses. In the adult frog it consists of a mainly cartilaginous plate (known as the corpus of the hyobranchial apparatus) under the tongue, with a pair of horns curving dorsally to join the cranium (Fig. 8-4). Although these horns reach all the way to the cranium they seem to be homologous with the hyoid bars of fishes: the hyomandibular which should form the upper part of the hyoid arch has become the stapes. Other projections from the corpus seem to be rudiments of other gill arches. Various

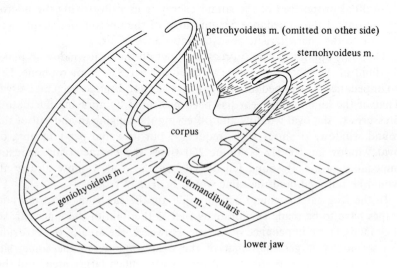

Fig. 8-4. The hyobranchial apparatus of a frog, and the muscles which move it.

muscles attach to the corpus. Sternohyoideus muscles run from it to the pectoral girdle. When they contract they pull the corpus posteriorly and ventrally, enlarging the mouth cavity much as the sternohyoideus muscles of fish do. The other muscles shown in Fig. 8-4 co-operate to reduce the mouth cavity by raising the plate and pulling it forward.

Professor Carl Gans and his colleagues have carried out a series of experiments to find out how frogs change the air in their lungs. They used pressure transducers to record the pressures in the lungs and in the mouth cavity, they identified the active muscles by electromyography and they took cinematograph films. Their conclusions are summarized in Fig. 8-5. The air is kept in the lungs under slight pressure, as in lungfish; it is prevented from escaping so long as the glottis is kept closed. To change the air, the mouth cavity is first enlarged by contraction of the sternohyoideus muscles, drawing air in through the open nostrils (Fig. 8-5a). Next the glottis is opened and the air from the lungs is driven out through the full mouth cavity, probably mainly by elastic recoil of the lungs and by contraction of their smooth muscles (Fig. 8-5b). Then the nostrils are closed and the mouth cavity is contracted by the other muscles indicated in Fig. 8-4, driving air from the mouth cavity into the lungs (Fig. 8-5c). Finally the glottis is closed and the nostrils opened again (Fig. 8-5d). This sequence of movements is very like the movements of a lungfish taking a breath of air, but it does not by any means change all the air in the lungs. The capacity

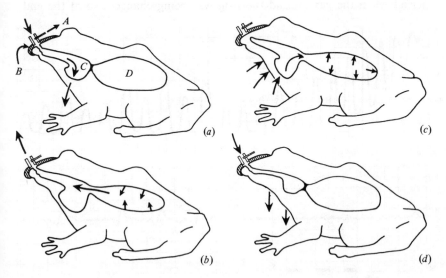

Fig. 8-5. Diagrams which are explained in the text, showing how frogs change the air in their lungs. *C,* Mouth cavity; *D,* lungs. From C. Gans, H. J. de Jongh & J. Farber (1969). *Science,* **163,** 1223–5. Copyright © 1969 by the American Association for the Advancement of Science.

of the expanded mouth cavity seems to be only about a quarter of the capacity of the lungs.

This sequence of movements is known as a ventilatory cycle. Between these cycles the frog keeps its glottis closed and its nostrils open and oscillates the floor of the mouth up and down. These oscillatory cycles move air in and out of the mouth cavity but do not affect the lungs. The constant rapid movement of the throat is easily seen on a living frog.

Two questions suggest themselves. What function have the oscillatory cycles? In the ventilatory cycles, does not the used air from the lungs get mixed with the fresh air waiting to enter the lungs (Fig. 8-5*b*)? Gans and his colleagues tried to answer both questions by experiments in which they used a mass spectrometer to record the composition of the gas passing in and out of the nostrils. The gas was sampled by means of the probe *A* (Fig. 8-5*a*) attached to a mask *B* made of a quick-setting rubber compound.

The most revealing experiments were ones in which the frog was kept for a while in a mixture of 80% argon and 20% oxygen, which was then changed rapidly to air (80% nitrogen and 20% oxygen). Fig. 8-6 shows a record obtained in one of these experiments. The lower line shows the pressure in the mouth cavity, recorded by a pressure transducer. The very small pressure fluctuations mark oscillatory cycles and the large ones (1, 2, 3, 4) ventilatory cycles. The upper line shows the argon content of air passing through the nostrils. The frog made no ventilatory cycles during the period when the gas around the frog was being changed, so at the end

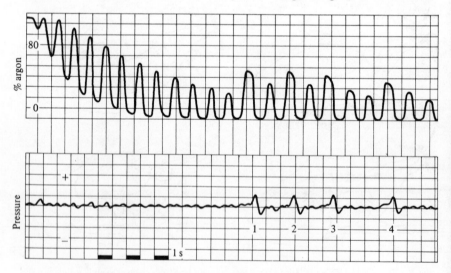

Fig. 8-6. Record of an experiment on frog respiration which is explained in the text. Source as Fig. 8-5.

of this period the frog was in air containing virtually no argon while its lungs still contained about 80% argon. After the change was completed the record shows alternation between virtually argon-free gas entering through the nostrils and gas with an appreciable argon content passing out through them. The argon content of the gas driven out during the first ventilatory cycle is about twice as high as in the preceding oscillatory cycle, so this gas must have contained a substantial proportion of gas from the lungs. The gas expired in the next oscillatory cycle contains considerably less argon, but nevertheless more than in the oscillatory cycle preceding the ventilatory one. This means that the gas from the lung was not all passed out through the nostrils in the ventilatory cycle, but some of it got mixed with the mouth contents and was only flushed out by subsequent oscillatory cycles. The record after ventilatory cycles 3 and 4 show gas from the lungs being flushed out of the mouth by a series of oscillatory cycles: note how the percentage of argon in successive expirations diminishes.

It used to be thought that the lining of the mouth served as an accessory air-breathing organ, and that the function of the oscillatory cycles was to ventilate it. However, the lining of the mouth is supplied with arterial blood which is already well oxygenated, and it has not a particularly rich blood supply. It seems probable that little or no oxygen is taken up through it and that the function of the oscillatory cycles is simply to flush out the air from the lungs left in the mouth after each ventilatory cycle.

Amphibians take up oxygen and lose carbon dioxide through the skin as well as in the lungs. Fig. 8-7 shows how the parts played in respiration

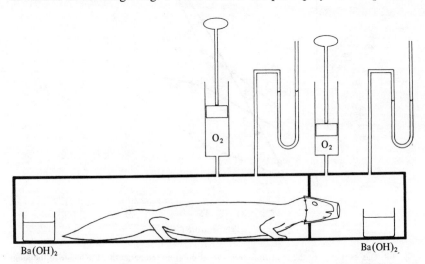

Fig. 8-7. Apparatus for separate investigation of gas exchange in the lungs and through the skin of amphibians.

by skin and lungs have been measured separately. A mask formed from a short piece of plastic tube was attached by a few stitches to the skin of the animal's face. The animal was put into a divided airtight container with the mask fitted into a hole in the dividing wall. An airtight seal was made round the mask. All air for the lungs had then to come from the front compartment while nearly all the skin was exposed to the air in the back one. Carbon dioxide given off in each compartment was absorbed by barium hydroxide which was titrated afterwards to find out how much carbon dioxide it had taken up. As oxygen was used up in each compartment the pressure fell, disturbing the manometer. It was replaced by a measured quantity of oxygen from the syringe, just sufficient to return the pressure to atmospheric pressure.

The relative importance of the parts played by skin and lungs varies with temperature. It also varies between species, but the results for toads shown in Fig. 8-8 are reasonably typical of amphibians in general. At

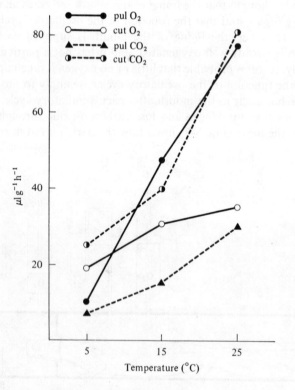

Fig. 8-8. Gas exchange of the toad *Bufo terrestris* at different temperatures. Exchange in the lungs (pul) and through the skin (cut) have been measured separately by the method illustrated in Fig. 8-7. From V. H. Hutchison, W. G. Whitford & M. Kohl (1968). *Physiol. Zool.* **41**, 65–85.

low temperatures more oxygen is taken up through the skin than in the lungs, but most of the extra oxygen needed at higher temperatures to maintain the higher metabolic rate comes from the lungs. Most loss of carbon dioxide is through the skin, at all temperatures. It seems that the main function of the lungs is to take up extra oxygen when it is needed at high temperatures or, presumably, in activity.

Why can the skin get rid of carbon dioxide so much faster than it can take up oxygen? The permeability constant (see p. 48) for carbon dioxide diffusing through water is about thirty times the value for oxygen. If the permeability constants for diffusion through skin are in the same ratio (a reasonable assumption) and if the difference in partial pressure of oxygen between the blood and the atmosphere is 0.2 atm (the maximum possible value), carbon dioxide can be lost through the skin three times as fast as oxygen is taken up with a difference in partial pressure of carbon dioxide of only 0.02 atm. Carbon dioxide could be eliminated very rapidly by the lungs if the air in the lungs was changed often enough, but this would require much more frequent changes than are needed for oxygen uptake. However much oxygen is removed from the air taken into the lungs at each breath, no more carbon dioxide can be lost there than will make the partial pressure of carbon dioxide in the lungs equal its low value in the blood.

Fig. 8-9 shows the structure of the skin, through which so much respiratory exchange occurs. It consists, as in fishes, of an inner dermis containing the collagen fibres that give the skin its strength and an outer epidermis. As in fishes the epidermis is formed by cell division at its inner surface and

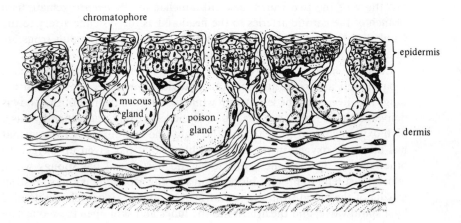

Fig. 8-9. A section through the skin of a frog (*Rana pipiens*). From C. J. & O. B. Goin (1971). *Introduction to Herpetology*, 2nd edit. Freeman, San Francisco. Copyright © 1971.

the young cells there are much less flattened than the older cells near the outer surface. However, amphibians and other tetrapods differ from fishes in that keratin accumulates in the epidermal cells. The outer layers of the epidermis consist of keratin-filled cells, which form a thin horny protective covering for the animal. The outermost layer of dead horny cells is shed from time to time as a whole or in pieces. Dead cells are lost individually from the surface of fish skin.

In reptiles, birds and mammals the horny layer is a very effective barrier against loss of water through the skin. It has been shown in experiments on man that the main barrier is not the outermost dead layer (which has air spaces in which diffusion would be rapid) but deeper layers where keratin and its precursors are being formed. A barrier to loss of water is valuable to a terrestrial animal unless it always has a plentiful water supply, but a horny epidermis is a barrier to diffusion of gases as well as to water. A horny skin cannot be both an effective waterproof layer and an important respiratory organ. Modern amphibians have only a thin horny layer, which allows the skin to play its important part in respiration but gives little protection against drying up. It has been found that water evaporates from the skin of amphibians in still, dry air about as fast as it would evaporate from a free water surface of the same area.

Fig. 8-10 shows how the principal blood vessels are arranged in a tadpole and in an adult frog. The arteries to the gills in the tadpole are arranged very much as in fish, but when the tadpole metamorphoses to an adult frog the gills and some of the arteries are lost. As in lungfish the lungs are supplied by the arteries of the last gill arch on each side, but the skin is also supplied from these arteries. The artery of the next gill arch disappears but those of the two which lack gill lamellae in *Protopterus* remain as a branch of the carotid arteries to the head and as the systemic artery to the trunk and limbs. Some of the arteries that are missing in adult frogs are retained in adult newts and salamanders.

Lungfish have partially divided atria and ventricles but frogs have separate left and right atria and an undivided ventricle. Blood from the lungs is delivered to the left atrium while blood from the rest of the body is delivered to the right one. The blood in the left atrium thus generally has a high oxygen content while that in the right one is a mixture of blood that has taken up oxygen in the skin and blood that has given up oxygen to other tissues. Though the ventricle has no partition it has spongy walls, which must tend to prevent blood from swirling around in it and so restrict mixing of the blood from the two atria. The conus is incompletely divided by a spiral valve, and the arteries to the lungs and skin open from one side of the partition while the systemic and carotid arteries open from the other (cf. Fig. 7-8). This arrangement is extremely similar to the arrangement

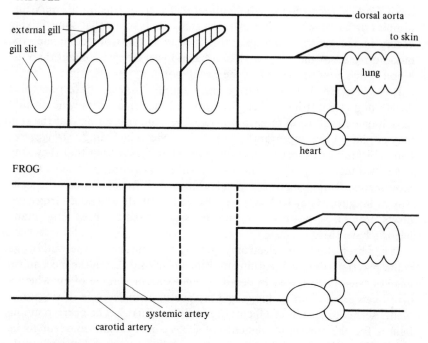

Fig. 8-10. Diagrams showing the arrangement of some of the principal blood vessels in a frog tadpole and an adult frog.

in lungfish and can be expected to make well oxygenated blood from the lungs tend to go through the carotid and systemic arteries to the general body circulation, while the rest of the blood, which is generally less well oxygenated, tends to be directed to the lungs and skin. If this separation occurs, it has the seeming disadvantage that blood from the skin where it has lost carbon dioxide and taken up oxygen tends to be directed to the skin again, or to the lungs.

The extent to which the blood entering the left and right atria is kept separate in its passage through the ventricle and conus has been investigated in many experiments. Dyes have been injected into veins and observations made of the arteries in which they appear after passing through the heart. Similar experiments have been made by injecting material opaque to X-rays and making X-ray observations of its passage through the heart. In still other experiments samples of blood have been withdrawn from various blood vessels and analysed for oxygen content, as in the experiments of lungfish. In all these experiments there is the difficulty that material must be added to the blood or blood withdrawn, and either may disturb the flow of blood. This danger was minimized in experiments on giant toads (*Bufo paracnemis*) weighing about 450 g, from which only very tiny samples of blood were withdrawn for analysis. It seems that some mixing occurs in the ventricle but that this mixing is far from complete: the blood directed to the lungs and skin has a lower oxygen content than the blood which goes to the rest of the circulation.

It is suspected that most of the early amphibians may have resembled lizards more than frogs in their respiration. Many had scales which would have hindered diffusion of gases through the skin, suggesting that the skin may not have been important in respiration. Most, including *Ichthyostega*, had substantial ribs largely encircling the chest, suggesting that they may have filled their lungs by rib movements (like lizards) rather than by throat movements. Rhipidistians presumably filled their lungs in much the same way as lungfish. They had short ribs or no ribs at all, and so do frogs and newts. Long ribs, however, are not necessarily used for breathing: many teleosts have them.

It has been shown by electromyography that muscles in the wall of the trunk play a part in the breathing movements of lizards, but the mechanism has not been investigated in detail. The maximum volume of air which a lizard can expel from its lungs and replace in a single cycle of rib movements seems to be about 60 cm^3 (kg body weight)$^{-1}$. The corresponding figures for the throat movements of frogs and *Protopterus* seem to be about 50 cm^3 kg^{-1} and 30 cm^3 kg^{-1}, respectively. (Frogs have relatively broader heads than lungfish and so can take larger mouthfuls of air.)

It has been suggested that use of the skin and of rib movements are alternative methods used by tetrapods to get the extra oxygen needed to meet the demands of life on land. They are probably better regarded as alternative solutions to the problem of getting rid of carbon dioxide into air. If a terrestrial animal cannot get rid of carbon dioxide through its skin like a modern amphibian, air taken into the lungs must have the partial pressure of carbon dioxide in it increased about as much as the partial pressure of oxygen is decreased (see p. 214). High partial pressures of carbon dioxide in the lungs, and so in the blood, can only be avoided by changing the air in the lungs frequently and removing only a small proportion of the oxygen from it. Lizards (*Lacerta* spp.) have been found to extract only 8% of the oxygen from the air which they breathe.

It is in any case doubtful whether life on land necessarily demands more oxygen than aquatic life. It has been found that at 20 °C a 1 kg Brook trout (*Salvelinus*) and a 1 kg lizard (*Iguana*) both use about 0.8 cm^3 oxygen min^{-1} while resting and a maximum of about 2.5 cm^3 min^{-1} in violent activity. However, the comparison is possibly a little unfair. Terrestrial animals are more likely than aquatic ones to be heated by solar radiation to high temperatures at which they use oxygen faster. If a 1 kg *Iguana* is heated to 30 °C it can run nearly three times as fast as at 20 °C, using 7 cm³ oxygen min⁻¹.

LOCOMOTION ON LAND

Early amphibians such as *Ichthyostega* were essentially newt-like in shape and might reasonably be expected to have walked on land like newts (Fig. 8-11). Happily we do not have to rely on such comparisons, for there have been preserved in Palaeozoic rocks tracks left by amphibians or reptiles walking over mud. (It is not always possible to decide whether a particular track was made by a reptile or an amphibian.) These indicate newt-like walking with feet well out to the side and generally with the toes pointing forward, though one particularly early track shows the toes pointing laterally. There are seldom marks of the body being dragged along the ground. Early tetrapods apparently normally walked with the belly held clear of the ground like the newt shown in Fig. 8-11. Newts do, however, sometimes crawl with the belly sliding on the ground, for instance on a wet surface.

Amphibians have four legs but they can be stable, like a three-legged stool, when standing on only three. The condition for stability is that a

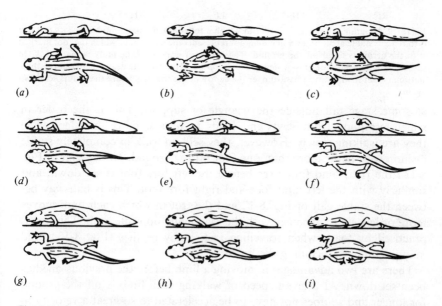

(a)   (b)   (c)

(d)   (e)   (f)

(g)   (h)   (i)

Fig. 8-11. Outlines traced from successive frames of a cinematograph film of a newt (*Triturus pyrogaster*) walking. A mirror was used to obtain dorsal and lateral views simultaneously. From B. Schaeffer (1941). *Bull. Amer. Mus. nat. Hist.* **78**, 395–472.

vertical line through the centre of gravity must pass through the triangle of which the corners are the points of contact of the three feet with the ground. If this vertical passes outside the triangle, the animal will topple over. Since three feet are enough for stability an amphibian can move its feet one at a time and always be stably supported by the other three, but the feet must be moved in a particular order (Fig. 8-12). If they were moved in any other order the vertical through the centre of gravity would

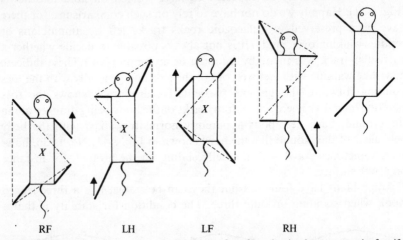

RF   LH   LF   RH

Fig. 8-12. Diagrams showing the order in which a four-footed animal must move its feet if it is to remain stable at all times. Each diagram represents the moment at which the foot marked with the arrow is lifted off the ground. The centre of gravity of the body is marked X. In cases where the centre of gravity lies above one edge of the triangle drawn between the three stationary feet it will move forward over the centre of the triangle before the foot is set down again.

at some stage fall outside the triangle of support. This is the order in which tetrapods, from newts and toads to horses, move their feet when they are walking slowly. However, they seldom walk in completely stable fashion with never more than one foot off the ground. Notice that in Fig. 8-13a the right hind foot rises before the left fore foot is set down, and similarly with the left hind foot and right fore foot. This is half-way between the stable gait of Fig. 8-12 and the trot in which each limb moves more or less simultaneously with the diagonally opposite limb. The trot is practised by toads when travelling fast, many reptiles (Fig. 8-13c) and many mammals, including the horse.

There are two advantages in moving a limb before the previous one has been set down. At a given speed of walking each limb is off the ground for longer and so does not have to be accelerated to so great a velocity to bring it forward to the position where it is to be set down. Less kinetic energy has to be given to the limb so the muscles need do less work. Also,

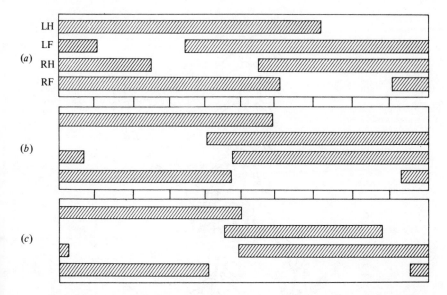

Fig. 8-13. Diagrams illustrating gaits of (a) of the newt *Triturus vulgaris*, and (b) and (c) the lizard *Lacerta viridis*. The lizard is running twice as fast in (c) as in (b). The hatched bars represent the period that each foot is on the ground. LH, LF, left hind and fore feet; RH, RF, right hind and fore feet. From S. Daan & T. Belterman (1968). *Proc. K. ned. Akad. Wet.* C, **71**, 245–66.

longer steps can be taken because the body travels further during the time that the foot is off the ground. A higher speed can be achieved with a given power output from the muscles.

Newts and most lizards do not run with the body straight, but bend it from side to side (Fig. 8-11). The movement suggests at first sight the swimming movements of a fish, but is fundamentally different. The waves of a swimming fish are travelling waves: each bend travels posteriorly along the body. The waves of the newt are standing waves, like the vibrations of a stretched string. The bending lengthens the newt's stride by making each shoulder or hip point backwards as the foot is raised and swinging it forwards before the foot is set down again (Fig. 8-14).

What changes were needed in the evolution of amphibians from fish to make walking possible? Foremost was the evolution of legs from pectoral and pelvic fins. No intermediate stage is known between a typical crossopterygian fin (Fig. 7-3) and the fully-developed limb. It seems fairly obvious that the humerus, radius and ulna are homologous with the pterygiophores indicated in Fig. 7-3, and similar homologies are apparent between hind leg and pelvic fin. It is not at all clear how the remaining bones at the distal ends of the limbs evolved.

The form of the primitive amphibian limb is obviously suited to the

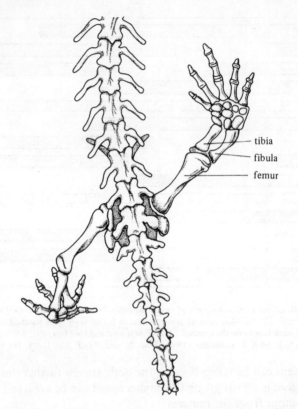

Fig. 8-14. Dorsal view of part of the skeleton of a newt showing how the tibia and fibula are parallel when a foot is set down on the ground in walking but become crossed before it is raised again. From B. Schaeffer (1941). *Bull. Amer. Mus. nat. Hist.* **78**, 395–472.

type of locomotion practised by newts. The large number of bones in the foot enables it to accommodate itself to the shape and angle of the surface the animal is walking on and to take advantage of irregularities to obtain a grip. Fig. 8-14 shows how the tibia and fibula can cross over each other and so keep the toes pointing forward all the time the foot is on the ground, instead of letting the foot swivel round with the femur (if the foot did swivel round it would be abraded, and work would have to be done against friction to turn it). The radius and ulna cross similarly in the fore limbs of newts. It is of course possible for an animal without a separate radius and ulna to walk in amphibian fashion and yet not swivel the foot, provided the elbow is not a simple hinge joint but allows sufficient variety of movement: this is the situation in frogs and toads.

The muscles of the trunk and tail do most of the work when a typical fish swims fast, but the muscles of the limbs do most of the work when a tetrapod runs. There is an appropriate difference in the relative sizes of the muscles. The limb muscles of tetrapods generally make up a very much larger proportion of the weight of the body than do the fin muscles of fishes, and the trunk and tail muscles are correspondingly smaller.

Not only did the origin of the tetrapods involve evolution of limbs from fins and enlargement of their muscles, but it involved modification of the limb girdles. The pectoral girdle of bony fishes consists mainly of a series of dermal bones stiffening the posterior wall of the opercular cavity (Fig. 8-15a). The two halves of the girdle meet ventrally and are attached to the cranium dorsally. The girdle provides an anterior attachment for

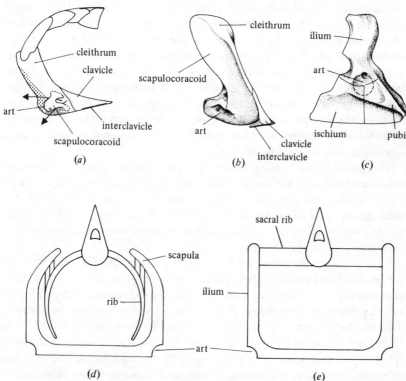

Fig. 8-15. (*a*) Median view of the pectoral girdle of a rhipidistian (*Eusthenopteron*). The probable area of origin of the fin muscles is stippled. Redrawn from S. M. Andrews & T. S. Westoll (1970). *Trans. Roy. Soc. Edinb.* **68**, 207–329.

(*b*) (*c*) Lateral views of the pectoral and pelvic girdles of a Palaeozoic amphibian (*Eryops*). Redrawn from A. S. Romer (1966). *Vertebrate Palaeontology*, 3rd edit. University of Chicago Press, Chicago. Copyright © 1966.

In (*a*), (*b*) and (*c*) the anterior part of the girdle is to the right.

(*d*) (*e*) Diagrammatic transverse sections through the pectoral and pelvic girdles, respectively, of a typical amphibian or reptile, showing how each is attached to the axial skeleton.

art, surface for articulation of fin or limb.

the trunk muscles and also an origin for the sternohyoideus muscle, which is used in feeding and respiration (Fig. 5-2). Most of the girdle is dermal bone and only a relatively small part attached to its inner side, the scapulo-coracoid, is sclerotome bone. It is with this that the fin articulates. The largest forces acting on the fins, in sculling with the fins or in braking, tend to act posteriorly or anteriorly. The main fin muscles originate partly on the scapulocoracoid and partly on the inner surface of the dermal part of the girdle (Fig. 8-15a).

Amphibians have no opercular cavity, so no reinforcement is needed for its posterior wall. The dermal part of the pectoral girdle is reduced but the scapulocoracoid is enlarged (Fig. 8-15b). Many of the limb muscles originate from it. The weight of the animal can best be supported without compressing the viscera and lungs if the pectoral girdle is not simply embedded in the body wall but attached more or less firmly to the axial skeleton. Its attachment to the skull in fishes is not well arranged for this because it is anterior to the fin and linked to it by a series of flexibly-connected bones. It has been replaced in tetrapods by an entirely new connection: the rib cage is suspended from the large scapula by short muscles (Fig. 8-15d). Detachment of the girdle from the skull makes the head more freely movable relative to the trunk.

The pelvic girdle in fishes is merely a pair of small plates of sclerotome origin embedded in the ventral body wall. In tetrapods it is larger, giving origin to large muscles and extending dorsally to make a firm attachment with the vertebral column (Fig. 8-15c, e). Fig. 8-15e shows how it is sutured to the ends of short stout ribs, which in turn are sutured to one or more vertebrae (the sacral vertebrae).

Now consider the trunk of an early amphibian or of a newt, between the pectoral and pelvic girdles. There are stresses that act on it in walking which are quite unlike the stresses of fish-like swimming. The trunk must not sag between the girdles, if the belly is not to trail on the ground. This can be prevented by tension in longitudinal muscles. It must also be resistant to twisting about its long axis, because during walking the body is supported at times only by two diagonally opposite limbs. The vertebrae of rhipidistians did not articulate with each other and any resistance to twisting would have to be provided by the notochord running through them, or by ligaments (Fig. 8-16a). The vertebrae of *Ichthyostega* were basically similar but the neural arch of each articulated with its neighbours, by articulating processes of which the faces were arranged radially with respect to the long axis of the column (Fig. 8-16b). Twisting would press the articulating processes of successive vertebrae together, and would be quickly stopped, but bending would not be prevented. Not all ancient amphibians had vertebrae as like rhipidistian ones as *Ichthyostega* did,

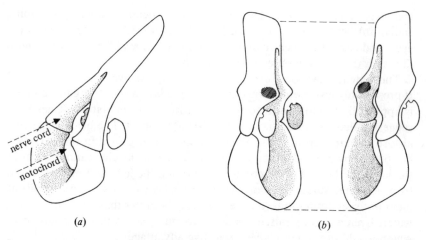

(a)   *(b)*

Fig. 8-16. Vertebrae of (*a*) the rhipidistian *Eusthenopteron*, and (*b*) the early amphibian *Ichthyostega*. The articulating processes are hatched.

and modern amphibians have very different vertebrae, but they have these articulating processes.

REPRODUCTION

Most modern amphibians lay numerous fairly small eggs enclosed in a layer of jelly. These are normally laid in water and develop into aquatic larvae (tadpoles) with external gills very like those of *Protopterus* and *Lepidosiren*, except that the fourth pair is rudimentary or absent. Traces of similar gills can be discerned on fossils of some immature early amphibians showing that they, too, had aquatic larvae.

The jelly is a protein gel. It is secreted by the oviducts as a thin, relatively concentrated layer which swells enormously when it has an opportunity to take up water. It presumably quickly absorbs any water that is present in the oviduct but appreciable swelling does not normally occur until the eggs are laid. The swollen jelly of the common frog (*Rana temporaria*) contains only 0.3% organic matter. Contact with water also makes the surface of the jelly sticky, and the eggs of *Rana* adhere to each other and sometimes to weed or hard surfaces when they are laid. They are usually laid in ponds or ditches, but some salamanders lay their eggs in streams and stick them individually to the undersides of stones so that they are not washed away.

Though the eggs of *Rana* stick together as a bunch of frogspawn they do not form a continuous mass of jelly. Each jelly capsule remains more or less spherical so that though they adhere at their points of contact, channels remain between them. (This can be demonstrated by dropping Indian ink on to submerged frogspawn.) Water can circulate between the eggs, and eggs at the centre of a bunch do not depend for their oxygen on diffusion through the outer parts of the bunch.

The hatched tadpoles do not eat the jelly, so what is its function? It presumably gives some protection from predators and its organic content is apparently too low for it to be itself attractive to predators. It has been suggested that its main function may be to trap heat. The black eggs must absorb radiation from the sun and sky well. Heat gained in this way cannot be lost by direct convection, since convection currents cannot occur in the jelly: the heat must be conducted through the jelly and only removed from its outer surface by convection. Frogspawn in its natural environment has been found to be on average 0.6 °C warmer than the surrounding water. Even a slightly raised temperature must lead to faster development which would presumably give a selective advantage.

Not all amphibians lay eggs in water. It can be an advantage to avoid having an aquatic larva because it tends to be vulnerable to predators such as insect larvae even if the eggs are protected by jelly. *Salamandra atra* does not lay eggs at all: the young are retained in the oviduct until after metamorphosis. The walls of the oviduct have a rich blood supply and the larvae have large external gills, so oxygen presumably diffuses from the blood of the mother to that of the young. Species of *Plethodon*, another genus of salamanders, lay eggs on land, sticking them to the walls of crannies in logs, or of caves. The larvae which develop in the eggs have large external gills but they do not hatch until after metamorphosis. The frog *Eleutherodactylus* also lays eggs on land which only hatch after metamorphosis. There are several species of which only some develop external gills, and the main respiratory organ of the larva in the eggs seems to be the broad, thin tail which has a rich blood supply. The males of another genus of frogs, *Rhinoderma*, carry the eggs in the vocal sac until after metamorphosis. (The vocal sac is an extension of the mouth cavity which plays a part in sound production, see p. 258.) *Salamandra* lives in Europe, *Plethodon* in North America, *Eleutherodactylus* in tropical America and *Rhinoderma* in Chile. Many other examples could be cited. Omission of a free larval stage means that the young cannot feed until the end of a long period of development. Enough food may be provided in a large yolky egg. For instance, *Eleutherodactylus guentheri* lays about 25 eggs, each containing about 50 mm^3 yolk, which hatch as adult frogs a month later. Another Brazilian frog of similar size lays about 1000 eggs, each containing 0.5 mm^3 yolk, which hatch as normal aquatic tadpoles after only one day. The eggs of *Eleutherodactylus* may be safer from predation, but fewer can be produced. *Plethodon* and *Rhinoderma* also lay a small number of large eggs. *Salamandra atra* eggs are not particularly large but the larva seems to obtain food as well as oxygen by diffusion from the mothers' blood.

Most frogs practise external fertilization. The male mounts on the female's back, holding on with his forelegs, and liberates the sperm while she liberates the eggs. Most salamanders however have internal fertilization. The male deposits gelatinous packets of sperm which the female picks up with the lips of the cloaca.

FEEDING

Modern amphibians are all carnivorous. Aquatic species feed on a wide variety of invertebrates, which they take in the same way as most fishes, sucking the prey in by enlarging the mouth cavity. This method of feeding will not work on land because prey which is similar in density to water is very much denser than air: the animal would have to produce an exceedingly fast current of air into the mouth if prey was to be sucked in. It is possible to catch prey simply by grabbing it between the teeth, as lizards do, but many terrestrial amphibians have sticky tongues which they use to catch and pick up the insects and other terrestrial invertebrates which they eat. The stickiness is due to mucus secreted by glands on the tongue. The tongue is attached to the anterior part of the floor of the mouth, and at rest is folded back in the mouth. When prey is to be caught it is flipped very rapidly forward, out of the mouth (Fig. 8-17a). The mechanism in salamanders with tongues of this sort is quite different from the mechanism of the tongue in frogs and toads, though the movements are similar.

MODERN AMPHIBIANS
Class Amphibia, subclass Lissamphibia

The modern Amphibia are divided into three orders, the Salientia (frogs and toads), Urodela (newts and salamanders) and Apoda (tropical limbless amphibians). Very different views have been expressed recently about their relationships to each other. At one extreme it has been held that Salientia show close similarity (especially in the structure of the snout) to one group of rhipidistian fishes, and the Urodela to another, and that this shows that they can have no common tetrapod ancestor: rather, limbs and other tetrapod features must have evolved separately in the ancestry of the two groups. At the other extreme it has been held that Salientia, Urodela and Apoda share features which distinguish them from all members of the subclasses Labyrinthodontia and Lepospondyli, and indicate that they have evolved from an unknown common ancester within one of those subclasses: they should therefore be grouped together in a third suborder, the Lissamphibia. I favour the latter view. The evidence that Salientia and Urodela evolved from separate groups of rhipidistians

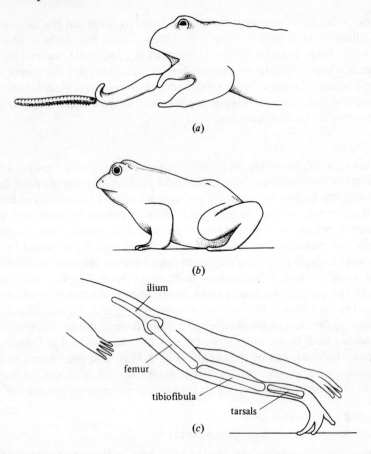

Fig. 8-17. (*a*) A toad (*Bufo calamita*) using its tongue to pick up a mealworm.
    (*b*) (*c*) Outlines traced from a cinematograph film of a frog (*Rana temporaria*) jumping. Some of the
bones are shown in position in (*c*).

is not convincing, because nearly all the characteristics of the snout on
which it depends seem to be consequences of having the nasal cavities
spaced far apart (in Urodela and one group of rhipidistians) or close together (in Salientia and the other group). A change in the spacing of the
nasal cavities is a basically simple one which might quite probably have
occurred several times in the course of evolution. Indeed it plainly has occurred several times, for there are Salientia with nasal cavities widely separated
and Urodela with them close together. The similarities between modern
amphibians include various features associated with their manner of
respiration: moist skin with a good blood supply and mucus glands, small
ribs and a more or less flattened skull (giving a large throat to pump air

into the lungs). There are other points of similarity which have no apparent functional relationship to respiration, among which the most striking are the presence of a small additional cartilage or bone (the operculum) at the proximal end of the stapes (Fig. 8-3*a*), and a peculiarity of the structure of the teeth. The operculum (which is, of course, nothing to do with the operculum of fishes) is present in Salientia and Urodela, and there is some embryological evidence that it may be present, though fused to the stapes, in Apoda. It has not been found in any labyrinthodont or lepospondyl, but it is not possible in most cases to be certain that it was absent, so it does not provide as convincing evidence as one would wish. More convincing is the peculiarity of the teeth, which is found in the great majority of modern amphibians of all three orders but not in any other known tetrapod living or fossil. Each tooth is in two parts, a base and a crown, both consisting mainly of dentine but separated by a layer of uncalcified tissue. These three characteristics, the manner of respiration (and all that goes with it), the operculum and the peculiar teeth are thus present in all or most modern amphibians but were probably not present in crossopterygians or the great majority of Palaeozoic amphibians. It seems unlikely that they should have been evolved independently by the three orders of modern Amphibia. It seems likely that they were all evolved in an unknown common ancestor of the three orders, which are therefore regarded as closely related to each other. There are other similarities of the three orders that support the argument, but these are the main ones.

FROGS AND TOADS
Subclass Lissamphibia, order Salientia

The frogs and toads are amphibians that have become quite extraordinarily modified for jumping and for swimming with their hind legs. Their most striking features are their big back legs and the lack of a tail. The jump is an effective means of escape because of the speed involved and because its direction is unpredictable, and it has been suggested that the ability to jump was evolved by the ancestors of modern frogs as a means of making a quick return to nearby water when danger threatened. The symmetrical two-footed kick which is used in jumping (Fig. 8-17*b*, *c*) is also used in swimming by frogs, and in the human breast stroke. No tail is therefore needed for swimming.

Long legs can push against the ground for longer when a frog jumps than short ones could. The velocity of take-off needed to jump a given distance is the same whatever the length of the legs and so is the work which must be done to accelerate a given mass of frog to this velocity, but if the legs are long the work can be done over a longer period and less power is required.

The bones are indicated in Fig. 8-17c to show how the leg has been lengthened. Not only are the femur and tibiofibula long, but changes in the pelvic girdle and ankle have effectively added two joints to the limb. The tarsal bones are long. The sacral vertebra is half-way along the trunk and the ilia are very long. Their joint with the sacral vertebra is movable: it is bent as the frog squats (Fig. 8-17b) and extends as the frog jumps (Fig. 8-17c).

Calling plays an important part in the life of frogs and toads. The functions of the calls are similar to those of the more familiar calls of birds. A single species may have distinct mating, territorial and alarm calls. The mating calls of different species living in the same locality are distinct.

An oscillogram of a short but otherwise typical call is shown in Fig. 8-18a. It consists of a series of pulses of high frequency sound. It is produced by expelling the air from the lungs into vocal sacs, which are extensions of the mouth cavity (Fig. 8-18b, c). The air is returned to the lungs for the next call. Sound production depends on the possession of a larynx and vocal chords.

The larynx is the upper end of the windpipe which is stiffened by cartilages in its wall. The two vocal chords are membranes each extending half-way across the windpipe, leaving only a narrow slit between them. When air is driven forcefully out of the lungs it sets the vocal chords vibrating, producing sound at the same frequency. In addition a pair of arytenoid cartilages in the larynx may be set vibrating, opening and closing the larynx and letting the air out in a series of puffs. This produces amplitude modulation of the sound. In the call shown in Fig. 8-18a the frequency of 4000 Hz is due to the vocal chords and the modulation at about 300 Hz to the cartilages. The vocal chords vibrate and sound is produced even when air is driven through the larynx of a decapitated frog. High-speed cine films have been taken of such experiments, and show the opening and closing of the larynx as the arytenoid cartilages vibrate. The frequency of vibration of the cartilages (and so the number of sound pulses per second) can be altered by means of muscles. In a series of experiments with decapitated frogs electrical stimuli were applied to the nerves serving these muscles while air was driven through the larynx under constant pressure. As the frequency of stimulation increased the muscles tightened, and the number of sound pulses per second increased. The sound pulses were more numerous than stimuli throughout this experiment: there is no question of each stimulus producing a pulse of sound.

The pulses of sound produced by intact frogs are generally made up of a fundamental tone (often around 1000 Hz) with harmonics. The frequency depends on the vocal chords, but the efficiency of sound production is greatly increased by resonance of the vocal sacs, if they are tuned to the

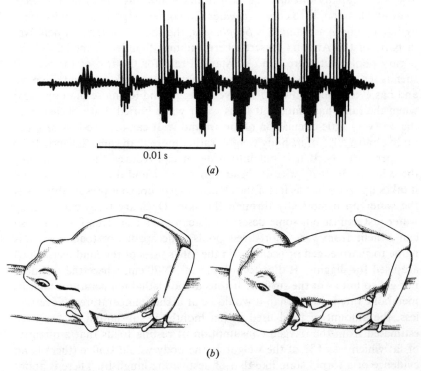

Fig. 8-18. (*a*) Oscillogram of a call by a tree frog, *Hyla a. arborea*. From H. Scheider (1967). *Z. vergl. Physiol.* **57**, 174–89.

(*b*) (*c*) Outlines traced from photographs of another tree frog, *Hyla a. meridionalis*, between calls and immediately after a call. Air is transferred during a call from the lungs to the vocal sacs. After H. Schneider (1968). *Z. vergl. Physiol.* **61**, 369–85.

same frequency. A vocal sac resonates at a very much higher frequency than a swimbladder of similar size because the resonance is of a different kind. A swimbladder pulsates, expanding and being compressed as the water around it vibrates (p. 172) but the vocal sac is not submerged during calling. It seems to act as a Helmholtz resonator, with air vibrating in and out through its opening into the mouth cavity.

Xenopus, the clawed toad, has already been referred to as an aquatic member of the Salientia that retains lateral line sense organs as an adult. The ilia are not hinged to the sacrum as in frogs but have a sliding joint with the sacrum. The whole pelvic girdle slides anteriorly and posteriorly as it swims.

In marked contrast to *Xenopus*, several species of Salientia have become adapted to life in deserts. This seems a strange habitat for animals which lose water rapidly through the skin in dry conditions. They do not seem to be any better able to conserve water than other amphibians, and survive by burrowing. An example is *Scaphiopus*, the spadefoot toad, which lives in parts of the Arizona Desert where summer thunderstorms leave temporary pools, but there is no substantial rain for the rest of the year. It spends ten months underground between one summer's rain and the next, and has even been known to survive underground for more than one year when the rains have failed. It loses water while buried, but not fast since there is very little circulation of air around it. It can survive loss of water up to almost 50% of its body weight, rather more than some Salientia from damper habitats. If it is put into water it can recover this loss in only three hours. It floats with its head above water and does not drink, and it takes up water just as fast if the cloaca is sewn up, so it presumably takes the water up osmotically through the skin. Ordinary frogs can take up water just as fast but some desert toads can take it up even faster and so can benefit from very short-lived pools. The spadefoot toad owes this name to sharp-edged projections on the inner sides of the hind feet, which are used for digging. It digs to a depth of 30–70 cm, where the soil does not get so hot as at the surface. It thus avoids lethal temperatures, and its metabolic rate is lower than it would be at higher temperatures. Nevertheless, the amount of food used in ten months is substantial. It has been estimated from the oxygen consumption of resting toads that a quantity of fat which was 5% of the weight of the body would suffice (there is no evidence of a torpid state like that of aestivating lungfish). There is about this amount of fat in well fed laboratory specimens, in the fat bodies in the body cavity which are a peculiarity of amphibians, but toads emerging from the winter fast have only small remnants of the fat bodies. Metabolism of fat yields more than twice as much energy as metabolism of equal weights of carbon or protein. Some protein must also be metabolized, for urea accumulates in the body. This urea, and the increased concentration of salts due to loss of water, may double the osmotic concentration of the blood (Fig. 8-19*a*).

At least two species of Salientia live in brackish water, including the crab-eating frog, *Rana cancrivora*, of Southeast Asian mangrove swamps. It can survive in 80% seawater. When it is put into water of high salinity it passes very little urine and urea accumulates in its blood, helping to keep its osmotic pressure slightly above that of the water. This prevents osmotic loss of water in just the same way as the urea in selachians and *Latimeria* does. The changes in the blood of a crab-eating frog in saline water are very like those of a hibernating spadefoot toad (Fig. 8-19*b*).

The tadpoles of frogs and toads differ markedly from their parents not only in being aquatic, but also in their feeding habits. They feed on algae,

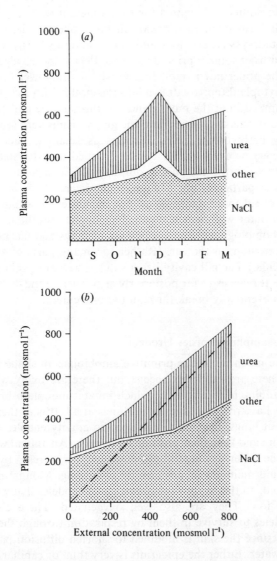

Fig. 8-19. Two graphs of total osmolarity of blood plasma, showing the contributions of sodium chloride and urea to the total.

(*a*) is a graph of plasma osmolarity against time for spadefoot toads (*Scaphiopus couchi*) hibernating in a laboratory experiment. Redrawn from L. McClanahan (1967). *Comp. Biochem. Physiol.* **20**, 73–99.

(*b*) is a graph of plasma osmolarity against osmolarity of the external medium, for crab-eating frogs (*Rana cancrivora*) acclimatized to various dilutions of seawater. Redrawn from M. S. Gordon, K. Schmidt-Nielsen & H. M. Kelly (1961). *J. exp. Biol.* **38**, 659–78.

while their parents eat insects and other terrestrial invertebrates. Since tadpoles and adults are adapted for such different ways of life transition to the adult form involves a drastic metamorphosis. The skin covering the jaws of tadpoles is very thick and horny, and this horny mouth is used like the mouths of some cyprinoid fishes (p. 189) to loosen algae from solid surfaces. The upper jaw is movable, as well as the lower one. Loosened algae, or phytoplankton, are drawn into the mouth with a current of water which is pumped in at the mouth and out through the gill slits. They are caught up in mucus and passed into the gut, which is very long and coiled as in many herbivorous fishes. Particles as small as bacteria (around 1 μm diameter) or even smaller can be trapped and digested. In experiments like those performed on sea squirts (p. 27) *Xenopus* tadpoles cleared suspensions of particles only 0.13 μm in diameter.

The gills of salientian tadpoles do not remain external but become enclosed in an outer wall, which is called an operculum though it does not seem to be homologous with the opercula of fishes and has nothing to do with the operculum in the ear. (Rather too many parts of amphibia are called opercula.) The gill cavity has a single external opening, usually on the left side. It reaches so far posteriorly as to enclose the developing forelimbs which eventually break through the operculum.

NEWTS
Subclass Lissamphibia, order Urodela

Urodeles are much more like primitive amphibians in shape than are the Salientia. There are no desert species but there are many which remain aquatic as adults, including some which inhabit mountain brooks. Many such species have reduced lungs or no lungs at all. It is to their advantage to be without lungs, because a high specific gravity enables them to rest on the bottom and keep their position by friction. An animal with the same density as the water would tend. to be washed downstream by the current unless it could maintain its position by swimming. Mountain brooks are generally cold, so the oxygen requirements of urodeles living in them will tend to be low. They are also well oxygenated. These characteristics enable urodeles to survive in them by respiration through the skin alone, particularly since the skin is modified to shorten diffusion paths between blood and water. Either the epidermis is very thin, or capillaries penetrate the epidermis and run close to its outer surface. The distance from the blood to the outer surface of the skin may be as little as 10 μm. When lungs are absent there seems no point in retaining a double circulation. The partition between the atria of the heart is incomplete and there is no spiral valve. There are also lungless terrestrial salamanders, but they seem to have evolved from stream-living ancestors.

A proportion of urodele species are neotenous. They fail to metamorphose or they metamorphose incompletely. The adults retain external gills and lack eyelids (which normally only develop at metamorphosis), and there may be no keratin in the epidermis. These are aquatic species, and they have come to retain in the adult form some of the aquatic features of larval urodeles. Among them the axolotl, *Ambystoma tigrinum*, is particularly interesting because it metamorphoses in some localities but not in others. Specimens of races which would not normally metamorphose can be made to do so by treatment with thyroid hormone.

Urodeles have no eardrum and often no stapes. The operculum (the cartilage referred to on p. 257) is generally connected to the pectoral girdle by a muscle (Fig. 8-20). This seems to provide a mechanism for detecting vibrations of the ground. When the trunk is made to vibrate by ground vibrations the head will vibrate less, because of its inertia and because its connection to the trunk is not absolutely rigid. However, the direct muscular attachment to the pectoral girdle will make the operculum vibrate more than the rest of the head. It will move like a piston in and out of the oval window, driving endolymph backwards and forwards in the slender canal which contains the cupula of the amphibian papilla. (The basilar papilla is small and probably unimportant in urodeles.) These movements are detected by the cupula. Since the cross-sectional area of the canal is much less than the area of the oval window, the amplitude of vibration in the canal is much greater than the amplitude of vibration of the operculum relative to the cranium. This improves the sensitivity of the system.

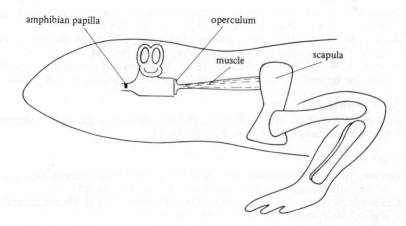

Fig. 8-20. A diagram illustrating the connection between the pectoral girdle and the ear in urodeles.

264 *Amphibians*

FURTHER READING

General

Goin, C. J. & O. B. Goin (1971). *Introduction to herpetology*, 2nd edit. Freeman, San Francisco.
Moore, J. A. (ed.) (1964). *Physiology of Amphibia*. Academic Press, New York.
Noble, G. K. (1931). *The biology of the Amphibia*. McGraw-Hill, New York.
Porter, K. S. (1972). *Herpetology*. Saunders, Philadelphia.
Smith, M. (1931). *The British amphibians and reptiles*. Collins, London.

Early amphibians

Schmalhausen, I. I. (1968). *The origin of terrestrial vertebrates*. Academic Press, New York.
Thomson, K. S. & K. H. Bossy (1970). Adaptive trends and relationships in early Amphibia. *Forma et functio*, **3**, 7–31.

Hearing on land

Geisler, C. D., W. A. van Bergeijk & L. S. Frishkopf (1964). The inner ear of the bullfrog. *J. Morph.* **114**, 43–58.

Respiration

Cox, C. B. (1967). Cutaneous respiration and the origin of the modern Amphibia. *Proc. Linn. Soc. Lond.* **178**, 37–47.
Gans, C., H. J. de Jongh & J. Farber (1969). Bullfrog (*Rana catesbiana*) ventilation: how does the frog breathe? *Science*, **163**, 1223–5.
Hutchison, V. H., W. C. Whitford & M. Kohl (1968). Relation of body size and surface area to gas exchange in anurans. *Physiol. Zool.* **41**, 65–85.
Whitford, W. C. & V. H. Hutchison (1963). Cutaneous and pulmonary gas exchange in the spotted salamander *Ambystoma maculata*. *Biol. Bull.* **124**, 344–54.

Locomotion on land

Daan, S. & T. Belterman (1968). Lateral bending in locomotion of some lower tetrapods, I & II. *Proc. K. ned. Akad. Wet.* C, **71**, 245–66.
Walker, W. F. (1971). A structural and functional analysis of walking in the turtle, *Chrysemys picta marginata*. *J. Morph.* **134**, 195–214.
Warren, J. W. & N. A. Wakefield (1972). Trackways of tetrapod vertebrates from the upper Devonian of Victoria, Australia. *Nature, Lond.* **238**, 469–70.

Modern amphibians

Parsons, T. S. & E. E. Williams (1963). The relationships of the modern Amphibia: a re-examination. *Q. Rev. Biol.* **38**, 26–53.

Frogs and toads

Gordon, M. S., K. Schmidt-Nielsen & H. M. Kelly (1961). Osmotic regulation in the crab-eating frog (*Rana cancrivora*). *J. exp. Biol.* **38**, 659–78.
Gradwell, N. (1972). Gill irrigation in *Rana catesbiana*. Part II. On the musculo-skeletal mechanism. *Can. J. Zool.* **50**, 510–21.
McClanahan, L. (1967). Adaptations of the spadefoot toad, *Scaphiopus couchi*, to desert environments. *Comp. Biochem. Physiol.* **20**, 73–99.

Martin, W. F. & C. Gans (1972). Muscular control of the vocal tract during release signalling in the toad *Bufo valiceps*. *J. Morph.* **137**, 1–28.

Wasserzug, R. (1972). The mechanism of ultraplanktonic entrapment in anuran larvae. *J. Morph.* **137**, 279–87.

Whiting, H. P. (1961). Pelvic girdle in amphibian locomotion. *Symp. Soc. exp. Biol.* **5**, 43–58.

Newts

Smith, J. B. (1968). Hearing in terrestrial urodeles: a vibration-sensitive mechanism in the ear. *J. exp. Biol.* **48**, 191–206.

Spotila, J. R. (1971). Role of temperature and water in the ecology of lungless salamanders. *Ecol. Monogr.* **42**, 95–125.

9

Reptiles in general

This chapter is about the distinctive features of the reptiles and about some of the earliest reptiles. A survey of the major groups of reptiles is left to Chapter 10.

EGGS OF REPTILES AND BIRDS

The most important differences between amphibians and reptiles concern their reproduction. The ova of amphibians (that is, the eggs excluding the jelly) are relatively small. The eggs of reptiles are generally much larger, even if only the ovum (yolk) is considered. For instance, typical frogs lay ova of about 2 mm diameter while lizards of similar weight lay eggs of nearly 1 cm diameter (with yolks of perhaps 7 mm diameter). Larger reptiles and birds lay larger eggs and ostrich (*Struthio*) eggs are about 14 cm in diameter. There is, however, a little overlap between amphibian and reptile egg sizes: the exceptionally large terrestrial ova of some amphibians (see p. 254), may be as much as 8 mm in diameter, while the tiny lizard *Sphaerodactylus*, only 5 cm long, lays eggs of diameter 4.5 mm. Amphibian eggs are usually laid in water and hatch as aquatic larvae. Reptile eggs are laid on land (if they are laid at all, for some reptiles are viviparous) and there is no aquatic larva. Amphibian eggs have no shell but reptile eggs have shells.

A large egg out of water needs a shell (or at least a tough membrane) to maintain its shape. That this is true of hen's eggs will be realized by anyone who has broken one open to fry it. The yolk is more or less spherical in an intact egg (or in a hard-boiled egg) and greatly flattened in an egg broken onto a plate (or in a fried egg). However, if the egg is broken into a jar of water the yolk is largely supported by buoyancy and remains more or less spherical. Consider a spherical egg of radius r and density ρ, enclosed in a taut but flexible membrane. A tension T acting in the membrane will set up a pressure difference $2T/r$ between the egg and its surroundings. There is a hydrostatic pressure difference inside the egg, between the top and the bottom, of $2r\rho g$ and if the egg were immersed in a fluid of density ρ there would be an equal hydrostatic pressure difference between the same levels in the fluid. If the egg is in air this is not the case, and it will only remain reasonably nearly spherical if

$$2T/r \gg 2r\rho g,$$
$$T \gg r^2\rho g.$$

To maintain a particular near-spherical shape as size increased, the tension would have to increase in proportion to the square of the radius. Consequently, a large terrestrial egg would need a relatively thicker membrane than a small one – or a rigid shell. Another important function which can be served by a shell on a terrestrial egg is to restrict water loss by evaporation. This will be discussed later.

Birds, tortoises and crocodilians lay eggs with stiff, brittle shells. Turtles, most lizards and snakes lay eggs with flexible leathery shells. The brittle egg shell of the hen contains only 3% organic matter, with 95% inorganic salts (mainly calcium carbonate) and 2% water. The shell is pierced by pores of about 20 μm diameter, through which oxygen diffuses into the egg while carbon dioxide and water vapour diffuse out. The whole shell is covered by a cuticle 5 μm thick which seems to cover the mouths of the pores. It is too thin to be a serious barrier to diffusion, but it may prevent micro-organisms from entering the pores. Within the shell are two porous shell membranes, each consisting of felted protein fibres. At the blunt end of the egg is the air cell, between the two shell membranes (Fig. 9-1). The stiff shell cannot contract as water evaporates from the egg, but the air cell gets larger. Crocodilian eggs have air cells, but reptile eggs with flexible shells do not.

The white or albumen of the hen's egg contains about 88% water, 11% protein and small quantities of carbohydrate and inorganic salts. Notice that its organic content is immensely greater than that of the jelly of frog eggs (p. 253). The twisted cords of less fluid albumen known as chalazae, which run lengthwise in bird eggs (Fig. 9-1), are not found in reptile eggs. The yolk of the hen's egg contains only about 50% water, with 16% protein and 32% fat. Only the ovum (yolk) is formed in the ovary. The white, shell membranes and shell are successively laid down around it as it passes down the oviduct, which takes about twenty-four hours.

As in other large ova such as those of hagfishes and selachians, only part of the ovum of a bird or reptile divides into cells. The embryo develops on the surface of the yolk, which becomes almost completely enclosed by cellular membranes (Fig. 9-2a). An inner membrane which is continuous with the wall of the gut (and is formed of endoderm and mesoderm) invests the yolk closely. It is known as the yolk sac. There is in addition an outer membrane continuous with the body wall (and formed of ectoderm and mesoderm). In selachians this outer membrane is an integral part of the yolk sac but in reptiles and birds it rises in folds around the embryo. (Fig. 9-2b). The folds fuse together over the embryo so that it comes to be covered by two membranes, the amnion and chorion (Fig. 9-2c). Mammals also develop an amnion, and the reptiles, birds and mammals are sometimes referred to as the amniotes, to distinguish them from the fish and

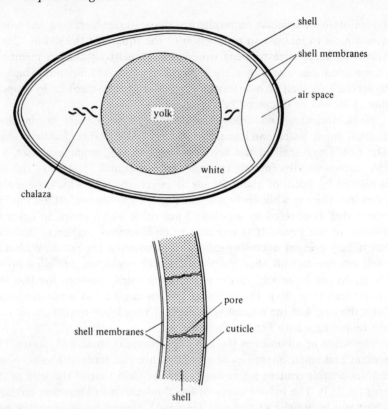

Fig. 9-1. Diagrammatic longitudinal section of a newly laid bird egg and (below) a section of the shell and shell membranes at a higher magnification.

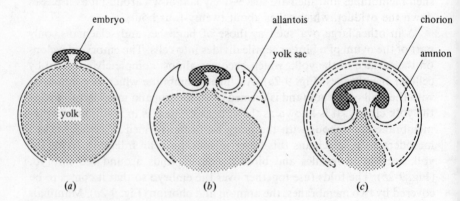

Fig. 9-2. Diagrammatic sections of successive stages in the development of a bird embryo, showing how the embryonic membranes are formed.

amphibians which have no amnion. The allantois is yet another membrane (formed of endoderm and mesoderm) which develops in amniotes as an outgrowth of the hind gut. Its main function in birds and reptiles is as an organ of respiration, and it grows so as to cover the whole inner surface of the chorion, to which it becomes attached. A network of blood capillaries develops in the allantochorion so formed, receiving blood from the dorsal aorta and returning it to the heart. Since this network is immediately inside the shell membranes, the blood receives oxygen and gives up carbon dioxide as it passes through. It is very delicate, with a thickness of less than 1 μm of tissue (in chick embryos) separating the blood from the inner shell membrane.

Various suggestions have been made as to why the amnion evolved. The most important reason was probably to allow the allantois to surround the embryo completely, so that the whole of the egg surface could be used for respiration. Another plausible suggestion is that it holds the embryo slightly clear of the shell membranes: temperature fluctuations are likely to be smaller deep in the egg than near its surface. A remark which is often repeated, that the amnion encloses the embryo in a 'private pond', is a pretty metaphor but does not explain anything.

While protecting the egg, the shell and shell membranes must be porous enough to allow respiration. They must allow diffusion of oxygen and carbon dioxide at the rates required for metabolism of the embryo, while the differences of partial pressure of these gases across the shell are reasonably low. The partial pressure of oxygen within the shell must remain high enough to support the respiration of the embryo, and the partial pressure of carbon dioxide must not rise to harmful levels. These partial pressures can be assessed by analysing samples of gas from the air cell in the egg. The partial pressure of oxygen is lowest (about 0.14 atm) and that of carbon dioxide highest (about 0.05 atm) towards the end of incubation when the metabolic rate of the embryo is highest. These partial pressures are close to the partial pressures in the air expired by adult fowl, which suggests that a less permeable shell might be harmful. Since the partial pressure of oxygen in air is 0.21 atm, the difference in partial pressure of oxygen between the air outside the egg and the air inside is about 0.07 atm.

A shell restricts evaporation from an egg but cannot altogether prevent evaporation, since it is permeable. A 60 g hen's egg incubated at 38 °C in air of relative humidity 60% loses about 7.7 g water in the 21 days of incubation. This is 0.015 g water h^{-1} or 20 cm^3 water vapour h^{-1}. Towards the end of incubation it uses about 25 cm^3 oxygen h^{-1}. Does the ability to take up 25 cm^3 oxygen h^{-1} make the loss of 20 cm^3 water vapour h^{-1} (in these conditions) inevitable? At 38 °C and 60% relative humidity, the partial pressure of water vapour is 0.026 atm less than in saturated air at the same temperature. Diffusion of water vapour through the egg shell would be driven by this partial pressure difference, which is 0.026/0.07 of the difference which is needed (as was seen above) to drive the diffusion of oxygen. The permeability constants (see p. 48) for diffusion of gases in air are inversely proportional to the square roots of their molecular weights, so the constant for water vapour (molecular weight 18) is $\sqrt{(32/18)}$ times the constant for oxygen (molecular weight 32). Hence, if oxygen and water vapour are diffusing (in opposite directions) through the same shell, the vapour should diffuse at $(0.026/0.07) \sqrt{(32/18)} = 0.5$ times the rate of the oxygen. Only 12.5 cm³ vapour h^{-1} need be lost. Actually, as we have seen, 20 cm³ h^{-1} is lost. The discrepancy is due to the oxygen having to diffuse along a longer path than the water vapour. The shell membranes are moist, and contain about 40% water even at the end of incubation, so water evaporates from their outer surface and has only to diffuse through the shell. Oxygen, however, has to diffuse through both shell and shell membranes.

This interpretation has been confirmed by experiments using the apparatus shown in Fig. 9-3, in which diffusion rates of gases through eggshells, with and without shell membranes, have been measured. Half an eggshell

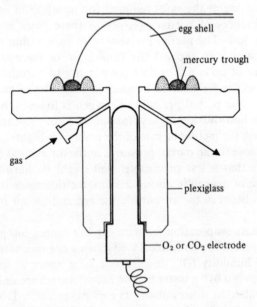

Fig. 9-3. Apparatus used to determine the permeability of egg shells to the diffusion of oxygen and of carbon dioxide. From H. Kutchai & J. B. Steen (1971). *Respir. Physiol.* **11**, 265–78.

was used, held with its edges immersed in an annular trough of mercury so that gas could only move between the atmosphere and its interior by diffusion through it. The interior was filled with a gas mixture, and subsequent changes of partial pressure of oxygen (or carbon dioxide) were monitored with the oxygen (or carbon-dioxide) electrode. It was found that in fertilized eggs which had been incubated for 5 days or more, the shell membranes offered about as much resistance to diffusion of oxygen as did the shell. Removing the shell membranes about doubled the rate of diffusion. Unfertilized eggs, or fertilized ones in the first few days of incubation, have shell membranes with a much bigger resistance to diffusion: indeed, diffusion of oxygen at the rate which occurs in fertilized eggs towards the end of incubation would require an impossible partial pressure difference of about 1.8 atm oxygen. The explanation seems to be that the pores in the shell membranes are initially all filled with water, so that diffusion through them is slow. In the first few days of incubation of a fertilized egg some of this water is lost (perhaps removed osmotically, due to an increase in the colloid osmotic pressure of the white of the egg which occurs at this time) and some of the pores become air-filled. They then allow much faster diffusion of oxygen and carbon dioxide, for a given difference in partial pressure. The rate at which water is lost by evaporation is about the same, whether the egg is fertilized or not.

The eggs of fishes and amphibians are laid in water and absorb water osmotically from their surroundings: for instance, an axolotl (*Ambystoma*) ovum increases its weight by about 75% by absorbing water after being laid. The eggs of birds are laid in generally dry places, in nests or on the ground, and lose water by evaporation. Reptile eggs are never laid in water, but some absorb water from damp soil or perhaps from moist air. These eggs are ones with leathery shells which allow some swelling. Some turtles, such as the loggerhead turtle, *Caretta*, bury their eggs in damp sand, from which they may absorb enough water to increase their weight by 50%. A terrapin, *Chrysemys*, buries its eggs in initially dry ground which it moistens with (very dilute) urine. The grass snake, *Tropidonotus*, lays in damp earth or rotting vegetation (such as in compost heaps) where its eggs take up water, but if the eggs are kept in drier conditions they lose water and the embryos die. Many lizards and snakes lay their eggs in more or less dry places, such as under stones or logs, and these eggs may lose water rather than gain it during development.

The amounts of fat, protein and carbohydrate consumed during the development of the embryo can be determined by analysing new-laid eggs and ones which are almost ready to hatch. Such analyses show that protein is the most important source of energy for fish and amphibian embryos. For instance, the material metabolized by frog (*Rana*) embryos has been found to be about 71% protein, 22% fat and 7% carbohydrate. Since fat metabolism uses 2–2.5 times as much oxygen as metabolism of the same weight of protein or carbohydrate, this implies that about 60% of the oxygen used during development is used for the metabolism of protein, and 35% for the metabolism of fat.

Protein metabolism produces nitrogenous waste products. In embryos of fishes and amphibians, as in adults, most of this waste is produced as ammonia or urea. These are soluble materials which diffuse out of the egg into the surrounding water. They diffuse out of turtle eggs which are laid in damp ' sand, but cannot diffuse in solution from bird eggs, or from reptile eggs which are laid in dry places. Ammonia could diffuse out of such eggs as gas, but not nearly fast enough to prevent a toxic concentration building up if a substantial proportion of the metabolism was of protein, releasing ammonia. If urea were produced and accumulated in the egg until it hatched, would it be likely to reach harmful concentrations?

In the course of development a hen's egg uses about 5 l of oxygen. If as high a proportion of protein were metabolized as in the frog egg, about 3 l of this would be used in protein metabolism. A mole of gas occupies 22.4 l (at STP) so this is 3/22.4 mol. A mole of urea results from protein metabolism using about six moles oxygen (see the chemical equation on p. 226), so if all the nitrogenous waste were produced as urea, about 0.022 mol would be produced. The contents of a hen's egg, excluding the large air space which is present at the end of incubation, occupy about 45 cm^3. 0.022 mol urea dissolved in 45 cm^3 fluid would have a concentration of about 0.5 mol l^{-1}. This is higher even than the concentrations found in selachians (p. 116), aestivating lungfish and spadefoot toads (pp. 226 and 260) and crab-eating frogs (p. 260). It would increase very substantially the osmotic pressure of the egg contents.

Analyses show that chick and turtle embryos use a much lower proportion of protein; in the case of the chick, only 6% of the material used in metabolism is protein, 91% is fat and 3% is carbohydrate. (The yolks of the eggs of birds and reptiles are rich in fat.) Only about 3% of the oxygen used by the chick embryo during incubation, and not the 60% supposed above, can actually be used in protein metabolism. If all the nitrogenous waste were retained in the shell as urea, the resulting urea concentration would be only 30 mmol l^{-1}.

Even this concentration is not accumulated, since most of the waste is produced not as urea but as uric acid, which is a purine closely related chemically to guanine (see pp. 145 and 160). Uric acid and its salts are too insoluble to contribute appreciably to the osmotic pressure of the egg contents, even when present as a saturated solution. They are deposited as a precipitate in the cavity of the allantois. Reptile embryos, like bird ones, include a proportion of uric acid in their nitrogenous waste, but the proportion varies greatly between species.

It will be seen later in this chapter that many adult reptiles excrete mainly uric acid or urate crystals, rather than a solution of urea, and that this can enable them to make useful savings of water when water is in short supply.

Internal fertilization is necessary for the production of the shelled, terrestrial eggs of reptiles. *Sphenodon* (the tuatara) has no intromittent organ but the males of other modern reptiles do. The tortoises and turtles and the crocodilians have a grooved penis formed from the ventral wall of the cloaca. It is erected for copulation by engorgement with blood, which also makes the edges of the groove meet to form a tube for the sperm to pass along. A male tortoise places his fore feet on the female's back and curls his tail under hers in copulation. Male lizards and snakes have paired hemipenes which at rest are diverticula of the cloacal cavity. They are erected and protruded from the cloaca by turning inside out: this involves both contraction of a muscle, and engorgement with blood. Only one of the pair of hemipenes is inserted into the cloaca of the female.

Various reptiles are viviparous, but it is probably no coincidence that they include many of the most northerly species. Examples are the Viper (*Vipera berus*) and Viviparous lizard (*Lacerta vivipara*), which reach latitudes 67° N. and 70° N. in Scandinavia. A viviparous species which basks in the sun can keep its body much warmer, during the day, than most hiding places where eggs could be laid. Temperature regulation by basking will be discussed later in this chapter. Sea snakes are also viviparous, for a different reason. Reptile eggs are not suitable for laying in water because oxygen could not diffuse in fast enough if the shell were waterlogged. Turtles and crocodilians lay their eggs on land but viviparity enables sea snakes to breed without leaving the water.

SKIN, WATER AND SALTS

The epidermis of lizards and snakes is very different from that of amphibians. Its horny layer is double (Fig. 9-4a). There is an inner sublayer which resembles the whole horny layer of amphibians, consisting of flattened but distinct keratinized cells. There is in addition an outer sublayer in which cellular outlines generally cannot be distinguished, though electron micrography reveals scattered fragments of plasma membrane. The outer surface is finely ridged. There is another difference between the two layers. The proteins classed as keratins occur in two distinct forms: the molecules are coiled in α-keratins and straighter in β-keratins, and this difference can be demonstrated by X-ray diffraction. The horny layer of the epidermis of amphibians and mammals contains only α-keratin, as

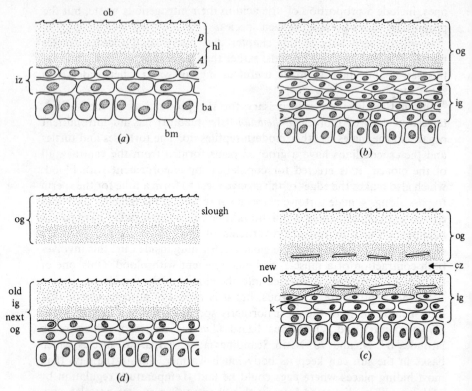

Fig. 9-4. Diagrammatic sections through the epidermis of a snake, showing four stages in the sloughing cycle. A, B, α- and β- keratin layers; ba, basal cell layer; bm, basement membrane; cz, cleavage zone; hl, horny layer; iz, intermediate zone of cells; ob, ridged surface of epidermis; ig, og, inner and outer epidermal generations. From A. Bellairs (1969). *The Life of Reptiles.* Weidenfeld & Nicolson, London.

does the inner sublayer of lizards and snakes, but the outer sublayer is of β-keratin.

The epidermis of crocodiles and of tortoises has a single horny layer with distinct cell outlines, as in amphibians, but in crocodiles at least this layer contains β- as well as α-keratin.

The horny layer is generally thicker in reptiles than in amphibians, and it is not interrupted by glandular openings. The horny scales are quite different in nature from the scales of fishes. They are merely thickenings of the continuous epidermis, which is folded where they overlap. The folds make it possible for the skin to stretch, although the horny layer is inextensible. Spectacular stretching is necessary when snakes swallow large prey. Often the horny scales overlie plates of bone in the dermis, which are more closely comparable to the scales of fishes.

The horny layer of the epidermis of crocodiles and tortoises is apparently added to from within, as in amphibians. It is not shed as a whole but small flakes are lost from its outer surface. The double horny layer of lizards and snakes could not be maintained in this way, as the outer sublayer would be worn away but only the inner one could be added to. Instead, it is shed as a whole, usually several times a year, and replaced by a complete new double layer. Lizards usually shed it in large flakes but snakes usually shed it complete. The process is illustrated in Fig. 9-4. The resting stage (*a*) has already been described. The process of shedding starts by division of the basal layer of the epidermis to form numerous layers of flattened cells (the inner epidermal generation, Fig. 9-4*b*). Keratin is laid down in these cells, forming a new double horny layer, while the innermost cells of the outer epidermal generation break down and the old horny layer splits free (Fig. 9-4*c*, *d*).

Most reptiles lose water far more slowly than amphibians, when exposed to dry air. This is well illustrated by an observation that a 17 g garter snake (*Thamnophis*) kept in a dessicator lost 13% of its weight in seven days while a frog (*Rana*) of similar weight in the same conditions lost the same percentage of its weight by evaporation in two to four hours. The difference is presumably due to the thick horny layer of the reptilian epidermis. The outer sublayer of lizards and snakes seems likely to be particularly effective in retaining water.

Evaporation of water cannot be cut down below a certain minimum rate, no matter how impermeable the skin. This is because water is also lost by evaporation from the moist surfaces of the lungs and respiratory tract. Air saturated with water vapour at 20 °C contains about 20 mg water l^{-1} and at 40 °C about 60 mg l^{-1}. Air contains about 200 cm^3 O_2 l^{-1}. If an animal with a body temperature of 20 °C breathes dry air, removes all the oxygen from it (replacing it by carbon dioxide) and then breathes it out again at 20 °C saturated with water vapour from the lungs, it will lose 0.1 g water for every litre of oxygen used. Similarly with a body temperature of 40 °C it would lose 0.3 g water per litre oxygen. In fact reptiles only remove a proportion of the oxygen from the air they breathe and must lose more water than this.

The loss is partially compensated by water formed in metabolism. For instance, when polysaccharide is metabolized according to the equation

$$(C_6H_{10}O_5)_n + 6n\ O_2 = 6n\ CO_2 + 5n\ H_2O$$

5 mol (90 g) water is formed for every 6 mol (134 l) oxygen used. That is, water is produced at the rate of 0.7 g l^{-1} oxygen. Fairly similar amounts are produced in fat and protein metabolism. This water production could compensate for the loss by evaporation from the lungs, if the evaporation

were kept reasonably near the theoretical minimum calculated in the previous paragraph.

Water loss from the respiratory tract and through the skin have been measured separately for a few reptiles, in experiments in which oxygen consumption was also measured. The animal was weighed and put into a chamber through which dry air was passed slowly. Oxygen consumption was calculated from the difference in oxygen content between the incoming and outgoing air. The animal was weighed again after a period in the chamber, and the total weight of water lost by evaporation was taken to be the loss of weight, minus a correction for the weight of carbon calculated (from the oxygen consumption) to have been lost as carbon dioxide. Urine and faeces were either collected and weighed, in which case their weight was allowed for in the calculations, or they were retained in the body by closing the cloaca with adhesive tape. The experiment was repeated with all but the head of the reptile enclosed by a plastic bag which was fastened closely round its neck. Water could then only be lost from the respiratory tract and from the skin of the head. The loss per unit area through the skin of the head was assumed to be the same as for the rest of the body, and so the respiratory loss could be calculated. Some of the results are displayed in Table 9-1. Note that these results are all for animals of similar size, so that it is reasonable to make direct comparisons between them.

There are several points of interest in the results. First, the three species lost water at 23 °C at very different rates. The crocodilian *Caiman*, which lives in and near water, lost water at very roughly one third of the rate which would be expected of an amphibian of similar size. The two lizards lost water much more slowly, but the desert lizard *Sauromalus* lost it even more slowly than the forest lizard *Iguana*. Secondly, *Sauromalus* lost water very much faster at 40 °C than at 23 °C. Thirdly, although one tends to think of reptile skin as highly impermeable, loss through the skin represented a large proportion of the total loss in every case. Finally, respiratory losses were always considerably above the minimum values of 0.1 g l^{-1} oxygen (for 20 °C) and 0.3 g l^{-1} (for 40 °C) calculated above. Probably only *Sauromalus* at 23 °C formed water fast enough by metabolism to compensate for evaporation from the lungs.

Reptiles lose water by excretion as well as by evaporation. Some turtles excrete most of their waste nitrogen as a solution of ammonia and urea, but most reptiles excrete mainly urates. Reptiles seem incapable of producing urine of higher osmotic concentration than their blood, so a certain minimum volume of water is required to excrete a given quantity of ammonia or urea. The solubility of sodium urate is only about 7 mmol l^{-1} (which is much less than the osmotic concentration of the blood) so urate can be precipitated and excreted in very little water as a paste (the white droppings of birds are similar urate pastes). How much water can be saved in this way?

A lizard in a warm climate might use about 3 l oxygen (kg body weight)$^{-1}$ day^{-1} (see Fig. 9-5a). Since most lizards feed mainly on insects and other small animals, containing a high proportion of protein, at least 1 l oxygen kg^{-1} day^{-1} would probably be used in protein metabolism. 1 l oxygen is about 0.04 mol, and protein metabolism using it would produce about 0.007 mol urea, if this were the nitrogenous end-product (see p. 226). The molar concentration of lizard blood plasma is typically about 0.35 mol l^{-1}, and we will suppose that the urea would be excreted as a solution of this concentration. The volume of water required would be 0.007/0.35 l or about 20 cm^3. The lizard could be expected to lose 20 cm^3 water kg^{-1} day^{-1} getting rid of nitrogenous waste. This would be a substantial loss, comparable to losses by evaporation (Table 9-1). It could be avoided by excretion of a urate paste.

TABLE 9-1.   *Data from experiments described in the text, on evaporative water loss from reptiles weighing about 0.13 kg.*

Some of the data is presented in a slightly different form from that in which it originally appeared, in Bentley, P. J. & K. Schmidt-Nielsen (1966). *Science*, **151**, 1547-9

	Temperature	Loss from skin	Respiratory loss	
	(°C)	(g kg^{-1} day^{-1})	(g kg^{-1} day^{-1})	(g l^{-1} oxygen)
Caiman	23	63.8	8.8	4.8
Iguana	23	9.6	2.3	0.9
Sauromalus	23	2.6	0.6	0.5
	40	6.8	8.1	1.5

Reptiles also lose water excreting excess salts. Salts must be excreted at the same rate as they are taken in with food. They are normally excreted in the urine, at concentrations not more than the osmotic concentration of the blood. A carnivorous reptile which ate food of about the same salt concentration as its blood would have to use about as much water excreting the salts, as was contained in the food. A herbivorous reptile eating terrestrial plants of lower salt concentration than the blood, on the other hand, need only excrete some of the water from its food and could retain the rest to help compensate for losses by evaporation. If urea were excreted, water needed for this would be additional to the water needed to excrete salts: the sum of the concentrations of urea and salts in the urine would presumably be limited by the concentration of the blood.

Many lizards live in hot, dry places where water lost by evaporation cannot easily be replaced by drinking. Drinking water may be sparse and seldom available. An example is *Amphibolurus ornatus* which lives on bare granite outcrops in Southwest Australia and feeds mainly on ants. It is apparently unable to maintain both salt and water balance in summer, when the only available drinking water is provided by infrequent thunderstorms. It maintains a constant proportion of water in its body, but the concentration of salts increases. The concentration of sodium in the blood plasma was 10% higher in specimens collected at mid-summer than in ones collected in later spring. (Very much larger increases have been observed in a prolonged drought.) There was a heavy rainstorm later in the same summer, and the lizards were seen catching raindrops and drinking from puddles. Specimens caught 10 h after the storm had as low a concentration of sodium in the plasma as in spring, and the same proportion of water in the body as in both the earlier samples. The excess salts had apparently been excreted in the few hours after the rain. No urine was found in the cloacas of the specimens taken during the mid-summer drought, but large quantities were found in all the specimens taken after the storm.

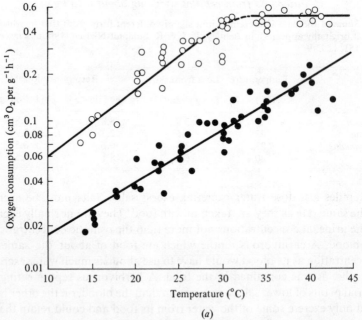

(a)

Fig. 9-5. (a) Resting (●) and maximum active (○) metabolic rates at different temperatures of the lizard *Iguana iguana*. The measurements were made on specimens weighing 0.37 to 1.22 kg.

(b) [*opposite*] Apparatus used to measure the metabolic rates of lizards running at various speeds.

From W. R. Moberly (1968). *Comp. Biochem. Physiol.* **27**, 1–20, 21–32.

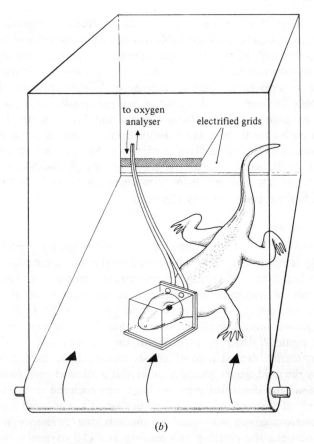

to oxygen
analyser

electrified grids

(b)

Amphibolurus seems able to excrete salts only in its urine, but some other lizards have glands in their nasal cavities which excrete salts. One of them is *Amblyrhynchus*, which lives in the surf of the Galapagos Islands and feeds on seaweed. Fluid collected from the nostrils of freshly caught specimens has been found to be a concentrated solution of salts, often three times as concentrated as seawater. One cannot be sure that it is secreted at this concentration, since it must tend to be concentrated by evaporation. The secretion is blown out of the nostrils, but an incrustation of salts tends to accumulate around them. There are other herbivorous lizards which have nasal salt glands, including *Sauromalus*, which lives in North American deserts and eats succulent plants.

Of the few marine reptiles all but the Marine crocodile (*Crocodilus porosus*) seem to have salt glands of one sort or another. It is clearly an advantage to them to be able to excrete a salt solution more concentrated than the blood plasma. Like other reptiles, they lose water by evaporation from the lungs. Fresh water which could be drunk to replace it is not available, except on land. Of available foods, fishes have salt concentrations similar to those of the reptiles, while marine invertebrates and algae have higher salt concentrations, about the same as seawater. Marine turtles leave the sea only to lay eggs on the shore. As they crawl up the beach they appear to weep. The tears, which are presumably also produced in the sea, have been collected from *Caretta* and *Lepidochelys*. They were found in each case to be a salt solution considerably more concentrated than seawater. They come from a gland in the orbit which apparently functions in the same way as the nasal glands of *Amblyrhynchus*. Sea snakes secrete salt solutions from somewhere on the head, but the gland involved has not been certainly identified.

TEMPERATURE

In reptiles as in fishes (p. 197) and other animals, metabolic rates increase with body temperature. This is illustrated by the measurements on *Iguana* shown in Fig. 9-5*a*. The resting metabolic rate increases with temperature but so does the maximum rate that can be achieved in activity, and the difference between the two is, in this case, greatest at 32 °C. This implies that the animal should be able to be most active at this temperature. This was investigated further by experiments on the treadmill shown in Fig. 9.5*b*. The animal ran on a continuously moving belt, discouraged from resting by the mild electric shock it suffered if it allowed itself to be carried back against the electrified grid. Its head was enclosed in a transparent plastic helmet through which air was passed at a measured rate, and a paramagnetic analyser was used to measure the difference in oxygen content between the entering and leaving air. The oxygen consumption could thus be calculated. It was found that the cost (in terms of oxygen consumption) of running at any given speed was the same at all temperatures between 20 °C and 40 °C, and that the maximum speed which could be sustained for several minutes was nearly three times as high at 30–35 °C as at 20 °C. This confirms what would be expected from Fig. 9-5*a*, that a rise in temperature, at least up to 30 °C, enables the lizard to become more active.

Though a reasonably high temperature may be advantageous, too high a one would be lethal. Reptiles become incapable of co-ordinated movement when their body temperatures reach a limit that varies between species but seems nearly always to be between 39 °C and 49 °C. They die at slightly higher temperatures. Some enzymes become inactive at temperatures in this range, apparently because of the disordering of molecular structure which is known as denaturation. The harmful effects of high temperatures are probably largely due to this but it has been suggested that disorganization of cell membranes due to melting of lipids may also be involved.

Terrestrial animals can warm themselves by basking in hot sun, and so obtain the advantages of a limited increase in temperature. Frogs of the genus *Rana* bask in the sun and may become up to about 7 °C warmer than the air, but terrestrial salamanders seem to avoid the sun. Amphibian body temperatures, in natural conditions, rarely exceed 35 °C and are generally much lower. Reptiles, with their less permeable skins, can bask with far less danger of desiccation. They make much more use of the sun and may attain remarkably high body temperatures, as will be seen. While taking advantage of the sun they have to avoid the danger of overheating.

The problem of temperature regulation is perhaps best appreciated by considering a hypothetical example such as the one illustrated in Fig. 9-6.

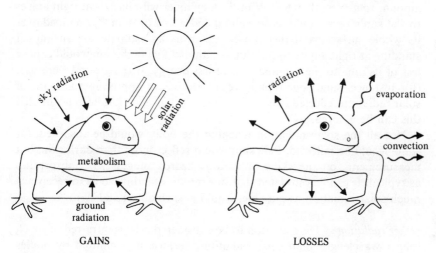

GAINS   LOSSES

Fig. 9-6. Diagrams illustrating the heat balance of a terrestrial animal.

The reptile is supposed to be standing on level, dark-coloured ground on a clear sunny day, with the sun high in the sky. The air temperature is 30 °C but the surface of the ground is at 50 °C because it is being heated by the sun. The reptile's body temperature is 37 °C. All these suppositions are reasonable ones, as we shall see when we go on to consider a real example. The body temperature of 37 °C may seem surprisingly high, for this would be a normal temperature for a mammal, but the bodies of many tropical and subtropical reptiles do, in fact, rise to such temperatures on sunny days. Our reptile exchanges heat with the environment by radiation and convection, heat is produced in its body by metabolism and heat is lost by the evaporation of water. The contributions which these make to its heat balance will be estimated in turn. It will be convenient to treat solar radiation and other radiation as separate categories.

Solar radiation.   The intensity of the radiation from the sun which reaches the surface of the earth depends on how high the sun is in the sky, because of absorption by the atmosphere. When the sun is near the horizon the radiation takes a much longer path through the atmosphere than when it is directly overhead, and more is absorbed before it reaches the earth. When the sun is high and the sky is clear solar radiation on a surface set at right angles to the sun's rays amounts to about 850 W m^{-2}. (This includes radiation scattered by the sky as well as direct rays from the sun.) Only part of the surface of an animal's body can be at right angles to the sun, and the solar radiation averaged over the whole surface of the body will amount to far less than 850 W m^{-2}. A cylinder with its axis at right angles to the sun's rays would receive about $850/\pi = 270$ W m^{-2}, averaged over its whole surface, and this can be taken as an estimate for an animal standing at right angles to the sun. An animal facing the sun would expose less of its area to the sun and receive less energy. If the ground were pale (for instance, sand) our reptile would receive an appreciable amount of solar radiation reflected from it, but since dark ground has been postulated this can be ignored.

Not all the solar radiation reaching the body would be absorbed, for some would be reflected. Pale skin would reflect more than dark skin, but measurements on the skins of various lizards suggest 75% absorption as typical. It can be estimated that our reptile, standing so as to receive as much solar radiation as possible, would absorb about 200 W m^{-2}.

Other radiation.   The radiation to be considered now is infra-red of much longer wavelength than solar radiation, because it is emitted by bodies which are much cooler than the sun. Though the surfaces of ground and of animals may be pale in colour and reflect much of the light which falls on them, they generally reflect very little of the long-wave radiation being considered now. In this range of wavelengths they are more or less perfect absorbers and emitters of radiation; in the language of physics, they behave as 'black bodies'.

All bodies emit radiation, at rates depending on the absolute temperatures of their surfaces. A black body at TK emits $5.7 \times 10^{-8}T^4$ W m^{-2}, so our reptile at 37 °C (310 K) should emit $5.7 \times 10^{-8} \times (310)^4 = 530$ W m^{-2}. It will also receive long-wave radiation from the ground and from the atmosphere (in addition to the short-wave solar radiation reflected from the ground and scattered by the atmosphere which was considered under the previous heading). Ground at 50 °C would emit 620 W m^{-2}. Different parts of the atmosphere are at different temperatures (the outer parts at very low temperatures indeed) and it does not seem possible to estimate atmospheric long-wave radiation from first principles, but it is found in practice that this radiation generally amounts to about 400 W m^{-2} when air temperatures near the ground are about 30 °C. The upper half of the reptile's body can thus be expected to receive about 400 W m^{-2} long-wave radiation, and the lower half 620 W m^{-2}, giving an average for the whole surface of the body of 510 W m^{-2}. Since radiation is being emitted at an estimated 530 W m^{-2}, exchange of long-wave radiation between the reptile and its environment results in an estimated net loss of 20 W m^{-2}.

Convection. In normal outdoor conditions wind is far more important than convection currents, and it is 'forced convection' due to wind that will be considered. Its rate (in W m^{-2}) is greatest if the temperature difference between the body and the air is large, if the wind is fast and if the diameter of the body is small. For a cylinder of diameter d m with its axis at right angles to a wind of velocity u m s^{-1}, the rate is $0.9\ \Delta T(u/d^2)^{\frac{1}{2}}$ W m^{-2} when the temperature difference is ΔT K.

In the case being considered ΔT is 7 K. A light wind might have a speed (at the appropriate height above the ground) of 1 m s^{-1} and a moderate-sized lizard might be regarded as approximating to a cylinder of diameter 2×10^{-2} m. With these values the formula gives as the rate of loss of heat by convection $0.9 \times 7(1/4 \times 10^{-4})^{\frac{1}{2}} = 90$ W m^{-2}.

Metabolism. The lizard *Sauromalus*, in the experiment at 40 °C referred to in Table 9-1, used about 4 cm^{3}O$_2$ (kg body weight)$^{-1}$ min^{-1}. Metabolism involving 1 cm^3 oxygen produces about 20 J heat, whatever food is being metabolized, so the lizards must have been producing about 80 J kg^{-1} min^{-1} or 1.3 W kg^{-1}. The specimens weighed 0.12 kg and their surface areas must have been about 0.025 m^2, so if the heat production is expressed in terms of area to correspond with the estimates for radiation and convection a value of 6.5 W m^2 is obtained.

Evaporation. The *Sauromalus* just considered lost a total of 15 g water kg^{-1} day^{-1}, or 0.17 mg kg^{-1} s^{-1}. The latent heat of vaporization of water is about 2.5 J mg^{-1}, so this represents a heat loss of about 0.43 W kg^{-1} or 2 W m^{-2}. Some other reptiles would lose water faster, but water loss would have to be very much faster to affect the heat balance of the reptile substantially. Lizards in laboratory experiments sometimes pant at high temperatures, and so increase evaporative losses, but this does not seem to be a frequent feature of behaviour in nature.

If the body temperature is to be kept constant, heat gains must balance heat losses. The hypothetical reptile we are considering can vary the amount of solar radiation it receives within wide limits, by changing its position. How much solar radiation would be needed to keep its body temperature constant? It has been estimated that exchange of long-wave radiation with the environment results in a net loss from the body of about 20 W m^{-2}. Convection causes a loss of about 90 W m^{-2}. Metabolism gives a gain of about 7 W m^{-2} and evaporation a loss of about 2 W m^{-2}. Hence the solar radiation needed is about 100 W m^{-2}. About twice as much would be received if the reptile arranged its body ar right angles to the sun's rays. To keep its temperature at about 37 °C it would have either to adopt a position in which it received less solar radiation, or spend some of its time in shade.

The ways in which reptiles control their body temperatures are well illustrated by observations on the Australian lizard *Amphibolurus ornatus*. It lives among barren granite outcrops where the only shelter from the sun is provided by the rocks themselves. The surfaces of the rocks become very hot in the midday sun, but very cold at night when they are exchanging long-wave radiation with a cold, clear sky. Fissures in the rocks are exposed neither to the sun nor to the night sky, and vary much less in temperature (Fig. 9-7).

When *Amphibolurus* is kept in a temperature gradient in laboratory experiments, it chooses its position so as to maintain its body temperature at about 37 °C. Body temperatures in the field have been investigated by inserting a thermistor in the rectum, with the leads fastened by adhesive tape to the tail. The wires continued through an overhead support to recording apparatus 50 m away which was also connected to thermistors registering air and rock surface temperatures. The lizards were free to run about, and were watched through binoculars. They behaved very much like lizards which had not been fitted with thermistors, except that their wires occasionally got caught in the rocks. Observations on a large monitor lizard (*Varanus*) with a thermistor connected to a miniature radio transmitter attached to its body (so that no wires were needed) gave similar results.

Fig. 9-8 summarizes observations made on a hot summer day. The night was spent in rock fissures, where night temperatures were higher than on the surface. The lizards emerged in the morning with body temperatures of about 25 °C, and basked in the sun, retiring again for a while if a cloud passed over the sun or if there was a gust of cold wind. They basked with their bellies in contact with the warm rock, in positions where their bodies were well exposed to the sun's rays (Fig. 9-8a). Their body temperatures rose at rates up to 1 deg C min^{-1}. When they reached about 37 °C they

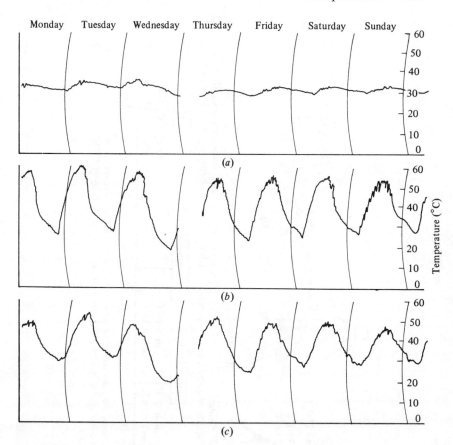

Fig. 9-7. Temperature records for a week in summer, in an area in SW Australia inhabited by the lizard *Amphibolurus ornatus*. Record (*a*) shows the temperature under a large fixed slab of rock, (*b*) on top of the same slab, and (*c*) under a small movable rock. From S. D. Bradshaw & A. R. Main (1968). *J. Zool., Lond.* **154**, 193–221.

became active – feeding, courting and defending their territories (Fig. 9-8*b*). The rock surfaces warmed up more slowly than the lizards at first, but eventually became much hotter. When they reached about 50 °C the lizards spent part of their time in the shade, or stood inactive, facing the sun with the belly and often the tail held clear of the ground. Facing the sun reduces the area exposed to it, as has been explained. Holding the body clear of the ground allows free circulation of cooling air around it, and minimizes heat gain by conduction from the rocks. In the hottest part of the day, with air temperatures around 40 °C and the rock surfaces over 53 °C, the lizards retired to the fissures (Fig. 9-8*c*), but they emerged again for a while in the evening before retiring finally for the night.

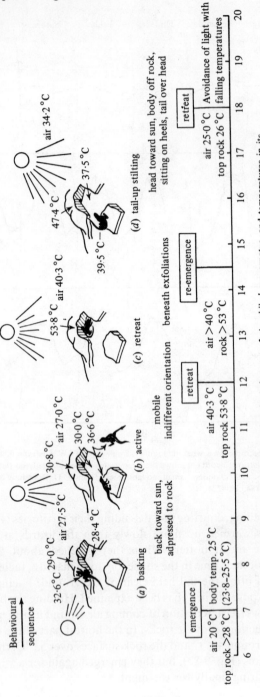

Fig. 9-8. A diagram illustrating the behaviour of *Amphibolurus ornatus*, and temperatures in its environment, at different times of day in summer. From S. D. Bradshaw & A. R. Main (1968). *J. Zool., Lond.* **154**, 193–221.

Many lizards that live in hot sandy regions behave in much the same way as *Amphibolurus ornatus*, using burrows made by themselves or by rodents instead of rock fissures. Some other reptiles make little use of solar radiation and may be active at much lower body temperatures. For instance, the lizard *Anolis allogus* lives in dense forest in Cuba where hardly any sunlight reaches the ground. It lives in small trees but never climbs high and so has virtually no opportunity to bask. Specimens caught during the day while air temperatures were 26–32 °C had body temperatures which were on average about 1 °C below air temperature. The legless burrowing lizard *Anniella* lives underground, where its body temperature was found, in a series of measurements in California, to average 21 °C. Its temperature must follow closely that of the soil, and high temperatures are avoided by digging deeply in summer; it is found mainly at depths of 0.3 m or less in spring, but at around 1.5 m at mid-summer. *Sphenodon*, the tuatara, spends much of the day in burrows but is active on the surface at night, when body temperatures of 6–13 °C have been measured. These temperatures are exceptionally low for an active reptile.

Lizards seem able to influence their rate of heating or cooling, by controlling the blood supply to the skin. In one series of experiments *Amphibolurus barbatus* was heated and cooled between 20 °C and 40 °C in an ordinary refrigerator. The average body temperature of active wild specimens of this species is 35 °C, so it might be expected to favour warming from 20 °C to 40 °C and resist cooling from 40 °C to 20 °C. It was found that heating was slightly faster than cooling while the animals were alive, but that dead animals heated and cooled at equal (lower) rates. These results could be due to restriction of blood flow through the skin of live animals during cooling, and absence of blood flow after death. The skin forms an insulating layer between the body and the air, thin enough to have been ignored in the very rough calculations presented above (p. 281). Restriction of blood flow increases the effective thickness of the insulating layer and so slightly reduces heat exchange. The effect seems to be small in reptiles, but it is interesting because control of blood flow in the skin plays an important part in the temperature regulation of mammals.

BLOOD CIRCULATION

Fig. 9-9a shows the arrangement of the major arteries of reptiles. The blood leaves the heart through three separate arteries that are twisted around each other, not through a conus and spiral valve as in amphibians. These three arteries are the pulmonary artery, which goes to the lungs, and the left and right systemic arteries, which serve the rest of the body. As in amphibians, the systemic arteries join together to form a median dorsal aorta. The carotid arteries, which serve the head, branch from the right systemic artery.

The hearts of crocodilians are rather different from those of other reptiles but will be described first because they are more easily understood. They have separate left and right atria as in frogs but they also have a partition separating left and right ventricles (Fig. 9-9*b*). The pulmonary and left systemic arteries open out of the right ventricle, and the right systemic from the left ventricle. (The twisting together of the arteries brings about this rather confusing situation in which the one that goes to the left arises on the right, and vice versa.) Where the left and right systemic arteries are in contact as they leave the ventricle, there is an opening between them known as the foramen Panizzae.

Small samples of blood from the main blood vessels of crocodiles and caimans have been analysed. It has been found as would be expected that

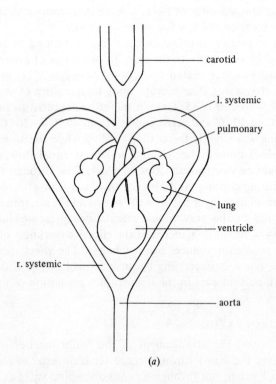

(*a*)

Fig. 9-9. (*a*) A diagram showing the arrangement of the heart and principal arteries of reptiles. (*b*) [*opposite*] A diagram showing the normal path of blood through the heart of a crocodilian. From K. Johansen (1971). In A. J. Waterman, *Chordate Structure and Function.* Macmillan, New York. © Macmillan Publishing Co. Inc., 1971.

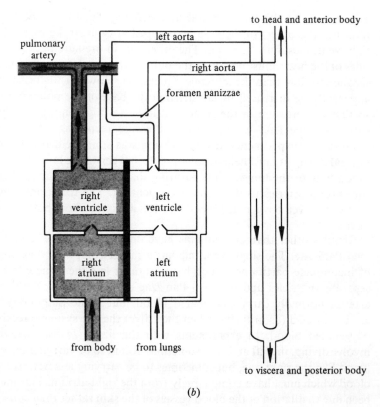

(*b*)

the pulmonary artery carries deoxygenated blood, with about the same oxygen content as the venous blood returned from the body to the sinus venosus. Also, the right systemic artery carries oxygenated blood with about the same oxygen content as the blood returning from the lungs in the pulmonary vein. Less predictably, the left systemic normally carries oxygenated blood of similar oxygen content. It apparently does not receive its blood from the right ventricle, but mainly or entirely through the foramen Panizzae from the left ventricle (Fig. 9-9*b*).

An explanation for this has been found by using pressure transducers to record pressures in the heart and major arteries of alligators. Pressures in both systemic arteries are generally considerably higher than those in the right ventricle and pulmonary artery. This pressure difference must prevent the valve from the right ventricle to the left systemic from opening, and ensure that this artery receives blood only through the foramen Panizzae.

It thus seems that all the oxygenated blood from the left atrium goes to the systemic circulation, and all the deoxygenated blood from the right atrium goes to the lungs. Blood driven from one side of the heart will return to the other, so the volume pumped at each stroke by the left and right ventricles must be equal. The difference in pressure between the two sides of the heart must be due to the pulmonary circulation offering less resistance to flow than the systemic circulation. However, its resistance can apparently be increased by constricting the base of the pulmonary artery so that the pressure in the right ventricle rises high enough to open the valve into the left systemic artery. Pressure recordings show that this happened in experiments in which diving was simulated. If it happens in normal diving, it must help to conserve the oxygen in the lungs (by reducing blood flow to the lungs) while ensuring that the blood supplied through the carotid arteries to the brain is oxygenated. The trunk and limbs will receive less well oxygenated blood which is a mixture from both sides of the heart.

Reptiles other than crocodilians have only one ventricle, and no foramen Panizzae. The single ventricle has a rather complicated arrangement of incomplete partitions, which have more or less the same effect as the separate ventricles and foramen Panizzae of crocodilians. Both systemic arteries normally carry oxygenated blood, but if the pulmonary vessels are constricted (as in turtles when they dive) the left systemic receives deoxygenated blood. In experiments with the lizard *Iguana* that did not involve diving, the left arch was sometimes found to be carrying oxygenated blood like the right one, but sometimes to be carrying less well oxygenated blood which must have come largely from the right atrium. This may have been due to dilation of the blood vessels of the skin rather than constriction of those of the lungs; it was explained on p. 287 that control of blood flow through the skin may be used as a mechanism of temperature regulation.

PRIMITIVE REPTILES

Class Reptilia, subclass Anapsida, order Cotylosauria

The reptiles seem to have appeared about the middle of the Carboniferous period, but it is not always easy to decide whether a particular Palaeozoic fossil is in fact an amphibian or a reptile. This is because the difference in their eggs is regarded as the fundamental difference between the classes, and it is generally impossible to find out what sort of eggs a fossil laid. Fossil eggs of reptilian type from the Permian period seem to be the earliest known, but fossil eggs are rare (which is not surprising, in view of their fragility) so this is not convincing evidence against an earlier origin for the group. Various features of the skeleton have been suggested as means of distinguishing reptiles from amphibians. Reptiles generally have two or more sacral vertebrae, each bearing a pair of sacral ribs, while amphibians have only one. The intercentrum, which is the main component of the body of each vertebra in most Palaeozoic amphibians, is reduced in reptiles to a small crescent of bone. The pleurocentra, which are generally a pair of small pieces of bone in amphibians, are enlarged and fused together in reptiles and form the main body of the vertebra. Reptiles generally have more phalanges than amphibians, and fewer tarsal bones. The seymouriamorphs are a group of Permian fossils which in these features are more like reptiles than amphibians, and they have sometimes been classed as reptiles. However, fossil larvae of seymouriamorphs have been found in which external gills can be distinguished. This establishes beyond doubt that the seymouriamorphas should be classed as amphibians.

There are other skeletal features which may perhaps be more reliable as a means of distinguishing early fossil reptiles from amphibians. Reptiles lack the otic notch which housed the eardrum in amphibians (Fig. 9-10). Primitive reptiles (and lizards) have down-turned transverse flanges on their pterygoid bones. Both these features seem to be associated with differences in jaw musculature. Fig. 9-10 shows how the jaw muscles are arranged in modern lizards, and how they are believed to have been arranged in some fossils. The jaw muscles of rhipidistians must have lain between the pterygoid and the dermal bones of the side of the head, emerging through the hole in the palate immediately anterior to the quadrate (Fig. 7-15). The shape of the space indicates that they must have run as shown in Fig. 9-10*a*, with the posterior fibres running roughly vertically and the anterior ones curving forward under the eye. Modern lizards (Fig. 9-10*f*) have posterior adductors which run more or less vertically to insert on the dorsal edge of the lower jaw, and pterygoideus muscles which run nearly horizontally from the transverse flange on the pterygoid to the inner face of the jaw. The pterygoideus muscle may be homologous with the anterior part of the jaw muscles of rhipidistians. The sequence of changes that probably occurred in the course of evolution of the reptiles from the rhipidistians is shown in Fig. 9-10*b–e*. *Ichthyostega* (*b*) is already familiar as a very early amphibian. *Palaeogyrinus* (*c*) and *Gephyrostegus* (*d*) are both Carboniferous amphibians, but the former is relatively primitive while the latter is probably very like the amphibians from which the reptiles evolved (it seems slightly too late to be an actual ancestor of the reptiles). *Paleothyris* (*e*), also from the Carboniferous, is believed to be a reptile, but it is an early and primitive one.

These changes can be interpreted as possible consequences of reduction in size and change of diet. The earliest known reptiles are small: for instance the length of *Paleothyris* (excluding the tail, which has not been found) is about 12 cm. They probably fed largely on insects, which were becoming common in the later Carboniferous, while early amphibians

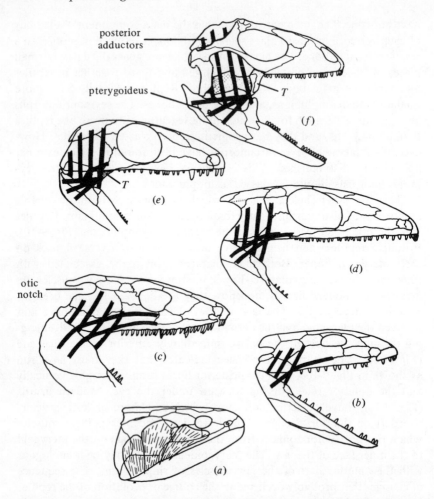

Fig. 9-10. (*a*)–(*e*) The probable arrangement of the jaw-closing muscles in various fossil vertebrates. (*a*) is the rhipidistian *Ectosteorhachis*; (*b*), (*c*) and (*d*) are the amphibians *Ichthyostega*, *Palaeogyrinus* and *Gephyrostegus*, respectively; (*e*) is the reptile *Paleothyris*; (*f*) the arrangement of jaw-closing muscles in a modern lizard. *T*, the flanges on the pterygoids. From R. L. Carroll (1969). *Biol. Rev.* **44**, 393–432.

probably fed largely on fishes. The prey of early reptiles probably tended not only to be absolutely smaller than that of early amphibians, but also to be smaller relative to the size of the predator. Small animals tend to have relatively large eyes, because a small eye cannot be made to resolve as fine detail as a large one: fewer sensory cells can be fitted into a small retina, and the resolving power of a lens depends on its diameter. In amphibians such as *Ichthyostega* and *Palaeogyrinus* (Fig. 9-10*b*, *c*) the jaw muscles probably extended forward below the eyes in much the same way as in rhipidistians. In smaller amphibians with relatively larger eyes, such as *Gephyrostegus* (Fig. 9-10*d*) there was less room for this. *Paleothyris* (Fig. 9-10*e*) was even smaller and in it and other early reptiles the flanges on the pterygoids provide an origin for the muscles for which there is no longer room under the eyes. These pterygoideus muscles must be short, but this need not matter if the food is too small to require a large gape. The flanges projected into the gullet, but would not hinder the passage of small prey. *Ichthyostega* and *Palaeogyrinus* have rather long jaws, with the jaw articulation posterior to the occiput (where the vertebral column joins the skull). Consequently the posterior jaw adductors do not run vertically, but slope posteriorly to the insertions. They pulled at right angles to the jaw, and presumably exerted their maximum moment about the articulation, only when the mouth was wide open. In *Paleothyris* the jaw articulations are further forward. This and the relatively large eyes must have made the posterior adductors run more vertically, so as to exert their maximum moment when the mouth was closed, or nearly closed as it would be when holding small prey. Running at this angle they would leave no room for the otic notch. The eardrums of early reptiles are believed to have been immediately posterior to the skull, in a more ventral position than in amphibians such as *Palaeogyrinus*. Modern reptiles including lizards (Fig. 9-10*f*) have their eardrums bordered by the curved edges of their tall quadrate bones, but the quadrates of early reptiles were much smaller.

In *Paleothyris*, as in early amphibians, the posterior jaw adductors were covered laterally by a continuous sheet of dermal bone. In many lines of reptile evolution dermal bone has been lost, so that the adductors are no longer completely covered (Fig. 9-11). In some cases the sheet of dermal bone has been emarginated from the posterior or ventral edge or both. In other cases it has been perforated by openings (temporal fenestrae). Skulls with no temporal fenestrae are described as anapsid, even if the dermal bone is emarginated. If there is one fenestra on each side of the head and it is dorsal in position (with the squamosal and postorbital bones meeting ventral to it) the skull is described as parapsid. If there is one fenestra which is ventral (with the squamosal and postorbital meeting dorsal to it) the skull is synapsid. If both dorsal and ventral fenestrae are present, it is diapsid. Reptiles having each of these conditions will be described in Chapters 10 and 12. The conditions have in the past been used as a basis for classification of the reptiles, but it would be unwise to assume without other evidence that two reptiles having, for instance, parapsid skulls were closely related. The variety of skull conditions indicates so widespread a tendency for bone covering the posterior adductors to be lost that it seems quite likely that

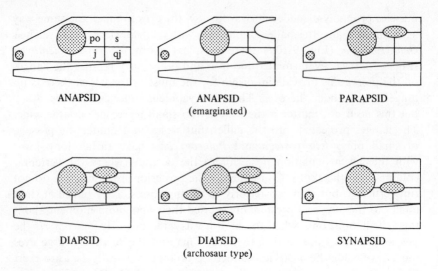

Fig. 9-11. Patterns of emargination and fenestration, in reptile skulls. j, jugal; po, postorbital; qj, quadratojugal; s, squamosal.

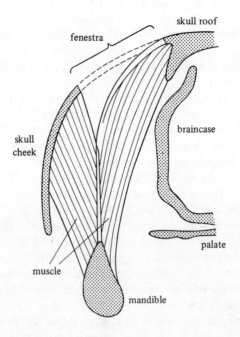

Fig. 9-12. Diagrammatic transverse section through the head of a tetrapod, showing the common pinnate arrangement of the posterior jaw adductor muscle. From T. H. Frazzetta (1968). *J. Morph.* **125**, 145–58.

each condition that has evolved from the anapsid one may have evolved more than once.

Closer study of the structure of the posterior jaw adductor may help to explain temporal fenestrae and emargination. In modern reptiles and presumably in extinct ones it is generally pinnate, with a central tendon inserting on the lower jaw (Fig. 9-12). In such cases there is no need for bone in the position of the fenestra shown in the figure. When the mouth closes the tendon and central parts of the adductor move dorsally, and skin covering a fenestra can bulge to allow this. A continuous sheet of bone could not bulge, and the bulging would have to occur from some more remote part of the head. This provides a possible explanation for the upper fenestra found in parapsid and diapsid skulls. The region labelled 'skull cheek' is the origin of the lateral part of the adductor, but it could still provide a firm origin if it were partially replaced by tough connective tissue, provided of course that its dorsal edge remained bone. This explains why the lower fenestra found in synapsid and diapsid skulls is possible, but does not explain why it evolved.

FURTHER READING

General

Bellairs, A. (1969). *The Life of Reptiles.* Weidenfeld & Nicolson, London.
Gans, C. (ed.) (1969–   ). *Biology of the Reptilia.* Academic Press, New York.
Grassé, P. P. (ed.) (1970). *Traité de Zoologie,* vol. 14, fasc. 2 & 3. *Reptiles.* Masson, Paris.
Romer, A. S. (1956). *Osteology of the Reptiles.* Chicago University Press, Chicago.
See also general works on herpetology in the list at the end of Chapter 8.

Eggs of reptiles and birds

Kutachai, H. & J. B. Steen (1971). Permeability of the shell and shell membranes of hens' eggs during development. *Respir. Physiol.* **11**, 265–78.
Needham, J. (1942). *Biochemistry and Morphogenesis.* Cambridge University Press, London.
Nelsen, O. E. (1953). *Comparative Embryology of the Vertebrates.* Blakiston, New York.
Romanoff, A. L. & A. J. Romanoff (1949). *The Avian Egg.* Wiley, New York.

Skin, water and salts

Bentley, P. J. & K. Schmidt-Nielsen (1966). Cutaneous water loss in reptiles. *Science,* **151**, 1547–9.
Bradshaw, S. D. & V. H. Shoemaker (1967). Aspects of water and electrolyte changes in a field population of *Amphibolurus* lizards. *Comp. Biochem. Physiol.* **20**, 855–65.
Cloudsley-Thompson, J. L. (1971). *The temperature and water relations of reptiles.* Merrow, Watford.

Dunson, W. A. (1969). Electrolyte excretion by the salt gland of the Galapagos marine iguana. *Am. J. Physiol.* **216**, 995–1002.
Krakauer, T., C. Gans & C. V. Paganelli (1968). Ecological correlation of water loss in burrowing reptiles. *Nature, Lond.* **218**, 659–60.
Maderson, P. F. A. (1965). Histological changes in the epidermis of snakes during the sloughing cycle. *J. Zool., Lond.* **146**, 98–113.
Schmidt-Nielsen, K. & R. Fänge (1958). Salt glands in marine reptiles. *Nature, Lond.* **182**, 783–5.

Temperature

Bradshaw, S. D. & A. R. Main (1968). Behavioural attitudes and regulation of temperature in *Amphibolurus* lizards. *J. Zool., Lond.* **154**, 193–221.
Moberley, W. R. (1968). The metabolic responses of the common iguana, *Iguana iguana* . . . (2 papers). *Comp. Biochem. Physiol.* **27**, 1–20.
Porter, W. P. & D. M. Gates (1969). Thermodynamic equilibria of animals with environment. *Ecol. Monogr.* **39**, 227–44.
Stebbins, R. C. & R. E. Barwick (1968). Radiotelemetric study of thermoregulation in a lace monitor. *Copeia* 1968, 541–7.
Whittow, G. C. (ed.) (1970). *Comparative Physiology of Thermoregulation*, vol. 1. Academic Press, New York.

Blood circulation

White, F. N. (1968). Functional anatomy of the heart of reptiles. *Am. Zoologist,* **8**, 211–19.

Primitive Reptiles

Carroll, R. L. (1969). Problems of the origin of reptiles. *Biol. Rev.* **44**, 393–432.
Frazzetta, T. H. (1968). Adaptive problems and possibilities in the temporal fenestration of tetrapod skulls. *J. Morph.* **125**, 145–58.

10

Various reptiles

The previous chapter was about reptiles in general and about their origin. This one is about interesting features of individual species or groups of reptiles. The synapsid reptiles from which the mammals evolved will be considered in Chapter 12.

TORTOISES AND TURTLES
Subclass Anapsida, order Chelonia

The tortoises and turtles are the only modern reptiles with anapsid skulls (Fig. 9-11; Many of them have skulls of the emarginated anapsid type). They are included with the primitive cotylosaurs in the subclass Anapsida, although they are so peculiar in structure that they might just as appropriately be given a subclass to themselves.

Their most peculiar feature is, of course, the shell. It consists of an inner layer of plates of bone and (usually) an outer one of plates of β-keratin. The sutures between the plates of bone generally do not coincide with the edges of the plates of keratin. Both the bone and the keratin are formed as part of the skin, the bone in the dermis and the keratin, like that of ordinary skin, in epidermis. The particularly thick keratin plates of the hawksbill turtle, *Eretmochelys*, are the tortoiseshell that used to be used for making small items such as combs and (with brass) for the marquetry on Boulle furniture.

The shell is open at its ends, with the head and forelimbs emerging in front and the hind limbs and short tail behind. Head, limbs and tail can be withdrawn into the shell by many of the chelonians, but not by all. This drives most of the air out of the lungs: the tortoise *Testudo* compresses its lungs to one-fifth of their initial volume when it retires into its shell. Box turtles such as *Terrapene* are particularly well protected when they withdraw, since the floor of the shell is hinged and can close the open ends. Marine turtles can withdraw neither head nor limbs.

Fig. 10-1 shows on its right-hand side how most of the trunk vertebrae and ribs are incorporated in the shells of typical chelonians. The dorsal processes (neural spines) of the vertebrae merge with median plates of dermal bone, and the ribs merge with more lateral plates. Vertebrae and ribs are of course cartilage bones, formed from sclerotome, but they grow so as to make contact with the dermis, where the plates of dermal bone develop as extensions of them. There are separate plates of dermal bone

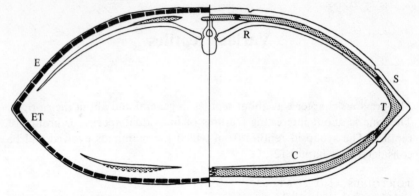

Fig. 10-1. Diagrammatic transverse sections of the shells of (left) *Dermochelys* and (right) a typical tortoise or turtle. C, clavicle; E, epidermis; ET (black), epithecal ossifications; S, keratin plates; T (stippled), thecal ossifications. From R. Zangerl (1969). In C. Gans (ed.), *Biology of the Reptilia*. Academic Press, London & New York.

around the edges of the shell, and in its floor. Three plates at the anterior end of the floor seem to be dermal bones of the pectoral girdle, the clavicles and interclavicle, that have become incorporated in the shell.

The leathery turtle *Dermochelys* (which reaches weights of well over 500 kg) has no keratin plates on its shell, but a covering of leathery skin with a mosaic of small plates of bone in the dermis. Deeper in the dermis are rudiments of the bony parts of the more normal shell that was presumably possessed by its ancestors; the arrangement is shown on the left-hand side of Fig. 10-1. The shells of the soft-shell turtles (*Trionyx*, etc.) are also covered with leathery skin instead of keratin plates, but are otherwise less peculiar.

The limb girdles of chelonians are enclosed within the shell (Fig. 10-2). The sacral vertebrae are almost immediately posterior to the vertebrae

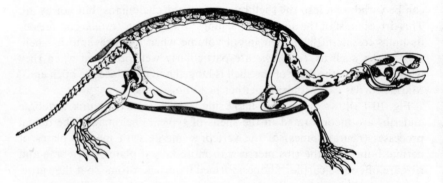

Fig. 10-2. Skeleton of the turtle *Emys*, with the nearer half of the shell cut away. From A. S. Romer (1956). *Osteology of the Reptiles*. University of Chicago Press, Chicago. Copyright © 1956.

which have their neural spines incorporated in the shell. The ilia are attached to the sacral vertebrae by sacral ribs, in the usual way (see Fig. 8-15*e*). In normal tetrapods the scapulae lie lateral to the anterior ribs (see, for instance, Fig. 8-1) but in chelonians they are inside the shell, with their dorsal ends attached to it close to the most anterior pair of ribs. It seems that as the chelonians evolved and the ribs moved to the skin, the scapulae were displaced to this position. The ventral ends of the scapulae are attached to the part of the floor of the shell, which is believed to derive from the pectoral girdle.

Most reptiles breathe by moving their ribs, but this is not feasible when the ribs are part of a rigid shell. The mechanism of breathing has been investigated in experiments with the snapping turtle, *Chelydra*. The animal was anaesthetized, a cannula was inserted into a lung or the body cavity and electrodes for electromyography were inserted in selected muscles. The cannulae and some of the electrodes were passed through holes in the shell and fixed by cement: turtles are particularly convenient for experiments like this. The animal was allowed to recover from the anaesthetic and walked or swam in a small enclosure, trailing the cannula and the electrode leads. The cannula was connected to a pressure transducer, and simultaneous records of pressure changes and muscle action potentials were obtained. Records for two experiments are shown in Fig. 10-3. The pressure records (labelled 1) show that in each breath pressure in the lungs first increased as air was driven out and then decreased as new air was drawn in. The electromyograms show that the muscles labelled 2 and 4 were active as the turtle breathed out, and 3 and 5 as it breathed in. These, with the muscles that open and close the glottis, seem to be the main muscles used in breathing. Fig. 10-4 shows how they work. Muscles 2 and 4 run across the body, under the viscera. When they contract they compress the lungs. Muscles 3 and 5 are just under the flexible skin which crosses the openings at the ends of the shell. When 2 and 4 contract, reducing the volume of the shell contents, 3 and 5 cave in. When 3 and 5 contract they flatten again, stretching 2 and 4 and expanding the lungs.

Similar experiments with the tortoise *Testudo* showed that it uses muscles 4 and 5 in the same way as the snapping turtle. However, the main pumping mechanism at the anterior end is movement in and out of the two halves of the pectoral girdle, which pivot about the dorsal and ventral articulations of the scapula with the shell. This makes the forelimbs move in and out as the tortoise breathes. The snapping turtle can make similar movements but makes little use of them in breathing.

The Nile turtle, *Trionyx*, often lies submerged, without visiting the surface to breathe, for very long periods. It has been known to remain submerged for as much as six hours. In experiments in which *Trionyx* were

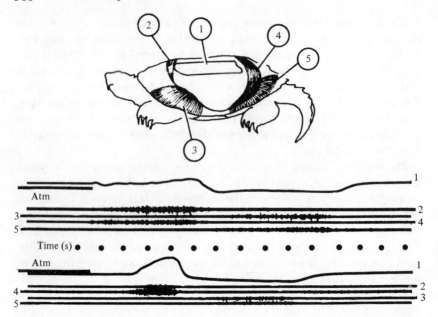

Fig. 10-3. Simultaneous records of pressure in the lungs of *Chelydra* (1) and of electrical activity in the muscles numbered (2) to (5) in the diagram. The set of records above the time scale is from a partially submerged turtle while the lower set is from a turtle out of water. From A. S. Gaunt & C. Gans (1969). *J. Morph.* **128**, 195–228.

prevented from breathing air, they took up oxygen from the water at a rate of only 0.12 cm^3 (kg body weight)$^{-1}$ min^{-1}, but this was apparently sufficient to keep them alive. The resting metabolic rate of iguanas of similar weight (about 1 kg) at the same temperature (24 °C) is about 0.8 cm^3 kg^{-1} min^{-1} (0.05 cm^3 g^{-1} h^{-1}, Fig. 9-5*a*). In further experiments, the *Trionyx* were submerged in a divided chamber so that oxygen uptake by the head and the rest of the body could be measured separately. It was found that only a small fraction of the total was taken up by the head. Blocking the cloaca made no difference to the rate of uptake. The oxygen is apparently taken up through the skin rather than by any special respiratory organ in mouth or rectum. *Trionyx* is one of the soft-shell turtles and

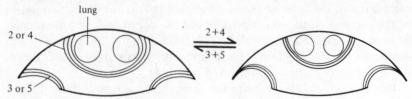

Fig. 10-4. Diagrammatic transverse sections through the shell of *Chelydra* showing how muscles (2) and (4) (see Fig. 10-3) compress the lungs, and how muscles (3) and (5) expand them.

the leathery skin which covers the shell probably plays its part in aquatic respiration.

The marine turtles such as *Eretmochelys* and *Dermochelys* have their limbs developed as flippers, which are used for swimming. They are not moved forward and back like oars, but up and down like birds' wings. They work as hydrofoils: their up-and-down movements produce a net forward force in essentially the same way as do the side-to-side movements of fishes' tails.

PRIMITIVE DIAPSIDS

Subclass Lepidosauria, orders Eosuchia and Rhynchocephalia

The eosuchians and rhynchocephalians are reptiles with skulls of the diapsid type shown in Fig. 9-11. The eosuchians are the more primitive group and are now extinct. They lived in the late Permian and early Trias. The rhynchocephalians are not very clearly distinct from them, but generally have beak-shaped premaxillae. They appeared in the Trias and survive as a single species, the tuatara (*Sphenodon*).

This is a lizard-like reptile weighing up to a kilogram which lives only on a few islands off New Zealand. These islands have large populations of sea birds, and are riddled with burrows dug by petrels. The tuatara spends much of the day in these burrows, or sometimes in burrows it has dug for itself. It is active mainly at night, and its low body temperature has already been remarked on (p. 287). It feeds largely on ground-living insects (crickets and beetles) and on snails. It grows very slowly and it is suspected that the largest wild specimens may be as much as a century old.

LIZARDS

Subclass Lepidosauria, order Squamata, suborder Lacertilia

The lizards, amphisbaenians and snakes are all included in one order, the Squamata, but they are given separate sections in this chapter. They are believed to have evolved from eosuchian ancestors with diapsid skulls but they no longer show the typical diapsid condition.

Lizards have lost the dermal bone ventral and posterior to the lower temporal fenestra (Fig. 10-5). This has made possible the evolution, in most lizards, of movable joints in the skull. The quadrate is no longer firmly attached to other bones all along its length, but has movable joints at its ends. There is an upper joint with the squamosal and supratemporal bones and a lower one with the pterygoid bone at the posterior end of the palate (in addition, of course, to the joint with the lower jaw). There is a flexible region which acts in effect as another joint further forward in the palate, and there is a hinge joint across the skull roof between the frontal and

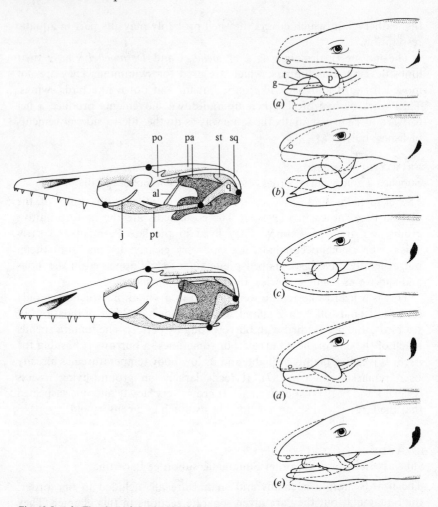

Fig. 10-5. (left) The skull of the lizard *Varanus* in two positions. The positions of the four movable joints described in the text are indicated. Based on an illustration in T. H. Frazetta (1962). *J. Morph.* **111**, 287–320.

(right) Tracings of consecutive frames of a cine film of the lizard *Gerrhonotus* feeding. Dotted outlines show the immediately following position. From the same paper by T. H. Frazzetta.

al, alisphenoid; g, gums; j, jugal; p, prey; pa, parietal; po, postorbital; pt, pterygoid; q, quadrate; sq, squamosal; st, supratemporal; t, tongue.

parietal bones. The positions of these joints are shown in Fig. 10-5, which also shows the movement they allow. The palate can be shifted anteriorly. pulling the ventral end of the quadrate forward and tilting up the snout. This movement is probably produced by a muscle which runs from the pterygoid bone to the base of the cranium. It is normally performed as the mouth is opened, though a different muscle, posterior to the quadrate, serves to open the mouth. The reverse action of lowering the snout seems to be performed automatically as the mouth is closed; the posterior jaw adductor runs at such an angle as to pull the jaw, and swing the quadrate. posteriorly (Fig. 9-10*f*). Skulls which can make movements like this are described as kinetic.

Fig. 10-5 also includes tracings from a film sequence showing a lizard feeding. The prey has been seized and is being squeezed repeatedly between the jaws. This kills it, and the holes made by the teeth may allow digestive enzymes to penetrate more quickly. Notice that the snout is raised when the mouth is open (*b*) and lowered when it is closed (*d*). Lowering of the snout probably makes it more difficult for squirming prey to escape from the front of the mouth. The upper and lower jaws converge in front of the prey as well as behind it, and indeed in (*d*) they seem to be meeting in front of the prey. The tongue is quite large and is used for manipulating food in the mouth. Notice in (*b*) that it is shifting the prey further back.

Most lizards feed on insects and other invertebrates, which they catch between their jaws and manipulate with their tongues. Chamaeleons have extraordinarily long tongues, which they use much as frogs do (p. 255) to catch small insects. They do not depend on their jaws for grasping prey and there would seem to be no advantage in having a kinetic skull: their skulls are not kinetic. Monitor lizards such as *Varanus* on the other hand eat rather large prey such as small mammals. Their skulls are kinetic with long jaws and large teeth. Their prey is too large to be manipulated by the tongue and indeed a large tongue might be in the way in swallowing. They have slender forked tongues, like snakes. They swallow their prey, after repeated squeezing between the jaws, by the method known as inertial feeding. They open the mouth momentarily, jerking the head forward as they do so. This moves the prey a little deeper into the mouth before the jaws close on it again. Yet other lizards, such as *Sauromalus*, are herbivorous.

Many lizards tongues, whether they are stout or slender, have forked tips. The tips are believed to convey material to a pair of accessory olfactory organs, known as Jacobson's organs. These develop as pockets of the nasal cavity but lose their connection with it and come to open directly into the mouth cavity, anterior to the internal nostrils. Though homologous organs are found in other reptiles they are most highly developed in Squamata, and it is only in them that they open into the mouth separately from the nostrils. Snakes and lizards with slender tongues are believed to pass particles to be sensed to Jacobson's organs, by inserting the tips of the tongue into the openings. Other lizards have ciliated grooves in the roof of the mouth which may carry forward to

Jacobson's organs particles placed in them by the tongue. Little is known about the physiology of Jacobson's organs, but when snakes move their tongues repeatedly in and out of their mouths they are probably collecting particles for investigation by the organs.

A few lizards can run on their hind legs, and run faster on their hind legs than on all fours. *Crotaphytus collaris* has been timed at the remarkable speed of 7 m s^{-1} (16 m.p.h.) on its hind legs, but at up to only 5 m s^{-1} on all fours. *Amphibolurus cristatus* achieved 5 m s^{-1} on its hind legs and 2 m s^{-1} on all fours. These, and the other species which run on their hind legs, have relatively long tails on which they depend for balance. If a third or more of the length of the tail is removed, they seem unable to make more than a very few steps on their hind legs. These species also have very much longer hind legs than forelegs. When they run on their hind legs, they take strides far longer than the short forelegs could make.

When a man walks, each foot is on the ground for 50% of the time. but when he runs each may be on the ground for less than 30% of the time and there are intervals when neither foot is on the ground. His run involves a succession of leaps. Similarly, an *Amphibolurus* running fast on its hind legs may have each hind foot on the ground for as little as 35% of the time. The advantages of incorporating leaps in a run are considered on p. 382. When a man or other mammal runs each femur swings forward and back in a vertical plane, but lizards running on their hind legs move these legs in the same sort of way as when they run on all fours, with the femurs pointing laterally. The points where the feet push against the ground are well to either side of the centre of gravity and the lizard runs with a rolling gait.

There are other lizards which have reduced limbs, or no limbs at all. Many of them live in deserts and 'swim' through loose sand much as eels swim through water, by passing waves of bending backwards along the body. A similar action serves limbless species for crawling over the surface of ground, and is referred to as serpentine movement because it is the most usual manner of crawling by snakes (Fig. 10-10*a*). The body forms bends which are generally rather irregular, positioned so as to take advantage of stones, tussocks of grass or mere irregularities of the surface of the ground. The bends are made to travel posteriorly along the length of the body but are kept stationary relative to the ground by the irregularities they rest against. The whole body moves forward, sliding past the irregularities without (if they are firmly fixed) displacing them. The process looks most mysterious, with every cross-section of the body following the same sinuous path.

When newts and lizards with normal legs run, they bend their bodies from side to side, forming stationary waves (i.e. the waves are stationary relative to the body. See p. 249.) Serpentine locomotion involves travelling waves (i.e. the waves move relative to the body, though not relative to the ground). The waves of bending involved in crawling by the lizard *Chalcides ocellatus*, which has very short legs, have been shown to be intermediate between the two types.

Not all limbless lizards live in deserts, as readers familiar with *Anguis*, the slow-worm, will realize. It is common in Britain, for instance amongst leaf litter.

AMPHISBAENIANS
Order Squamata, suborder Amphisbaenia

Amphisbaenians live mainly in Africa and South America. They have no limbs, except for one genus which has forelimbs. They do not swim through loose sand like so many limbless lizards but live in damper, compact soil in which they make systems of tunnels. They are mostly between 10 cm and 1 m long, and remarkably like earthworms in appearance. Their skin is pleated in rings which suggest earthworm segments. Many lack skin pigment and are consequently pink. Their eyes are small and inconspicuous, and the tail is rounded like the head, so that it is not immediately obvious which end is which. They can crawl backwards – and have to do so in their burrows since there is not room to turn. The genus *Amphisbaena* from which the group as a whole takes its name was named after the mythical amphisbaena, a serpent with a head at each end described by ancient and mediaeval writers.

Most of the peculiarities of amphisbaenians are obviously related to their life underground. If they had limbs, or if they were not so slender, they would have to make wider burrows. Eyes and skin pigment have no obvious function underground, but species which visit the surface regularly retain skin pigment. Their ears are modified in much the same way as in snakes (described in the next section of this chapter). The modifications probably increase their sensitivity to ground vibrations, at the expense of sensitivity to airborne sound.

Amphisbaenians make branching systems of tunnels and apparently patrol them, searching for earthworms, termites and other prey which may enter them. Really dry soils are generally either rock hard or so loose that tunnels would cave in, and amphisbaenians seem to live only in moist soils. Their skin is highly permeable to water (in dry air some lose water about as fast as amphibians of similar weight) but this is probably no disadvantage in moist soil.

Fig. 10-6a shows how a tunnel is extended. The snout is driven into the soil. The head is then raised or (in other groups of amphisbaenians) moved in other ways, to widen the hole. Since the head is used as a ram the skull

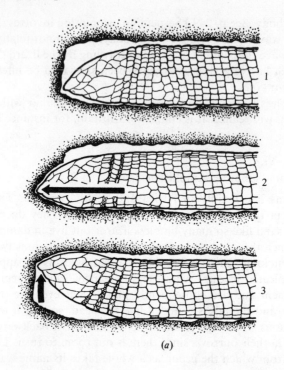

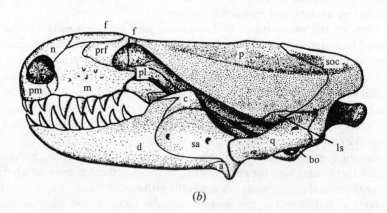

Fig. 10-6. (*a*) The burrowing technique used by a common group of amphisbaenians. The head is driven forward into the end wall of the tunnel, and then raised. Notice the separation of the pleats of the skin posterior to the head in 2 and 3. From C. Gans (1968). *Am. Nat.* **102**, 345–62. Copyright © University of Chicago Press, 1968.

(*b*) The skull of *Amphisbaena*. Abbreviations include: f, frontal; p, parietal; q, quadrate. From A. S. Romer (1956). *Osteology of the Reptiles*. University of Chicago Press, Chicago. Copyright © 1956.

must be strong with no weak bones in exposed positions. Amphisbaenians have presumably evolved from lizards but their skulls are much more compact, lacking the squamosal and usually the postorbital and jugal bones (Fig. 10-6b). The braincase is reinforced by downgrowths of the frontal and parietal bones, so that the brain is completely enclosed in bone: in typical lizards the side walls of the braincase consist partly of soft connective tissue. The skull is not kinetic.

Amphisbaenians crawl by a technique that involves forward and backward sliding of the skin over the underlying tissues, and looks rather like the crawling of earthworms.

SNAKES
Order Squamata, suborder Ophidia

Snakes are reptiles which are specialized for eating large prey, and have a number of features in common with amphisbaenians and burrowing lizards. It has been suggested that they may have evolved from burrowing ancestors, and indeed the modern snakes which seem to be most primitive are burrowers.

Snakes have no limbs or limb girdles, apart from rudiments of the pelvic girdle and hind limb in the more primitive groups. They may have evolved this condition as an adaptation to burrowing, but the size of prey they could swallow would be limited if it had to pass through a pectoral girdle. The braincase is extended forwards by downgrowths of the frontal and parietal bones (Fig. 10-7), which may have evolved to protect the brain in burrowing but also serve to protect it when very large prey is being forced down the throat. There are no jugal or squamosal bones, and though the postorbital (which is usually missing in amphisbaenians) remains it no longer forms the lower border of an upper temporal fenestra. (The bone labelled supratemporal in Fig. 10-7 has sometimes been identified as a squamosal, but it is attached to the parietal in exactly the same way as the supratemporals of lizards and almost certainly is in fact the supratemporal.)

Though the skull is strong, it can make an extraordinary range of movements, as Fig. 10-7 shows. The skull illustrated is that of a python, a typical non-poisonous snake. The snout can be tilted up as in lizards, by pulling the pterygoid bones forward. However, the joint in the skull roof which allows this is not between the frontals and parietal (which are incorporated in the rigid braincase) but more anteriorly, between the frontals on the one hand and the prefrontals and nasals on the other. There are also movable joints between prefrontal and maxilla, between maxilla and ectopterygoid, between pterygoid and quadrate, between quadrate and supratemporal, and elsewhere. These are not simply hinges, but allow rotation about more than one axis and the ligaments are in many places loose enough to allow some sliding of one bone over another. The left and right maxillae can be moved independently, as can the two halves of the lower jaw.

Pythons feed mainly on mammals and birds, including amazingly large ones. A leopard measuring 1.25 m from snout to rump has been found in a

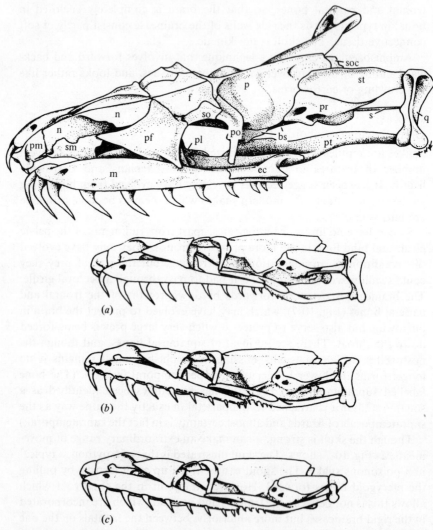

Fig. 10-7. The skull of the African rock python, *Python sebae*. Ventral [*opposite*] and lateral views in three positions, (*a*), (*b*) and (*c*), are included to illustrate the range of movement of the skull. bs, basisphenoid; ec, ectopterygoid; f, frontal; m, maxilla; n, nasal; p, parietal; pf, prefrontal; pl, palatine; pm, premaxilla; po, postorbital; pr, pro-otic; pt, pterygoid; q, quadrate; s, stapes; sm, septomaxilla; so, supraorbital; soc, supraoccipital; st, supratemporal. From T. H. Frazetta (1966). *J. Morph.* **118**, 217–96.

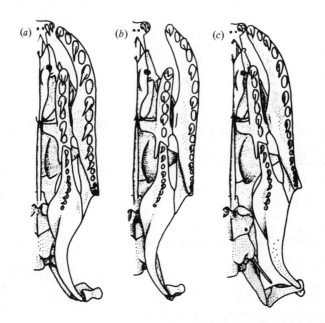

very large python, 5.5 m long. Prey is seized, once the python is near enough, by a sudden strike. The snake forms the anterior part of the body into an S-shape and then extends it very rapidly with the mouth open (Fig. 10-8, left). The snout is tilted up in the strike in position (*b*) of Fig. 10-7. The long recurved teeth penetrate the prey, and when the snake has got a firm grip in this way it coils its body round the prey and squeezes it. Breathing is prevented, and the prey eventually dies.

The extraordinary mobility of the head makes the swallowing of large prey possible. The two sides of the mouth are used rather as the left and right hands are used to pull in a rope hand over hand. The prey is held firmly in the left side of the mouth while the right teeth are released and moved forward to a new position, and vice versa. The right maxillary teeth can be released as the right half of the lower jaw is lowered, by tilting up the right side only of the snout. Fig. 10-8 (right) shows the two halves of the lower jaw moving independently. They can be separated widely, if necessary, when large prey is being swallowed. The glottis can be pushed forward out of the corner of the mouth so that it is not blocked, and breathing can continue during swallowing of prey.

Evolution of venom was a further adaptation for dealing with large and possibly formidable prey. Venom is injected in a quick strike and the

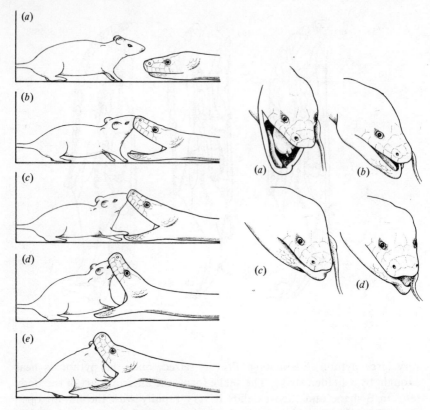

Fig. 10-8. (left) Tracings from a high-speed cine film of a *Python sebae* seizing a mouse. Note how the snout is tilted up.

(right). Tracings from a cine film of an Indian python, *Python molurus*, swallowing a mouse. Note the independent movement of the two halves of the lower jaw. Both after T. H. Frazzetta (1966). *J. Morph.* **118**, 217–96.

snake may then withdraw, avoiding the danger of being wounded by the dying prey. The snake returns later, using its tongue and Jacobson's organ if necessary to follow the trail of the victim to the place where it died.

Some venomous snakes have fangs with grooves down which the venom flows. In more advanced ones the edges of the grooves have met so that the fangs are tubular, like hypodermic needles. The duct of the venom gland enters the base of the tooth, and the venom emerges from its tip.

Some of the most dangerous snakes, including cobras, have quite short fangs and can close their mouths with the fangs erect. Vipers and rattlesnakes have longer fangs which have to be folded back when the mouth closes. This is possible because the maxillae (which bear the fangs) have become extremely short. Pythons move their palates forward as they strike, and so tilt their relatively long maxillae through a small angle (Fig. 10-7*b*). Vipers move their palates in the same way but the maxillae rotate through a much larger angle, because they are so short (Fig. 10-9).

Vipers keep their fangs folded down while swallowing prey, and use the teeth of the palate and lower jaw. The palatal teeth can be lowered for this purpose, below the level of the fangs (Fig. 10-9*c*). This is possible because the joint between the pterygoid and ectopterygoid, which is fixed in pythons, is movable in vipers.

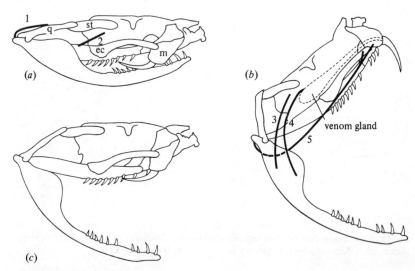

Fig. 10-9. The skull of a viperid snake, (*a*) with the mouth closed and fangs retracted, (*b*) with the mouth open and fangs erected, and (*c*) with the pterygoids lowered. Muscle (1) opens the mouth, (2) erects the fangs, (3) (the posterior jaw adductor) closes the mouth, (4) squeezes venom out of the gland and (5) retracts the fangs. Abbreviations as in Fig. 10-7. This diagram is based on sketches of *Agkistrodon* supplied by Dr K. V. Kardong.

Lizards and snakes have strips of glandular tissue under the skin of the lips. They discharge into the mouth, through numerous small openings, a mucus secretion which probably helps as a lubricant in swallowing. The venom glands of some snakes seem clearly to be modified parts of the gland in the upper lip, but in other cases their homology is less obvious.

The venoms of snakes are varied and complicated in composition. They include various polypeptides with toxic effects, including in at least many cases one with a curare-like action, which causes paralysis by putting neuromuscular junctions out of action. Paralysis of the respiratory muscles results and quickly causes death. Venoms also include a variety of enzymes, including proteases which aid digestion of the prey.

Some snakes such as *Boa* can crawl with their bodies straight, by a method similar to that of amphisbaenians (p. 307). It differs in that only the ventral skin is moved forward and back. The commonest method of crawling by snakes is serpentine locomotion, which has already been described (p. 304 and Fig. 10-10a). Sidewinding is a remarkable variant of serpentine locomotion practised by rattlesnakes (*Crotalus*) and some other snakes. They use it in crawling over sand and smooth surfaces lacking the projections needed to give a purchase for serpentine locomotion. The process is illustrated in Fig. 10-10b. Only parts of the body which are

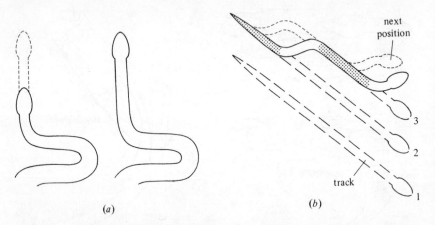

(a)   (b)

Fig. 10-10. (a) A diagram illustrating serpentine locomotion. Two successive positions of the body are shown. (b) A diagram, which is explained in the text, illustrating sidewinding.

stippled rest on the ground. They form on sand a series of parallel tracks (1, 2, etc.) which show when they have rested. The head is being lifted across from track 3 to start a new track 4 and the unstippled region in the middle of the body is being lifted from track 2 to track 3.

Snakes have no eardrums or air-filled middle ears. The distal end of the stapes articulates with the quadrate and is held in contact with it by a ligament. The quadrate, of course, articulates with the lower jaw, and when the head rests on the ground, vibrations of the ground should be transmitted to the ear. The arrangement might be supposed to make snakes particularly sensitive to ground vibrations, but rather insensitive to airborne sound.

These suppositions have been tested experimentally. Snakes were placed on a platform which could be vibrated by a piezoelectric transducer, and were also exposed to sound from loudspeakers. Sensitivity to the stimuli was investigated by electrical recording from the brain. It was found that they were being detected both by the ear and by receptors in the skin. The sensitivities of the two systems could be investigated separately by cutting the spinal cord (which isolated the skin receptors from the brain) or by destroying the inner ears.

The results of the experiments are summarized in Fig. 10-11. Thresholds for detection of stimuli by the ear are labelled 'auditory' and for the skin receptors 'somatic'. (The curve labelled 'snake skin' is for entirely separate experiments, in which action potentials were recorded from a vibration receptor in an isolated piece of snake skin.) The figure shows that for frequencies between about 150 and 500 Hz, the snake auditory system is more sensitive than the somatic system. Outside these limits, the somatic system is the more sensitive.

Fig. 10-11a shows that the snake auditory system is as sensitive to airborne sound as the frog ear, though over a much narrower range of frequencies. Both are considerably less sensitive than mammal ears. Fig. 10-11b shows that snakes are very much more sensitive than man to vibration. At the best frequency of 300 Hz, the snake auditory system can detect vibrations of only 1 Å amplitude.

Their sensitivity to vibrations may give snakes valuable information about approaching prey or predators. Other peculiar sense organs which apparently help in finding prey are the radiant heat receptors of the pit

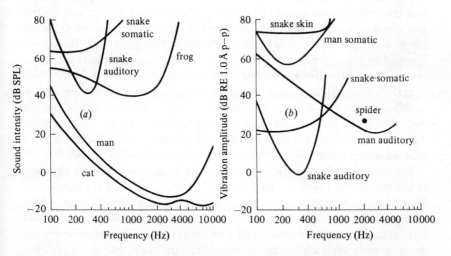

Fig. 10-11. (a) thresholds for detection of airborne sound of various frequencies (Hz) by the ear ('auditory') and skin receptors ('somatic') of snakes, and by the ears of a frog, the domestic cat and man. The thresholds given for snakes are typical values based on the results of experiments with several species.

(b) Thresholds for detection of vibration by the ear ('auditory') and skin receptors ('somatic') of snakes: by a single (apparently much less sensitive) receptor in an isolated piece of snake skin ('skin'), by the ear of man to skull vibration ('auditory'), and by the fingers to vibration of a rod ('somatic') and by the leg of a spider. From P. H. Hartline (1971). *J. exp. Biol.* **54**, 349–72.

vipers (which include the rattlesnakes) and of boas. These snakes feed largely on birds and mammals, which are generally warmer than their surroundings.

The most sensitive radiant heat receptors are the 'pits' of pit vipers. These are a pair of deep cavities in the snout, with forward-facing openings. Stretched across each cavity, under the opening, is a thin membrane served by a substantial sensory nerve. The branched endings of the axons are so tightly packed in the membrane as to comprise about half its thickness. Radiation entering the opening must tend to warm the membrane, and it has been demonstrated by electrical recording that the nerve endings in the membrane are very sensitive to changes of temperature. They are insensitive to light, but sensitive to infra-red radiation of the range of wavelengths emitted by bodies at temperatures around 300 K. The openings of the cavities are narrow, so radiation from different directions falls on different parts of the membranes. Blindfolded rattlesnakes can locate a rat at a distance of 50 cm accurately enough to strike at it.

DINOSAURS AND THEIR ANCESTORS
Subclass Archosauria, orders Thecodontia, Saurischia and Ornithischia

Archosaurs, like lepidosaurs, have diapsid skulls, but in addition to the two temporal fenestrae most have a fenestra between the eye and nostril, and another in the lower jaw (Fig. 9-11). The teeth have roots which are housed in sockets in the jaws, whereas lepidosaurs have their teeth fastened to the edges of the jaws, or in grooves. The archosaurs may have evolved from primitive lepidosaurs, but it seems as likely that they evolved directly from cotylosaurs and reached the diapsid condition independently. Another suggestion which has been made (and supported by evidence) is that they evolved from an ancestor among the pelycosaurs which are described in chapter 12. Pelycosaurs have synapsid skulls but could have evolved upper temporal fenestrae in addition to their lower ones.

The earliest archosaurs belong to the order Thecodontia, which appeared at the end of the Permian and flourished in the Trias. The reptiles known as dinosaurs belong to the orders Saurischia and Ornithischia, which evolved from the thecodonts in the Trias. They were the dominant terrestrial animals in the Jurassic and Cretaceous. The differences which distinguish early saurischians from thecodonts are slight, but there are clear differences between the saurischians and the ornithischians. The latter have peculiar pelvic girdles: the pubis is V-shaped, with one arm running parallel to the ischium (Fig. 10-12b). They also have an additional bone at the front of the lower jaw. Both differences can be seen by comparing the saurischian shown in Fig. 10-13 with the ornithischian shown in Fig. 10-14a.

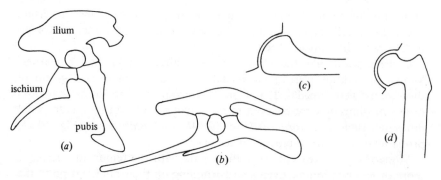

Fig. 10-12. (*a*), (*b*), lateral views of the right side of the pelvic girdle of (*a*) a saurischian and (*b*) an ornithischian dinosaur. (*c*) (*d*), diagrammatic posterior views of the proximal ends of the right femurs of (*c*) a typical amphibian or reptile and (*d*) a dinosaur, bird or mammal.

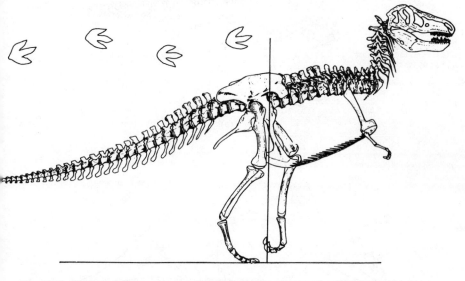

Fig. 10-13. Skeleton of *Tyrannosaurus*, a bipedal saurischian dinosaur, in what is believed to have been the normal walking position. Length about 8.5 m. From B. H. Newman (1970). *Biol. J. Linn. Soc.* 2, 119–23. (Inset) Footprints of a similar dinosaur. Redrawn from R. T. Bird (1944). *Natural History, N.Y.* **53**, 60–7.

Footprints have been found preserved in rock which show that some dinosaurs were bipedal, walking on their hind legs alone (Fig. 10-13). That many thecodonts and dinosaurs were bipedal is also indicated by their proportions. The hind limbs were long, in comparison both to the fore-limbs and to the distance from hip to shoulder. They were capable of so much longer a stride than the forelimbs, that running on all fours seems unlikely. Their tails seem long and heavy enough to balance the trunk about the hip joint, though the length of the tail is sometimes uncertain because of fossils being incomplete. The importance of tail length for bipedal lizards has already been noted (p. 304).

Considerations like these make it seem fairly certain that the earliest known thecodonts were quadrupedal, but that an important group of thecodonts were bipedal. It is from this group of thecodonts that both orders of dinosaurs are believed to have evolved. Obviously quadrupedal dinosaurs such as *Torosaurus* (Fig. 10-14b) and *Barosaurus* (Fig. 10-15) probably evolved from bipedal ancestors.

Modern reptiles (except crocodiles) walk with the femur and humerus more or less horizontal. Even lizards running on their hind legs point the

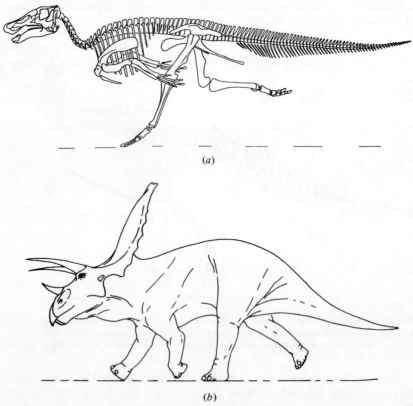

(a)

(b)

Fig. 10-14. (a) Skeleton of *Anatosaurus*, a bipedal ornithischian dinosaur, in what is believed to have been the normal running position. From P. M. Galton (1970). *J. Palaeontol.* **44**, 464–73.

(b) Reconstruction of *Torosaurus*, a quadrupedal ornithischian dinosaur. From R. T. Bakker (1968). *Discovery, Peabody Museum* 3, 11–22. Courtesy of the Peabody Museum of Natural History, Yale University.

Both animals were about 9 m long.

femur laterally. Birds and mammals walk with the femur and (in mammals) the humerus moving more or less in a vertical plane, and crocodiles sometimes do the same. The difference in posture is reflected by the shapes of the articulating surfaces of the bones, particularly at the hip joint. The socket in the pelvic girdle, where the femur articulates, points laterally. To fit into this socket, the femur needs an articulating surface on its end (if it is held horizontal) or on its inner side (if it moves in a vertical plane). The difference is shown in Fig. 10-12*c*, *d*. The structure of the joints in bipedal thecodonts and in dinosaurs indicates that humerus and femur were moved in vertical planes, as in mammals rather than modern reptiles. This is confirmed by footprints, both of bipedal dinosaurs and of quadrupedal ones, which show left and right footprints not far to either side of the mid-line (Figs. 10-13, 10-15). Moving the legs in this way makes a longer stride possible, and reduces the tendency to a rolling gait like that of bipedal lizards. Few fossil footprints are accompanied by signs of a trailing tail, so most dinosaurs (bipedal and quadrupedal) probably carried their tails clear of the ground.

When a dinosaur ran on its hind legs, the part of the vertebral column over the hips must have been held stiff to support the weight of trunk and tail. Some bipedal dinosaurs such as *Anatosaurus* (Fig. 10-14*a*) had this stiffened by ossified ligaments (which are preserved in fossils) joining the neural spines. Complete skeletons of these dinosaurs always have this part of the vertebral column straight.

Many dinosaurs were big, and the biggest were the quadrupedal saurischians known as brontosaurs. *Barosaurus* (Fig. 10-15) was one of them. With their long necks and tails, some were 30 m long. They could have raised their heads 12 m from the ground, a good deal higher than most houses and more than twice as high as a tall giraffe. It has been estimated that the largest, *Brachiosaurus*, weighed over 80 tons, or as much as a dozen large elephants. This estimate was obtained by making a scale model based on careful measurements of fossils, but showing how the animal probably looked when alive. The volume of the model was measured and that of the actual animal calculated from it. The weight was estimated by making the reasonable assumption that the density of the animal was about the same as that of water.

The question has often been asked whether such enormous animals could support their weight on land. If they waded in fairly deep water much of their weight would be supported by buoyancy. The bones in their legs were thick, but disproportionately thick legs are needed by very large terrestrial animals. This can be appreciated by considering stresses in animals of similar proportions but different size. If the larger of two such animals is twice as long as the smaller it will be eight times as heavy, but

Fig. 10-15. Reconstruction of *Barosaurus*, a quadrupedal saurischian dinosaur (brontosaur). Length (with neck horizontal) about 25 m. From R. T. Bakker (1968). *Discovery, Peabody Museum* 3, 11–22.

(Inset) Footprints of a similar dinosaur. Redrawn from R. T. Bird (1944). *Natural History, N.Y.* **53**, 60–7.

the cross-sectional areas of the bones and muscles which support it will be only four times as great as in the smaller animal. Stresses (force per unit area) in corresponding parts of these bones and muscles will be twice as high as in the smaller animal. More generally, cross-sectional areas are proportional to (body weight)$^{0.67}$ and stresses due to body weight are proportional to (body weight)$^{0.33}$, in animals with the same proportions. This is why elephants cannot have the same proportions as gazelles, but need relatively thicker legs. Were the legs of brontosaurs thick enough to support them on land?

Consider *Apatosaurus*, which is estimated to have weighed 30 tons. Most of its weight must have been carried by the hind feet, and when one of these was lifted to take a step on land the other would have to support about 20 tons. It would not be enough for it to support this weight when upright like a pillar: it would also have to support it when it was sloping at the beginning or end of a step, or with the knee bent. In these circumstances bending moments would act which would set up much larger stresses in the bones than when the leg was used as a pillar. Calculations using the dimensions of the bones, the tensile strength of bone (as measured on fresh bones of modern animals) and elementary engineering theory, lead to the conclusion that the leg bones of *Apatosaurus* were indeed strong enough for walking on land.

This, of course, does not prove that brontosaurs did walk on land, but there are other reasons for this belief. Their fossils are found in rocks believed on geological grounds to have been deposited on flood plains rather than lake bottoms, and they are found among fossils of other dinosaurs which are not suspected of being aquatic.

The long necks of brontosaurs provide another reason for suspecting that they lived on dry land. What was their advantage? Some plesiosaurs (which will be described in p. 326 and were undoubtedly aquatic) may have used their long necks to catch fish by darting at them, but brontosaurs have teeth which seem better suited to plucking vegetation than to catching fish. Their necks seem unnecessarily long for reaching plants, unless they fed like giraffes on the leaves of trees. It has been suggested recently that they did.

The evolution of large dinosaurs could hardly have occurred, if the typical reptile posture with humerus and femur horizontal had been retained. Large dinosaurs probably stood like horses and elephants, with their legs straight and vertical. Standing in the typical reptile position, or with legs bent, requires more energy, because of the tension that has to be maintained in muscles. The power required in a given position seems likely to be proportional to the weight supported multiplied by the length of muscle involved. Since muscle length (in animals of similar proportions) would be proportional to (body weight)$^{0.33}$, the power would be proportional to (body weight)$^{1.33}$. Metabolic rates of similar animals tend to be proportional to (body weight)$^{0.75}$ (Fig. 11-4). Standing with legs flexed or with humerus and femur horizontal would be very tiring for a brontosaur, even if its bones and muscles were strong enough.

It is often said that dinosaurs had very small brains. The brain itself is not preserved in fossils but the volume of the cavity for it in the skull can be measured, and comparison with modern reptiles suggests that the brain would have occupied about half this volume. Fig. 10-16 includes estimates made on this basis. It shows that though the brains of large dinosaurs were only a tiny proportion of body weight, their range of sizes was about what might be expected by extrapolation from data for modern reptiles. The notion that they were extraordinarily small is due to the false assumption that brain weight should be proportional to body weight, for

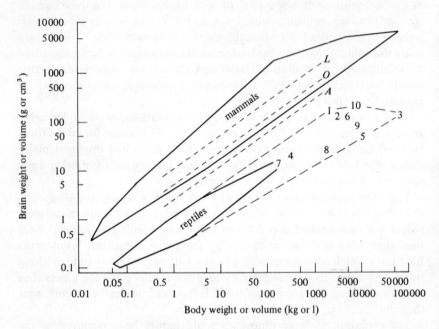

Fig. 10-16. Brain size and body size of reptiles and mammals. The continuous outlines indicate the ranges of variation among living reptiles and mammals. The numerals indicate estimated brain and body sizes of dinosaurs, including (2) *Anatosaurus*, (3) *Brachiosaurus*, (9) *Triceratops* and (10) *Tyrannosaurus*. The broken lines labelled *L*, *O* and *A* are regression lines for living (*L*) Oligocene (*O*) and Eocene and earlier (*A*) mammals. From H. J. Jerison (1969). *Amer. Nat.* **103**, 575–88. Copyright © University of Chicago Press, 1969.

related animals of different size. The figure shows that it is more nearly proportional to (body weight)$^{0.65}$. There was a good deal of variation in brain size between dinosaurs of similar size but different habits, but this variation seems to have been no greater than is found in mammals.

Many bipedal saurischians such as *Tyrannosaurus* (Fig. 10-13) and *Antrodemus* seem to have been carnivorous. Their sharp-edged, sabre-like teeth strongly suggest that they ate flesh, and brontosaur skeletons have been found with scratches on the bones which correspond in spacing to the teeth of *Antrodemus*. In one case a broken *Antrodemus* tooth was preserved nearby. It is not of course clear whether *Antrodemus* normally fed on brontosaurs, or even that it killed the ones it seems to have eaten. Smaller carnivorous dinosaurs must have attacked smaller prey. The brontosaurs and all ornithischian dinosaurs seem to have been herbivores, but what plants they ate is often a puzzle. Ornithischians had generally no front teeth, but the texture of the anterior parts of their jaws indicates a horny beak, like the beaks of birds and tortoises. Further back in the jaws were teeth which were sometimes extremely numerous. *Anatosaurus* (Fig. 10-14a) and its allies have beaks so shaped as to give them the name of duck-billed dinosaurs and numerous back teeth which seem adapted for grinding. The arrangement reminds one of the toothless horny front of the upper jaw, and the grinding teeth, of cattle. Each tooth had a series of replacements, ready to take over when it wore out. These dinosaurs presumably plucked and ground tough vegetation. Remains of plant food have been found in fossils, including conifer needles, twigs and seeds. The horned dinosaurs such as *Triceratops* and *Torosaurus* (Fig. 10-14b) had narrow hooked beaks, and the worn faces of their teeth are vertical, indicating a vertical chopping action rather than a grinding action. The teeth were closely fitted together to form a continuous cutting edge, and each tooth had two or three replacement teeth ready below it. The extraordinary frill at the back of the skull seems to have provided an origin for enormous jaw muscles, as well as protecting the neck. What food could have needed such extraordinary provision for chopping it up? Ordinary leaves, fruits and seeds can be broken up by crushing or grinding, but very fibrous plant tissue might be dealt with more effectively by chopping. Horned dinosaurs seem to have been the only ones equipped for chopping, rather than grinding, plant food and it has been suggested that they may have fed on the fibrous leaves of palms and cycads.

PTEROSAURS
Subclass Archosauria, order Pterosauria

The pterosaurs were reptiles with bat-like wings (Fig. 10-17a), living in the Jurassic and Cretaceous periods. They had diapsid skulls with a fenestra between eye and nostril, like other archosaurs. Those that had teeth had them in sockets, again like other archosaurs. The wings of bats are stiffened by four long fingers, but in pterosaurs only the last of the four digits was enlarged and involved in the wing. It formed the leading edge of the outer half of the wing, and a hinge joint between it and its metacarpal seems to have been the main joint for folding the wing. The outline of the wing can be distinguished in some fossils.

Nearly all pterosaur fossils have been found in sedimentary rocks that are believed to have been deposited in the sea. Most of them had long jaws. Some had pointed teeth which seem suitable for catching fish, while others had a toothless beak. Fossils of *Pteranodon* (which had a beak) have been found with remains of fish inside them.

Pteranodon had a wing span of over 8 m, more than any other known flying animal living or fossil. The wing span of the albatross *Diomedea exulans* is only 3.4 m. *Pteranodon* had a relatively small trunk and its bones

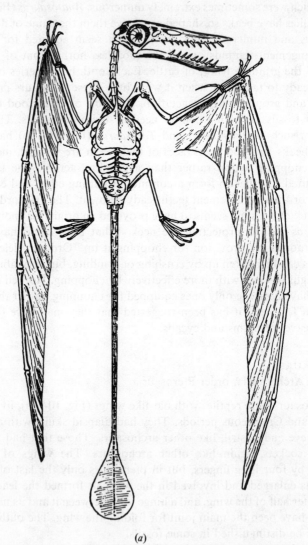

(a)

Fig. 10-17. (*a*) The pterosaur *Rhamphorhynchus*; (*b*) [*opposite*] an ichthyosaur. From A. S. Romer (1966). *Vertebrate Paleontology*, 3rd edit. University of Chicago Press, Chicago. Copyright © 1966.

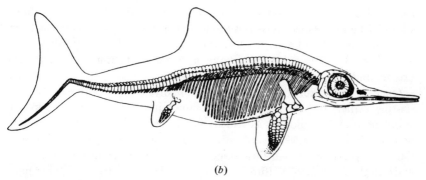

(b)

were lightly built, but even so it has been estimated to have weighed 18 kg, which is more than any flying bird. There is a keel on the sternum like the keel which separates the left and right flight muscles of birds (Fig. 11-6), but it is not particularly large and does not suggest particularly powerful flight muscles. How could *Pteranodon* have taken off?

Small birds can take off from level ground, in still air, without a run. They start by hovering, and gradually gather speed. The mechanics of hovering is discussed on p. 350, where it is explained that hovering requires too high a power output for large birds. It seems quite plain that *Pteranodon* could not have taken off in this way. Large birds take off by facing into a sufficiently strong wind, or by diving from a tree or cliff face, or after a taxiing run. These are all ways of getting the wings moving fast enough, relative to the air, to generate lift equal to the weight of the body. *Pteranodon* seems ill-suited to taking taxiing runs whether over land or, like a swan, over water. It would presumably not always have been convenient to land on cliffs. Could it have relied on taking off into the wind?

By equation 4. 2, the wind velocity u required to lift a *Pteranodon* of mass m and wing area A is given by

$$mg = \tfrac{1}{2}\rho A u^2 C_{L\,max}$$

$C_{L\,max}$ is the maximum lift coefficient, and would probably be at least 1.0 for a *Pteranodon* wing. The density ρ is that of air, 1.3 kg m^{-3}. By putting these values in the equation we get

$$u = 3.9\sqrt{(m/A)}$$

Notice that the velocity required is proportional to the square root of the wing loading, m/A. Since for animals of the same proportions m would be proportional to the cube of length and A only to the square, wing loading and take-off speed tend to be high for large flying animals, and for large aircraft. Large birds do not, of course, have the same proportions as small ones, but they do have substantially higher wing loadings. However, the huge wings of *Pteranodon* gave it a remarkably low wing loading for its size. Large albatrosses and vultures have wing loadings of about 14 and 7 kg m², respectively, but the wing loading of *Pteranodon* has been estimated as only 3 kg m^{-2}. This would give it a take off speed of only about $3.9\sqrt{3} = 7$ m s^{-1}. It could take off simply by facing into a moderate breeze, and spreading its wings.

Once it had taken off *Pteranodon* probably kept airborne mainly by soaring rather than by flapping flight. Since its wing loading was low it could presumably glide so as to lose height only very slowly in still air, and rise in very gentle upward currents. If it spent most of its time over the sea it presumably had little opportunity to gain height by using thermals, in the manner of vultures and man-made gliders (p. 352). Thermals are bodies of warm, rising air which develop over parts of the ground that are more strongly heated by the sun than their immediate surroundings. It seems more likely that *Pteranodon* depended on the upward currents that occur where the wind is deflected upward by waves and by steep shores. It seems less likely to have used the peculiar soaring technique of albatrosses, which depends on the ability to glide at high speeds.

Early pterosaurs such as *Rhamphorhynchus* (Fig. 10-17*a*) had long tails like most other reptiles. At least some had a fin-like flap at the end. This must have the same sort of effect as the feathers at the rear end of an arrow, which help to keep the arrow in line with its path through the air, giving it aerodynamic stability. Advanced pterosaurs such as *Pteranodon* had very short tails and must have been less stable. This would also have made them more manoeuvrable. A highly stable aircraft does not manoeuvre well, and the design of aircraft is a compromise between stability (which makes an aircraft easier to fly) and manoeuvrability.

Birds and bats have feathers or fur as heat insulation and are homoiothermic: that is, they maintain fairly constant high body temperatures. Modern reptiles do not, but traces of what seems to be fur have been found on *Rhamphorhynchus*, which suggests that pterosaurs may have been homoiothermic.

CROCODILES
Subclass Archosauria, order Crocodilia

Crocodilians are known as fossils from the Triassic period onwards, and survive to the present day. They have diapsid skulls with additional fenestrae in the lower jaw and (in early fossils) between eye and nostril. Their teeth have sockets. They should plainly be included in the Archosauria.

Crocodilians spend part of their time on land and part in water, where they swim by tail movements. Daily fluctuations of temperature are much less in the water of rivers and lakes, than they are on land. The Nile crocodile generally spends the night in water (where it is warmer than it would be on land) and spends the day on land (where its body temperature rises above that of the water). It sometimes returns to the water for the hottest part of the day.

The external nostrils of crocodiles are on top of a hump on the tip of the snout. Their eyes are also raised above the general level of the top of the head. Crocodiles often float in quiet water with little but the eyes and nostrils projecting above the surface. They have a view over the surface of the water and they can breathe, but they are inconspicuous. They sometimes lurk in this way off drinking places, submerging when a mammal comes to drink and swimming underwater to seize it unawares. Young crocodiles feed largely on insects but larger ones feed mainly on vertebrates, including fish, birds and mammals. Mouthfuls of flesh are torn from large prey, though the teeth are conical, without cutting edges.

The long jaws of crocodilians have evolved by elongation of the snout, and the part of the skull posterior to the eyes is relatively short. There is little room for the posterior jaw adductors, and the pterygoideus muscles (which extend forward ventral to the eye) are the main jaw-closing muscles.

Crocodilians have muscles which close their nostrils when they submerge. They also have an adaptation which enables them to breathe through the nostrils while these are above water, even when the mouth is open under water. They can thus hold prey under water to drown it, and go on breathing without getting water into their lungs. The main structure involved is the secondary palate. This is a 'false ceiling' to the mouth cavity, ventral to the primary palate. It is essentially similar to the secondary palate of mammals (described on p. 390) and consists of extensions of the premaxillae, maxillae, palatines and pterygoids. Because it is there, the internal nostrils no longer open into the anterior part of the mouth cavity, but much further back, close to the glottis. Air is conveyed to these openings, in the space between the primary and secondary palates. The passages from the external to the internal nostrils are indicated in Fig. 10-18. Flaps of tissue descending from the internal nostrils and rising from the glottis, fit together to close off the air passage from the mouth cavity (though they have to separate when the animal swallows).

Unlike lizards, crocodilians do not depend on rib movements alone to pump air in and out of their lungs. A sheet of connective tissue crosses the body cavity posterior to the lungs, in the position of the mammalian diaphragm. The liver is attached to the posterior face of this sheet, and is moved backwards and forward by muscles shown in Fig. 10-18. The diaphragmaticus muscle pulls it posteriorly, drawing air into the lungs, and at the same time making the abdomen swell. The transverse muscles of the

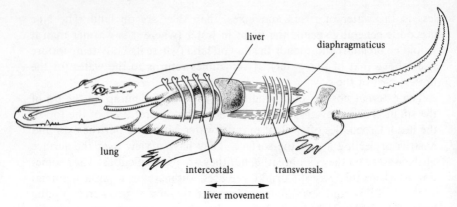

Fig. 10-18. A diagram showing structures concerned with breathing in a crocodilian (*Caiman*). After C. Gans (1970). *Evolution* **24**, 723–34.

abdominal wall have the reverse effect, driving the liver forward and emptying the lungs.

The peculiar structure of the hearts of crocodilians has already been described (p. 288).

EXTINCT MARINE REPTILES
Subclasses Euryapsida and Ichthyopterygia

The marine reptiles of the Mesozoic were almost as remarkable as the terrestrial dinosaurs. As well as marine turtles and the giant swimming lizards known as mosasaurs, there were the two subclasses of reptiles described in this section. Both had parapsid skulls (Fig. 9-11) but they do not seem to be at all closely related and probably evolved the parapsid condition independently.

The Euryapsida was quite a large and varied group, and its best known members are the plesiosaurs. They were up to 12 m long with short tails that could hardly have been used in swimming, but their limbs were flippers, like those of marine turtles. Indeed, the limb skeleton was even more strangely modified than in turtles. The flippers were presumably used as turtles use theirs, as hydrofoils rather than oars. The ventral parts of the limb girdles were very large, presumably for the attachment of large muscles used in the downstroke.

Plesiosaurs had slender, pointed teeth which seem well adapted for catching fish. Some had long jaws and short necks, and presumably relied on swimming speed to catch their prey. Others had small heads on extraordinarily long necks. They would have been badly streamlined if they swam with necks extended under water, and may perhaps have swum just below the surface, with the neck rising above the surface like the neck of a swan. They may have caught their prey by darting with the long neck at passing fishes, rather than by swimming after them.

Ichthyosaurs (Fig. 10-17*b*) apparently looked remarkably like sharks or tunnies. Their bodies seem to have been well streamlined, with no neck and with a tapering tail. The outline of the body can be distinguished in some fossils, and shows that they had a tail fluke which was arranged vertically like the tail fins of fishes, not horizontally like the flukes of dolphins. There was also a dorsal fluke, as in fishes and dolphins. The limbs were modified as flippers but were generally smaller than in plesiosaurs, and the limb girdles were much smaller. Ichthyosaurs presumably swam with their tails like fishes, and used their flippers mainly for steering and braking rather than for propulsion.

The extinct group of pelagic cephalopods known as belemnites were plentiful in the sea at the same time as ichthyosaurs. The very durable remains of their shells are often found inside ichthyosaurs, which had presumably eaten them. The long jaws and conical teeth of ichthyosaurs resemble those of dolphins which feed on fishes and cephalopods. Also found within ichthyosaur skeletons are remains of very small, apparently embryonic, ichthyosaurs. It seems that ichthyosaurs were viviparous like sea snakes and so, like them, did not have to go ashore to lay eggs.

FURTHER READING
General

See the list at the end of Chapter 9.

Tortoises and turtles
Gaunt, A. S. & C. Gans (1969). Mechanics of respiration in the snapping turtle, *Chelydra serpentina* (Linné). *J. Morph.* **128**, 195–228.
Girgis, S. (1961). Aquatic respiration in the common Nile turtle, *Trionyx triunguis* (Forskål). *Comp. Biochem. Physiol.* **3**, 206–17.

Primitive diapsids
Dawbin, W. H. (1961). The tuatara in its natural habitat. *Endeavour* **21**, 16–24.

Lizards
Daan, S. & T. Beltermann (1968). Lateral bending in locomotion of some lower tetrapods. *Proc. K. ned. Akad. Wet.* C, **71**, 245–66.
Frazzetta, T. H. (1962). A functional consideration of cranial kinesis in lizards. *J. Morph.* **111**, 287–320.
Snyder, R. C. (1962). Adaptations for bipedal locomotion of lizards. *Am. Zool.* **2**, 191–203.

Amphisbaenians
Gans, C. (1969). Amphisbaenians – reptiles specialized for a burrowing existence. *Endeavour*, **28**, 146–51.

Snakes

Boltt, R. E. & R. F. Ewer (1964). The functional anatomy of the head of the puff adder, *Bitis arietans* (Merr.). *J. Morph.* **114**, 83–106.

Frazzetta, T. H. (1966). Studies on the morphology and function of the skull in the Boidae (Serpentes). Part II. Morphology and function of the jaw apparatus in *Python sebae* and *Python molurus*. *J. Morph.* **118**, 217–96.

Gans, C. (1962). Terrestrial locomotion without limbs. *Am. Zool.* **2**, 167–82.

Gans, C. & W. B. Elliott (1968). Snake venoms: production, injection, action. *Adv. Oral Biol.* **3**, 45–81.

Hartline, P. H. (1971). Physiological basis for detection of sound and vibration in snakes. *J. exp. Biol.* **54**, 349–72.

Dinosaurs and their ancestors

Bakker, R. T. (1971). Ecology of the brontosaurs. *Nature, Lond.* **229**, 172–4.

Colbert, E. H. (1962). *Dinosaurs*. Hutchinson, London.

Galton, P. M. (1970). The posture of hadrosaurian dinosaurs. *J. Paleont.* **44**, 464–73.

Jerison, H. J. (1969). Brain evolution and dinosaur brains. *Am. Nat.* **103**, 575–88.

Kurten, B. (1968). *The Age of Dinosaurs*. Weidenfeld & Nicolson, London.

Ostrum, J. H. (1966). Functional morphology and evolution of the ceratopsian dinosaurs. *Evolution*, **20**, 290–308.

Reig, O. A. (1967). Archosaurian reptiles: a new hypothesis on their origins. *Science*, **157**, 565–8.

Swinton, W. E. (1934). *The dinosaurs*. Murby, London.

Crocodilians

Cott, H. B. (1961). Scientific results of an enquiry into the ecology and economic status of the Nile crocodile (*Crocodilus niloticus*) in Uganda and Northern Rhodesia. *Trans. zool. Soc., Lond.* **29**, 211–356.

Gans, C. (1970). Respiration in early tetrapods – the frog is a red herring. *Evolution*, **24**, 723–34.

Pterosaurs

Bramwell, C. D. (1971). Aerodynamics of *Pteranodon*. *Biol. J. Linn. Soc.* **3**, 313–328.

11

Birds

There are at the present time more species of birds than of reptiles, and their habits tend to make them more conspicuous. However, birds are much more uniform in structure. This is why they are allotted only a single chapter in this book, though two were devoted to reptiles. Their feathers are among their most distinctive features, and are described first.

FEATHERS AND BODY TEMPERATURE

Birds have keratin scales on their feet which are much like the horny scales of reptiles, but most of their body surface is covered by feathers. These are formed by the epidermis and consist of β-keratin. The structure of a typical feather is shown in Fig.11-1. The base of the shaft is housed in a follicle in the skin. The barbs are branches on either side of the shaft and they in turn have branches, the barbules, which are of two types. Those on the distal side of each barb have hooks (Fig. 11-1a) while those on the proximal side do not (Fig. 11-1b). The hooks of the distal barbules catch on the proximal barbules of the next barb (Fig. 11-1, circular inset) so that the barbs form a coherent lamina that is light but not fragile. Rough treatment is less likely to tear it, than to separate some of the barbs. Separated barbs can be interlocked again simply by re-arranging them side by side; this can be done by pulling a feather between the fingers, and birds do it with their bills when they preen.

Many feathers have a fluffy region near the base, where the barbules are all of the type shown in Fig. 11-1c. There are no arrangements for interlocking, and the barbs do not cohere. Chicks are covered with down feathers in which none of the barbs interlock and similar feathers are present (usually in concealed positions) on adult birds.

Barbs which get separated can be interlocked again, but a part which is broken off a feather cannot be replaced. Such damage can only be made good when the feathers are moulted and replaced by entirely new ones, which usually happens at least once a year. Feathers are formed by papillae at the bases of the follicles and as each new one is formed by the same papilla as its predecessor it is not formed until its predecessor has been shed. Moulting generally proceeds gradually so that the bird never has very much less than its full complement of feathers.

The part played by the feathers in flight is considered in a later section of this chapter. They have another function in providing heat insulation.

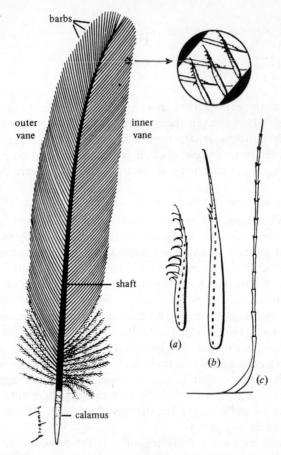

Fig. 11-1. One of the smaller feathers from the wing of a domestic fowl. The enlarged circular inset shows how the barbules interlock. (*a*) is a typical distal barbule, (*b*) a typical proximal barbule, and (*c*) a barbule from the fluffy region at the base of the feather. From M. E. Rawles (1960). In A. J. Marshall (ed.), *Biology and comparative physiology of birds*. Academic Press, New York.

They trap close to the body a layer of stationary air in which convection currents are effectively prevented. Little heat can pass across this layer except by conduction, and air is a very poor conductor of heat.

The advantage of a fairly high body temperature was considered in Chapter 9, as were the means whereby many reptiles achieve it. They rely on the sun to heat them above air temperature, and metabolism is relatively unimportant in their heat balance. Birds and mammals do not depend on solar radiation in this way, but maintain fairly constant high body temperatures by their metabolism. This is expressed by describing them as homoiothermic. The temperature is usually kept constant, day and night, within remarkably narrow limits: a very small deviation of human body temperature from normal is regarded as a symptom of illness. The body temperatures of birds are generally 40–43 °C, and of mammals 36–40 °C. Maintaining such temperatures in cooler surroundings would require much higher metabolic rates were it not for the insulating effect of the air trapped by feathers or (on mammals) fur.

Birds are almost bound to get wet from time to time, whether by entering water or simply being out in the rain. If water penetrated between their feathers and displaced the layer of trapped air it would spoil the heat insulation, since water conducts heat much better than air does. If it evaporated off, this would cool the bird. In fact, water does not easily penetrate between feathers, or remain on a bird until it evaporates: its tendency to run off a duck's back is proverbial.

How easily a liquid runs off a solid surface depends on the contact angle. If the contact angle is low, as between water and glass, drops tend to spread and they roll off slowly. If it is high, as between mercury and glass, much rounder drops are formed which roll off very easily indeed. Water on feathers forms drops with high contact angles, which roll off easily. This is partly because the feathers are thinly coated by an oily secretion produced by a gland at the base of the tail, and perhaps also by gland cells in other parts of the skin. Not quite all birds have the gland, but it is particularly large in water birds. Birds spread the secretion over their feathers when they preen. Its effect is greatly enhanced by the structure of the feathers, as Fig. 11-2*a, b* shows. Because the vane of a feather consists of parallel barbs with spaces between, the contact angles are much higher than if it were a continuous lamina. Water-repellent fabrics exploit the same principle.

Not only does water run easily off plumage, but it cannot easily penetrate between the barbs. This is because any pressure tending to force it

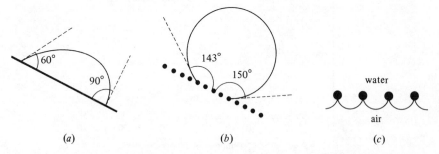

(*a*)   (*b*)   (*c*)

Fig. 11-2. (*a*), (*b*) Sections through drops of water moving towards the right on (*a*) the shaft and (*b*) the barbs of a duck's feather. Based on data in A. M. Rijke (1968). *J. exp. Biol.* **48**, 185–9.

(*c*) A diagrammatic section showing how pressure tending to force water between the barbs of a feather is resisted by surface tension.

through is resisted by surface tension in the air–water interface (Fig. 11-2c). This waterproofing (which is particularly important to swimming birds) depends largely on the even spacing which is imposed on the barbs by the interlocking barbules. If the barbs were free to clump together water would penetrate more easily.

Even a well insulated bird or mammal can only maintain its body temperature, when the temperature of its surroundings falls below a certain point, by increasing its metabolic rate. This can be done, for instance, by shivering. Fig. 11-3 shows measurements of the metabolic rates of small birds of the same species at various temperatures. The birds were resting, and since they had not recently fed they cannot have been using much energy for digestive processes. Each bird perched in a five-litre container in a constant-temperature cabinet. Air was passed slowly through the container, and oxygen consumption was calculated from its rate of flow and from the fall in its oxygen concentration as it passed through. At temperatures below 18 °C the rate of consumption of oxygen was more or

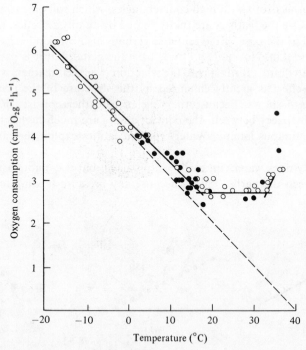

Fig. 11-3. A graph of oxygen consumption against environmental temperature for 22 cardinals (*Richmondena cardinalis*), weighing about 40 g each. ● Birds kept indoors and ○ birds kept out of doors in cold (−10 °C to +5 °C) weather. From W. R. Dawson (1958). *Physiol. Zool.* **31**, 37–48. Copyright © University of Chicago Press, 1958.

less proportional to the difference between body temperature (about 40 °C) and the temperature of the surroundings. This is as might be expected since rates of loss of heat are in general proportional to temperature differences. At temperatures between 18 °C and 33 °C, however, the metabolic rate remained constant at what is presumably the rate required for purposes other than temperature maintenance. This is known as the basal metabolic rate. If it is sufficient to maintain body temperature at 18 °C, it is more than sufficient at higher temperatures, at which the bird must make adjustments to avoid overheating. The feathers (which are fluffed out at low temperatures) are held flatter against the body so that the heat insulating layer is thinner, and more of the blood is directed to the un-insulated parts of the legs. At very high temperatures, above 33 °C, the metabolic rate rises again because the bird's body temperature rises and because it starts panting. Though panting involves muscular activity and so heat production, it results in a net loss of heat because it increases evaporation from the respiratory tract. It has been shown that birds and mammals tend to pant at the resonant frequency of the respiratory system; the energy required is less at this frequency than at higher or lower frequencies.

Fig. 11-4 shows how basal metabolic rate depends on body size, for lizards, birds and mammals. The order Passeriformes, which includes most small birds (see p. 368) is distinguished from the rest. In each group basal metabolic rate is more or less proportional to (body weight)$^{0.7}$ so basal metabolic rate per unit body weight (the quantity plotted on the graph) is more or less proportional to (body weight)$^{-0.3}$. The lizard rates are much lower than those of birds and mammals of similar weight, though measured with the lizards warmed to a typical mammalian body temperature. This shows that the high metabolic rates of birds and mammals cannot be explained as direct effects of high body temperature on chemical reaction rates. Some more fundamental metabolic change must have occurred when they evolved from reptiles. No similar change seems to have occurred when reptiles evolved from lower vertebrates: few fishes would survive heating to mammalian body temperatures, but at lower temperatures fishes and lizards of the same weight use oxygen at similar rates (see p. 246).

If animals of similar weight are compared, Passeriformes have basal metabolic rates almost twice as high as other birds and mammals, which in turn have basal rates about four times as high as reptiles.

Graphs such as Fig. 11-3 indicate that resting birds and mammals generally use metabolic energy either at the basal rate, or at the rate needed to maintain body temperature, whichever is the higher. An attempt will now be made to explain the rate needed to maintain body temperature,

334   Birds

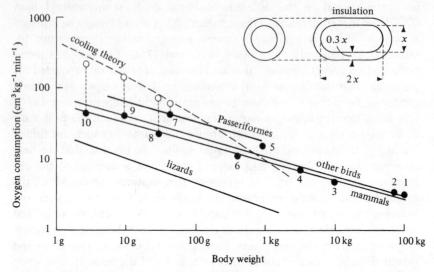

Fig. 11-4. Graphs on logarithmic co-ordinates of oxygen consumption per unit body weight, against body weight. The following regression lines are shown:

Lizards. Resting metabolic rates at 37 °C of 13 species. (J. R. Templeton (1970). In G. C. Whittow (ed.), *Comparative physiology of thermoregulation*, vol. 1. Academic Press, New York.)

Birds. Separate lines for basal metabolic rates of Passeriformes (35 species) and other birds (58 species). (R. C. Lasiewski & W. R. Dawson (1967). *Condor*, **69**, 13–23.)

Mammals. Basal metabolic rates of eutherian mammals (13 species). (S. Brody (1945). *Bioenergetics and growth*. Reinhold, New York.)

The broken line shows the expected oxygen consumptions of birds and mammals with skin temperatures of 37 °C in surroundings at 10 °C. It has been calculated from equations 11.1 and 11.2, which are based on the model shown as an inset.

●, Basal metabolic rates of a small sample of species and ○, metabolic rates at 10 °C. 1, Ostrich (*Struthio*); 2, man; 3, vulture (*Vultur*); 4, domestic cat (*Felis*); 5, raven (*Corvus*); 6, rat (*Rattus*); 7, cardinal (*Richmondena*); 8, deer mouse (*Peromyscus*); 9, harvest mouse (*Reithrodontomys*); 10, various hummingbirds. The raven and cardinal are Passeriformes; other birds listed here are not.

Points 1 to 6 are based on the same data as the regression lines; 7 on the data of Fig. 11-3; 8 on M. Muric (1961). *Ecology*, **42**, 723–40; 9 on O. P. Pearson (1960). *Physiol. Zool.* **33**, 152–60; 10 on R. C. Lasiewski (1963). *Physiol. Zool.* **36**, 122–40.

in terms of heat exchange theory. The basic physics of convection and radiation that will be used has been explained already, in the discussion of reptile heat balance on p. 281.

A simple model animal is needed as a basis for the calculations. The The model chosen is a cylinder with hemispherical ends, of the dimensions (in metres) shown in Fig. 11-4 (inset). The proportions are intended to be typical of small birds and mammals with their limbs pressed against the body. The thickness of the outer heat insulating layer is intended to correspond to that of the plumage or fur of well insulated birds and mammals (that is, in general, of small ones or ones from cold climates).

Thicker plumage or fur would be apt to get in the way of the limbs. It will be assumed that the body temperature is 310 K (37 °C) and that even the skin is kept at this temperature by the circulating blood. The air temperature will be taken as 283 K (10 °C), as a convenient example of a moderately low temperature. It is assumed that no sunlight falls on the animal, that all its surroundings are at the same temperature as the air and that the wind speed is very low, about 0.1 m s^{-1}. These assumptions are intended to correspond to conditions in the apparatus used to measure the actual metabolic rates, which will be compared with the calculated values.

When a steady state has been reached the outer surface of the animal's insulating layer will have a temperature T K, intermediate between the temperatures of the body and of the air. This surface will radiate heat at a rate $5.7 \times 10^{-8}T^4$ W m^{-2} and since its surroundings are at 283 K it will receive radiation from them at a rate $5.7 \times 10^{-8}(283)^4$ W m^{-2}. The net rate of loss of heat by radiation is thus $5.7 \times 10^{-4}(T^4 - 283^4)$ W m^{-2}. This can be shown by the binomial theorem to be approximately $5.2(T - 283)$ W m^{-2}. Heat will also be lost by convection. The formula for convection given on p. 283 will be used, although it applies strictly only to long cylinders: this will not cause serious error in a calculation as rough as this one. The diameter of the outer surface is $1.6x$ m and a wind speed of 0.1 m s^{-1} has been assumed, so loss by convection can be estimated as $0.9(T - 283)[0.1/1.6x)^2]^3 = 0.31(T - 283)x^{-2/3}$ W m^{-2}. The total rate of loss by radiation and convection is thus $(T - 283)(5.2 + 0.31x^{-2/3})$ W m^{-2}. The area of the outer surface can be calculated from the dimensions given in Fig. 11-5 to be $4.2\pi x^2$ m^2, and the rate at which heat is lost from this whole surface must equal the rate, M watts, at which it is being produced by metabolism. Thus

$$M = 4.2\pi x^2(T - 283)(5.2 + 0.31x^{-2/3}). \qquad (11.1)$$

Before it is lost by radiation and convection, this heat must be conducted across the insulating layer. The rate of conduction of heat is (conductivity × area × temperature gradient). The thermal conductivity of air is 0.025 W m^{-1} K^{-1} but measurements on fur have given a higher value of 0.05 W m^{-1} K^{-1}, presumably because some conduction occurs through the solid matter of the fur. The latter value will be used in this calculation. The heat is being conducted from the skin surface (area $2\pi x^2$ m^2) to the outer surface of the insulating layer (area $4.2\pi x^2$ m^2), so an intermediate area of $3\pi x^2$ m^2 will be used in the calculation. The temperature difference across the insulating layer is $(310 - T)$ K and the thickness of the layer is $0.3x$ m so the temperature gradient is $(310 - T)/0.3x$ K m^{-1}. The rate of conduction of heat across the layer equals the rate of heat production by metabolism so

$$M = 0.05 \times 3\pi x^2(310 - T)/0.3x$$
$$= 0.5x(310 - T). \tag{11.2}$$

The unknown T can be eliminated from 11.1 and 11.2, giving an equation for M in terms of x. The volume of the body (excluding insulation) can be calculated from the dimensions shown in Fig. 11-4 to be $0.42\pi x^3$ so if its density is 1000 kg m^{-3} (the same as for water) its mass is $420\pi x^3$. Hence M can be obtained in terms of body weight. The oxygen consumption corresponding to this rate of heat production can be estimated, since metabolism using 1 cm^3 oxygen yields about 20 J.

Metabolic rates estimated in this way are shown by the broken line in Fig. 11-4. They refer to an animal maintaining a body temperature (or more precisely, a skin temperature) of 37 °C, in surroundings at 10 °C. Also shown are some measured values of actual metabolic rates of small birds and mammals at 10 °C. They agree with the theory as well as could be expected, bearing in mind that the theory makes no allowance for differences of body shape, body temperature or relative thickness of insulation. Note that the metabolic rate per unit body weight, needed to maintain body temperature, falls quite steeply as body weight increases. This is because the ratio of surface area to volume is smaller for large animals, and also because large animals can have thicker insulation than small ones. The theoretical curve cuts the basal metabolic rate lines for birds and mammals at weights of 150 g (for Passeriformes) and about 1 kg (for other birds and mammals). Well insulated birds and mammals which are heavier than this can be expected to use oxygen at the basal rate even at 10 °C. The metabolic rates of Eskimo dog pups (weighing 9–15 kg) and and of the Arctic gull (*Larus hyperboreus*, 1.6 kg) have been found not to rise above their basal values even at −20 °C, but the basal rates of these Arctic animals are exceptionally high.

At the other extreme, it is found that a 3 g humming bird at 10 °C consumes oxygen at four times its basal rate. A 1 g bird or mammal at the same temperature would have to multiply the basal rate by an even larger factor. There must be a minimum weight below which it is no longer feasible to maintain a high body temperature by metabolism. Precisely what that minimum was would presumably depend on the lowest temperature likely to be experienced, but even in the tropics there are no birds or mammals lighter than about 2 g. This is the weight of both the smallest humming birds and the smallest shrews.

Some birds and mammals conserve energy by allowing the body temperature to fall during the part of the day when they are inactive. As they cool, they become torpid. Humming birds in captivity often become torpid at night, and torpid humming birds have sometimes been found in nature. Many bats which feed at night become torpid by day.

Adult birds and mammals may be large enough to be able to maintain their own body temperatures, but their embryos are not. Bird embryos are kept close to the adult body temperature by incubation of the eggs, and mammal embryos by being retained in the mother's body (and later in the pouch in marsupials). Even if it were feasible for isolated embryos to maintain a high temperature by their own metabolism it would be wasteful of energy; just as less energy per unit body weight is needed to maintain the body temperature of a large animal than of a small one, less is needed to maintain the temperature of embryos being incubated or in the uterus, than if they were separated from their parent.

The part of the parent bird's skin which rests against the eggs generally loses its feathers and develops a rich blood supply before or during incubation. This facilitates heat transfer to the eggs. Many nests must have quite good heat insulating properties. Nevertheless, egg temperatures during incubation are lower than adult body temperatures. 34 °C seems to be a typical egg temperature for Passeriformes.

There is a striking difference in maturity between the newly hatched young of different groups. Ducks and some other birds hatch in an advanced state, covered with down and moderately well able to maintain their own body temperature. They are very soon able to run about, and they quickly leave the nest. Passeriformes and some other birds, however, are naked and helpless when they hatch. They are for some days entirely unable to maintain their own body temperature. Meanwhile, their parents brood them.

Loss of water by evaporation from the respiratory surfaces of lizards was discussed on p. 275. Birds also lose water in this way and they might be expected to lose it rather rapidly in dry conditions on account of their high body temperature. With the water they would lose its latent heat of vaporization. How much water and heat is likely to be lost in this way?

Birds breathe in air containing 210 cm^3 oxygen per litre and breathe out air containing, typically, 140 cm^3 oxygen per litre. One litre of air breathed by a bird thus loses about 70 cm^3 oxygen which is used in metabolism yielding about 1400 J. If it left the body at 40 °C, saturated with water vapour, it would carry with it 60 mg water vapour. If the air had been cool (say, around 10 °C) and fairly dry when it was breathed in, nearly all of this water would have been obtained by evaporation within the bird. The latent heat of vaporization of water is about 2500 J g^{-1}, so the heat lost in this way would be about $0.06 \times 2500 = 150$ J, or around 10% of the metabolic heat production. If the metabolic rate had to be increased accordingly to maintain body temperature, this would be a substantial loss of energy.

In fact, neither birds nor mammals breathe air out at body temperature. This has been established by fitting microbead thermistors in their nostrils. These temperature-sensitive devices respond so quickly to changes of temperature that the temperatures of the air entering and leaving the nostrils as the animal breathes in and out are registered separately. Six species of bird breathing in air at 12 °C were found to breathe it out again at 14–21 °C, and only one (the domestic duck) breathed out at a higher temperature. A litre of air saturated with water at 14–21 °C contains only 14–21 mg water vapour, which is much less than the 60 mg at 40 °C, so the saving of water and heat is considerable. At higher air temperatures the expired air is warmer and the savings are less dramatic. Similar observations have been made on mammals.

The air must reach body temperature in the lungs but it is cooled again in the nasal cavities. Both in birds and in mammals these cavities contain turbinals, which are thin scrolls of bone or cartilage covered by epithelium. The spaces between the turbinals are narrow, and the total area of the epithelium is large. Incoming air cools the turbinals and is itself warmed. Outgoing air encounters the cooled turbinals, and is cooled again. The principle is the same as that of the counter-current heat exchanger in tuna muscles (p. 199), though flow in the opposite directions is not simultaneous.

WINGS AND FLIGHT

Fig. 11-5a shows the skeleton of a bird wing. The part which corresponds to the fore feet of other tetrapods is grossly modified. Its largest component is the carpometacarpus, which develops from three metacarpals and some carpals that fuse together as a single bone. The three metacarpals are those of the second, third and fourth digits, and a few phalanges of these digits articulate with them. The first digit (thumb) and fifth digit (little finger) can be found as vestiges in some embryos, but not in adults.

The radius and ulna cannot cross over each other in the manner illustrated in Fig. 8-14, but remain always parallel. If they could cross, muscles would have to be used to prevent twisting of the wing during flight. The elbow and wrist joints are so constructed that when the elbow is bent the wrist automatically bends as well, by a parallel-rule mechanism (Fig. 11-5b). These movements occur when the bird folds its wings.

The main flight feathers are the primaries and secondaries, which are attached along the posterior edges of the hand bones and of the ulna, respectively. The group of smaller feathers known as the alula or bastard wing is attached to the second digit (i.e. to the most anterior of the three

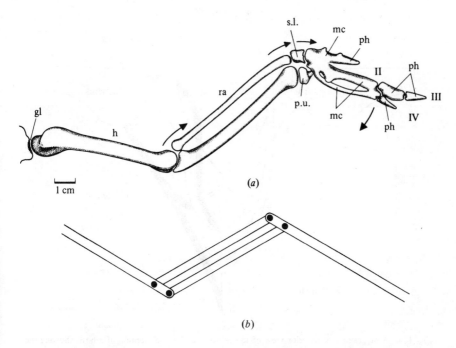

Fig. 11-5. (*a*) Right wing skeleton of domestic fowl. The arrows indicate movements which occur as the wing is folded. gl, glenoid; h, humerus; mc, carpometacarpus; ph, phalanx; p.u., s.l., carpals; ra, radius; ul, ulna; II, III, IV, digits. From A. D'A. Bellairs & C. R. Jenkin (1960). In A. J. Marshall (ed.), *Biology and comparative physiology of birds*, vol. 1. Academic Press, New York.
  (*b*) An equivalent parallel-rule mechanism.

digits). Other small feathers are arranged to give the wing a streamlined surface around the bones and muscles.

The pectoral girdle and sternum are almost as peculiar as the wing itself (Figs. 11-6, 11-7). To understand them, one must know how the wing muscles are arranged. The largest muscle is the pectoralis, which is responsible for the downstroke of the wings. In reptiles, which stand with the humerus more or less horizontal, pointing laterally, the pectoralis has an important weight-supporting function. It runs from the humerus to the sternum and interclavicle, and tension in it keeps the humerus horizontal and so keeps the animal's chest off the ground. The coracoid, running between the sternum and the shoulder joint, prevents the pectoralis from pulling the shoulders in towards the mid-line. In flying birds, the pectoralis is enormous and the sternum is greatly enlarged, with a deep keel, to provide most of its origin. The coracoids, which prevent the pectoralis from pulling sternum and shoulders towards each other, have become stout pillars of bone. The scapulae are long, but attached by muscle to the ribs as

340   *Birds*

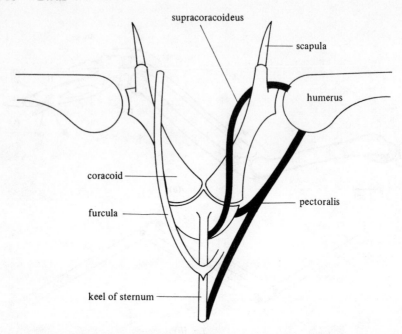

Fig. 11-6. A diagrammatic anterior view of the pectoral girdle and sternum of a bird, showing the arrangement of the main flight muscles.

in reptiles. The clavicles have apparently joined to form the furcula (wishbone) which, though generally a single bone, has separate halves in some parrots and owls.

The pectoralis muscle which is responsible for the downstroke of the wing runs directly from the sternum and furcula to the humerus. The supracoracoideus muscle, which is responsible for the upstroke, works less directly (Fig. 11-6). It also originates on the sternum but its tendon runs over a notch in the pectoral girdle which serves as a pulley, so that though the muscle lies ventral to the wings it serves to raise them. The surface of the notch is covered by cartilage, and the sliding tendon is presumably lubricated in the same way as joints between bones. In many birds that fly strongly including pigeons, budgerigars and hummingbirds (these examples are chosen because experiments on them are described later in this section) the pectoralis and supracoracoideus muscles together make up 25% or more of the weight of the body. In running birds such as the domestic fowl and waterfowl such as the coot (*Fulica atra*), they tend to be smaller, making up, in the coot, only 11% of the weight of the body. It will be shown later that the downstroke is the main power stroke in normal flight but that the upstroke produces important forces in hovering. In most

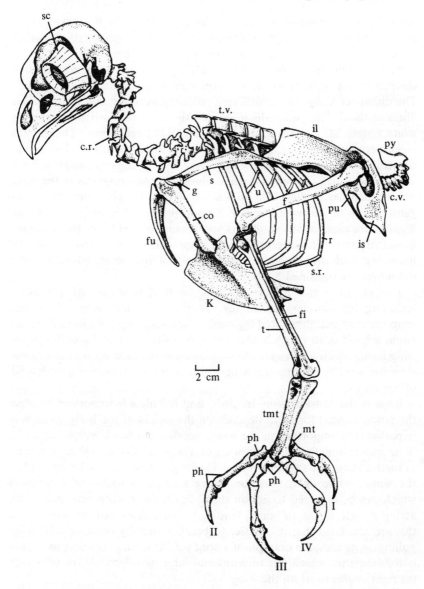

Fig. 11-7. Skeleton with wings removed of an eagle owl (*Bubo bubo*). co, coracoid; c.r., cervical rib; c.v., caudal vertebrae; f, femur; fi, fibula; fu, furcula; g, glenoid; il, ilium; is, ischium; k, keel on sternum; mt, first metatarsal; ph, phalanx; pu, pubis; py, pygostyle; r, rib; s, scapula; sc, scleral ossicles; s.r., sternal rib; t, tibiotarsus; tmt, tarsometatarsus; t.v., thoracic vertebrae; u, uncinate process; I–IV, digits. From A. d'A. Bellairs & C. R. Jenkin (1960). In A. J. Marshall (ed.), *Biology and comparative physiology of birds*, vol. 1. Academic Press, New York.

birds the supracoracoideus muscle which produces the upstroke is only around one tenth the weight of the pectoralis, but in hummingbirds (which hover a great deal) it is often half the weight of the pectoralis.

Anyone who has eaten chicken and pigeon will have noticed that chicken breast meat is white, while pigeon breast meat is much darker. The difference is due to the different proportions of red and white muscle fibres in them. The pectoralis muscle of the domestic fowl contains 67% white fibres, 22% intermediate fibres and 11% red fibres. That of the pigeon contains only 14% white fibres, but 86% red fibres. The supra-coracoideus muscles differ in the same way, though the percentages are not exactly the same. The red and white fibres differ in properties in the same way as in fishes (p. 94). The domestic fowl and its wild relatives do not generally fly far, so white muscle is appropriate for beating their wings. Pigeons are capable of long flights and depend on red fibres for sustained power output. The pectoralis and supracoracoideus muscles of the humming bird *Archilochus* consist entirely of red fibres, which provide the power for sustained hovering.

Though the breast muscle of domestic fowl is white, other muscles including leg muscles are darker. They presumably include a higher proportion of red fibres. Red leg muscles seem appropriate, since fowl run about a good deal. It is less easy to explain why the small muscles in the wing itself, which are responsible for adjustments of wing position rather than for actually beating the wings, also contain substantial proportions of red fibres.

Bone is the densest tissue in birds, and lightness is important because the power needed for flight depends on the weight of the body. Lightness is particularly important in the wings because the work which has to be done accelerating them in every wing beat is proportional to their moments of inertia about the shoulder joints. The long bones both of the legs and of the wings, while light, must be strong enough to withstand the forces which can be expected to act on them. Particularly dangerous are forces acting at right angles to bones, tending to bend them: readers who doubt this are reminded that it is easier to break a stick by bending it than by pulling on its ends. An example of a bone subject to large bending moments is the humerus, which has to withstand large aerodynamic forces acting (at right angles to it) on the wing.

If a light structure is required which will be strong enough to resist a given bending moment, a hollow tube is preferable to a solid cylindrical rod. A wide, thin-walled tube can be made lighter than a narrow, thick-walled one, so long as one does not go to such extremes as to make the wall of the tube liable to buckle or kink. Tubes are used extensively in engineering (for instance, in bicycle frames) and many of the long bones of tetrapods are hollow. The advantage in terms of lightness can be calculated, using engineering theory. For instance, the shaft of the human femur is a tube with its internal diameter about 0.5 of its external diameter. It can be calculated that a solid cylindrical shaft would have to be 1.3 times as heavy, to be equally strong in bending. The shafts of many bird bones are relatively wider, with thinner walls. For instance, the internal diameter of a swan humerus is 0.9 times the external diameter, and a solid shaft would have to be 2.5 times as heavy. So thin-walled a tube might be liable to collapse by kinking (like a bent drinking straw), and some bird bones are protected from kinking by internal struts.

The cavities inside the bones of other vertebrates house the bone marrow, where fat is stored and red blood corpuscles are formed. The relatively larger spaces in bird bones are in general only partly filled by marrow, and partly by air. There are air-filled cavities in the skull which are extensions of the nasal and middle-ear cavities, but the cavities in other bones are continuous with the air sacs (p. 354).

Fig. 11-8 shows the movements involved in ordinary horizontal flight. The wings beat up and down. During the downstroke (*c*) the primary feathers at the wing tip are bent upwards and forwards, indicating that an upward, forward force is acting on them. During the upstroke, they

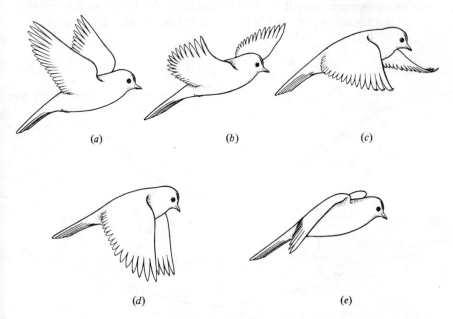

Fig. 11-8. Outlines traced from photographs taken at intervals of 0.01 s, of a Great tit (*Parus*) flying fast. From R. H. J. Brown (1963). *Biol. Rev.* **38**, 460–89.

are not obviously bent, indicating that forces on them are small. Fig. 11-9 shows how this might happen. Since the wings rise and fall while the bird moves forward, their path through the air is sinuous. The figure shows longitudinal sections through the wing, during the downstroke and during the upstroke. In the downstroke the outer part of the wing is set at a substantial angle of attack, so that lift and drag act as indicated. The resultant force on the wing has an upward component which helps to support the weight of the body and a forward (propulsive) component which counteracts drag on the body. These are the components of the force deduced from the bending of the feathers. In the upstroke the outer part of the wing is set at a different angle, so that it has little or no angle of attack and the forces which act on it are small. If it was set at an appreciable angle of attack the resultant force on it would act backwards as well as upwards, and slow the bird down. The inner parts of the wing move up and down less than the outer parts and probably produce an upward force in both strokes, but any forward component of the force which acts on them during the downstroke is probably cancelled out by a backward component of force during the upstroke.

So much can be learnt from films of birds flying past a camera. More detailed photographic analysis is possible and physiological measurements can be made more easily if the bird is stationary, flying against a jet of air in a wind tunnel at just such a speed as neither to gain ground nor be

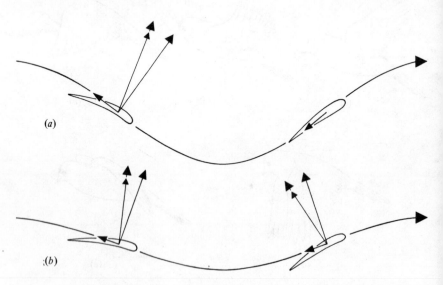

Fig. 11-9. A diagram showing the paths through the air of (*a*) the outer and (*b*) the inner part of a bird's wing in normal fast flight, and the forces which probably act on these parts of the wing.

blown back. If the wind tunnel is well designed airflow (in the absence of the bird) will be smooth and uniform throughout the jet. The movements of air relative to a bird flying and keeping station in such a wind tunnel against a wind velocity u will be the same as if the bird were flying at velocity u through still air. Several recent investigations have involved training birds to fly in wind tunnels. Fig. 11-10 shows a tunnel built for experiments with pigeons (*Columba*). Air is sucked in at the left and blown out at the right by the large fan c. The vanes e and the 'honeycomb' f stop the jet from swirling around. Any unevenness in air velocity in this part of the tunnel is swamped by the sudden uniform acceleration of the air as it passes through the contraction h. The bird flies in the position shown. The whole tunnel is suspended, pivoted at b, so that ascending and descending flight can be simulated.

Birds have been trained to fly in wind tunnels by various means. The experiment illustrated in Fig. 11-10 depended on an inducement of food. An end view of the working end of the tunnel is shown as an inset. A tube fixed diagonally across the opening has a teaspoon bowl soldered to its end. Maple peas were rolled down the tube into the bowl, and the bird could only get them by flying in the required position. In other experiments with other species the working sections of wind tunnels have been enclosed

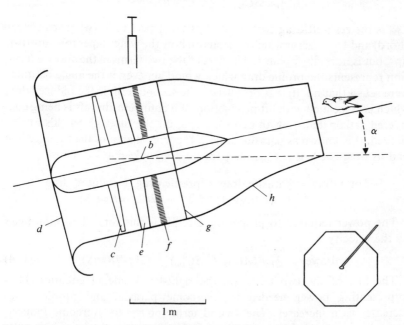

Fig. 11-10. A wind tunnel used for investigations of pigeon flight. It is explained in the text. From C. J. Pennycuick (1968). *J. exp. Biol.* **49**, 509–26.

by grids which could be electrified, and the birds learned that they would suffer a mild electric shock if they landed while an experiment was in progress.

Dr V. A. Tucker has measured the oxygen consumption of budgerigars (*Melopsittacus*) flying in a wind tunnel. Transparent plastic masks were used, as in the experiment with a lizard which is illustrated in Fig. 9-5*b*. The masks were rounded and fitted the head much more closely than the ones used on lizards, and probably did not interfere at all severely with streamlining. Results are illustrated in Fig. 11-11. They show as one would expect that oxygen is used more rapidly ascending than in descending flight, and also that in horizontal flight oxygen is used less rapidly at 35 km h^{-1} than at higher or lower flying speeds. In this, the budgerigar resembles man-made aircraft. There is a particular speed at which an aircraft needs least power for horizontal flight. If it flies faster or slower than this minimum-power speed, its engines must supply more power.

The explanation for this involves the distinction between profile drag and induced drag. The drag on an aerofoil of plan area S moving at velocity u relative to a fluid of density ρ and generating lift L is, by equation 4. 3

$$\tfrac{1}{2}\rho u^2 S C_{D0} + (2kL^2/\pi R\rho u^2 S).$$

C_{D0} is the zero-lift drag coefficient, R is the aspect ratio (wing span/mean chord) and k is a factor which depends on how the wings taper towards their tips, but is normally about 1. The first of the two terms in the above expression represents the profile drag which would act even if the angle of attack were set so that no lift was produced. The second term is the induced drag which is associated with lift production. If a complete aircraft is being considered rather than just an aerofoil, drag on the fuselage has also to be sidered. It is known as parasite drag. If the frontal area of the fuselage is A and its drag coefficient C_{DF}, the parasite drag is $\tfrac{1}{2}\rho u^2 A C_{DF}$. Thus

$$\begin{aligned} \text{Total drag} &= \text{parasite drag} + \text{profile drag} + \text{induced drag} \\ &= \tfrac{1}{2}\rho u^2 (AC_{DF} + SC_{D0}) + (2kL^2/\pi R\rho u^2 S). \end{aligned} \tag{11.3}$$

The power required to propel the aircraft is this total drag multiplied by the velocity

$$\text{Power} = \tfrac{1}{2}\rho u^3 (AC_{DF} + SC_{D0}) + (2kL^2/R\rho u S). \tag{11.4}$$

The first of the two terms on the right-hand side of equation 11. 4 (representing power needed to overcome parasite and profile drag) increases as u increases. The second term (power to overcome induced drag) decreases as u increases. There is therefore a particular value of the velocity u, at which the power required for flight is least.

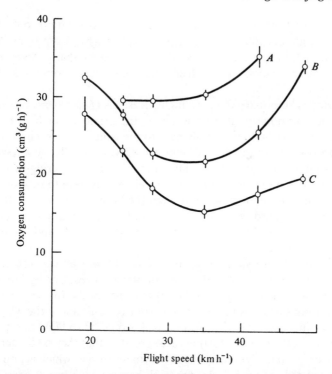

Fig. 11-11. Graphs of oxygen consumption against speed for budgerigars (*Melopsittacus*) flying in a wind tunnel. (A) refers to ascending flight at 5° to the horizontal, (B) to level flight and (C) to descending flight at 5°. From V. A. Tucker (1968). *J. exp. Biol.* **48**, 67–88.

Equation 11.4 applies to fixed-wing aircraft, with wings and fuselage moving at the same velocity. A bird's wings, beating up and down, move rather faster than the rest of the body, but the explanation of the minimum power speed is otherwise the same as for fixed-wing aircraft.

Though an aircraft uses least power at the minimum-power speed, a rather higher speed is the most economical in terms of miles per gallon of fuel. It is known as the maximum-range speed since it allows the longest possible flight without refuelling. Consider the oxygen consumption of budgerigars per kilometre travelled in level flight, taking data from Fig. 11-11. At the minimum-power speed, 35 km h, 22 cm^3 oxygen g^{-1} h^{-1} is used, or $22/35 = 0.63$ cm^3 g^{-1} km^{-1}. At 39 km h^{-1} the consumption would apparently be 23.5 cm^3 g^{-1} h^{-1} or 0.60 cm^3 g^{-1} km^{-1}, which would allow a slightly longer range.

Fat is the main fuel for long flights by birds, which often accumulate fat stores of more than 25% of the weight of the body. Metabolism of 1 g fat uses 2 l oxygen, so that the oxygen consumption of 0.6 cm^3 g^{-1} km^{-1}

which has just been calculated implies use of fat at a rate of only 3% of the body weight per 100 km. This helps to explain why birds can make long migrations. Migrating birds fly 1400 km across the Sahara Desert apparently without an opportunity to feed, and others make even longer flights over the sea.

Hovering is stationary flapping flight. Kestrels (*Falco tinnunculus*) keep more or less stationary 10 m or so above the ground, watching for prey on the ground. They do this by flying into the wind at just such a speed as neither to make headway nor to be blown downwind. The next few paragraphs are not about this sort of hovering but about hovering in still air, which is a much more strenuous activity. The wings have to be moved fast enough to produce enough lift, although the body is not moving through the air. Hummingbirds weighing 2–20 g can hover indefinitely but pigeons weighing 0.4 kg can hover only for a very few seconds and large birds cannot hover at all.

Birds hovering in still air keep their bodies more or less vertical and beat their wings horizontally. Two techniques are used, one by hummingbirds and one (illustrated in Fig. 11-12) by other birds. In the latter technique the wing surface is never very far from vertical, and if the wing were acting as a single aerofoil its angle of attack would be so high that it would stall. However, the primary feathers are spread so that they no longer overlap and are free to be twisted by the aerodynamic forces which act on them. They are twisted to lower angles of attack because each has its shaft much nearer the anterior than the posterior edge of the vane. Separation of the feathers so that they act as separate aerofoils increases the total aerofoil area by eliminating overlap, and so reduces the wing speed needed to produce sufficient lift.

Feather separation apparently prevents stalling at the wing tip. Further in on the wing, the tendency to stall may be reduced by the alula, which can be seen projecting as a separate aerofoil in Fig. 11-12*b, c*. This supposition is based on the principle of the slotted wing as used in aircraft, and on experiments with bird wings in wind tunnels. It has been found that at high angles of attack bird wings may produce up to 25% more lift with the alula in the extended position than with it fixed in the normal position flat against the main wing surface.

Hummingbirds hover in front of flowers, using their long slender tongues to take the nectar on which they feed. They play the same role as pollinators of some flowers, as nectar-feeding insects do of many others. They will also hover to take sugar solution from suitably designed bottles, and are fed in this way in aviaries. Oxygen consumption of hummingbirds has been measured as they hovered at bottles of this sort in small closed chambers. In one of these experiments a 3 g hummingbird hovered continuously

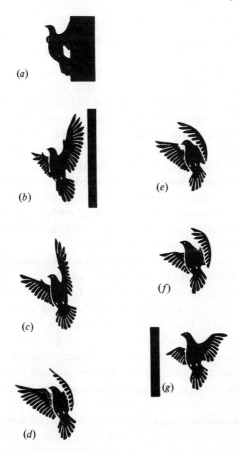

Fig. 11-12. Silhouettes traced from a cine film of a domestic pigeon hovering briefly before landing. The interval between successive silhouettes was about 0.017 s. From W. Nachtigall & B. Kempf (1971). *Z. vergl. Physiol.* **71**, 326–41.

for 35 min, using 2.1 cm³ oxygen min⁻¹. This is 42 cm³ (g body weight)⁻¹ h⁻¹, which is high in comparison to the rates for flying budgerigars (Fig. 11-11).

The power required for hovering can be estimated from engineering considerations. The wing beat is a reciprocating action: the wings are halted at the end of each beat and accelerated in the opposite direction. Each time they lose their kinetic energy, and have to be given kinetic energy afresh. Thus work is done against wing inertia at the beginning of each beat. Work is also done throughout the beat, against aerodynamic drag. Thus power is used both in overcoming inertia, and in overcoming drag. These two components have been estimated from measurements on a typical hummingbird, as 0.029 W (g body weight)$^{-1}$ (for inertia) and 0.026 W g^{-1} (for drag). The simple total of these components is 0.055 W g^{-1}, but the power actually needed is only 0.046 W g^{-1} because some of the kinetic energy of the wings is converted to work against drag at the end of the beat as drag helps to slow the wings down. Most of the rest of the kinetic energy is probably lost as heat in the muscles that also play a part in slowing the wings down. (There may however be some saving of energy due to tendon elasticity, which seems certainly to be important in human running. See p. 383.) Vertebrate striated muscles seem generally to work at efficiencies of about 20%, and metabolism using 1 cm^3 oxygen yields about 20 J, so the oxygen consumption required for hovering can be estimated as about $0.046 \times 5/20 = 0.012$ cm^3 g^{-1} s^{-1} or 43 cm^3 g^{-1} h^{-1}, which agrees very well with the measured value of 42 cm^3 g^{-1} h^{-1} (or 36 cm^3 g^{-1} h^{-1} if the resting metabolic rate is subtracted, as representing energy being used for purposes other than hovering).

It can be shown by further analysis that more power is needed for hovering than for forward flight at moderate speeds. Also, (power for hovering)/(body weight) can be shown to be larger for large birds than for small ones. As size increases, power required for hovering must increase faster than power available, so that birds above a certain size cannot hover. The largest hummingbirds weigh only about 20 g, and this may be about the limit of size for birds which can hover without incurring an oxygen debt. Larger birds hover more briefly: 400 g pigeons are apparently unable to hover for more than about a second, and substantially larger birds cannot hover at all.

Small and medium-sized birds often take off by hovering and then gradually gathering speed, changing gradually from the hovering action to the normal flying action. Birds too large to hover cannot do this, and can only take off in still air by diving from a high perch or by taking a taxying run. They must get up enough speed for their wings to lift them. The relationship between body size and take-off speed has already been discussed in connection with a very large pterosaur (p. 323). The largest birds which can fly, such as the Kori bustard (*Ardeotis kori*) weigh up to about 20 kg. Some flightless birds such as the ostrich (*Struthio*) and the Emperor penguin (*Aptenodytes*) are of course much heavier.

The gliding ability of birds has been investigated in wind tunnels including the one illustrated in Fig. 11-10. Dr C. J. Pennycuick trained pigeons to glide and fly in this tunnel. With the wind speed fixed he varied the angle at which the tunnel was tilted. He found the angle at which the bird could only just glide. It could glide more steeply at the same speed if it used its feet to brake but it could not glide at shallower angles and had to resort to flapping flight. Observations were made at different wind speeds, finding the shallowest gliding angle for each. The results of these experiments and of similar ones with other species are shown in Fig. 11-13. It is not the angle itself which is plotted but the sinking speed, which is the rate at which the bird would lose height in still air. The falcon and vulture were much better gliders than the pigeons.

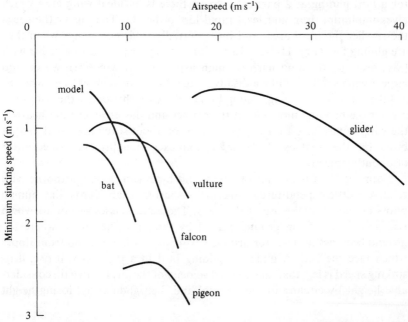

Fig. 11-13. Graphs of minimum sinking speed for gliding at various airspeeds for a glider (wing span 17 m), a model glider (wing span 1.1 m), a vulture (*Coragyps atratus*), a falcon (*Falco jugger*), the pigeon (*Columbia livia*), and a tomb bat (*Rousettus aegyptiacus*). Data from C. J. Pennycuick (1971). *J. exp. Biol.* **55**, 833–45; V. A. Tucker & G. C. Parrott (1970). *J. exp. Biol.* **52**, 345–68; and G. C. Parrott (1970). *J. exp. Biol.* **53**, 363–74.

Fig. 11-13 shows that for each species (and for a man-made glider) there is a particular gliding speed at which the sinking speed is least. This is easily explained. Fig. 11-14a shows the forces acting on a bird of weight m g which is gliding at an airspeed u and an angle θ, so that its sinking speed is $u \sin \theta$. The lift is L and drag D. At equilibrium

$$D = m \text{ g} \sin \theta$$
$$L = m \text{ g} \cos \theta$$
$$\simeq m \text{ g} \quad \text{if } \theta \text{ is small.}$$

Using these equations and equation 11.3

$$\text{Sinking speed} = u \sin \theta$$
$$\simeq Du/L$$
$$= (\rho u^3/2L)(AC_{DF} + SC_{D0}) + (2kL/\pi R\rho uS). \quad (11.5)$$

The first of the two terms on the right-hand side of the equation increases as u increases, and the second decreases. There is therefore a particular airspeed u at which the sinking speed is least.

Just as there is an ideal area of selachian pectoral fins which enables them to give the required lift at a particular speed with least drag (p. 100), so for a bird gliding at a particular speed there is an ideal wing area which gives minimum drag and least rapid loss of height. The higher the speed the smaller the ideal area, and birds partially fold their wings when they are gliding fast (Fig. 11-14*b*). Just as fins of high aspect ratio give lift with least drag, so do long narrow (high aspect ratio) wings. However, too large a wing span could be a disadvantage. Long wings might be awkward at take-off or when flying among branches. Also, the longer the wings the greater the bending moments at their bases and the stronger (and heavier) the wing bones have to be. Aspect ratios of about seven are usual among birds, but the albatross *Diomedea* has an exceptionallly high aspect ratio of about eighteen.

Flapping flight uses a lot of energy but birds can remain airborne with relatively little expenditure of energy by soaring. Many birds, like human pilots in gliders, make use of thermals. These are masses of hot air which rise periodically from ground heated by the sun. They form where the ground becomes hottest: for instance, from dry dark areas and from slopes which face the sun. A glider or gliding bird in a thermal will rise, if its sinking speed is less than the upward velocity of the air. It can gain considerable height by circling in the thermal and then glide away, losing height

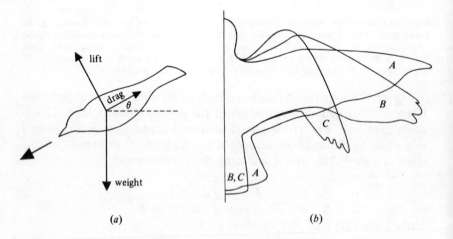

(a)   (b)

Fig. 11-14. (*a*) A diagram showing the forces which act on a gliding bird.
   (*b*) Outlines of a falcon gliding at (*A*) 6.6 m s^{-1}, (*B*) 8.5 m s^{-1} and (*C*) 14.3 m s^{-1}. From V. A. Tucker & G. C. Parrott (1970). *J. exp. Biol.* **52**, 345–67.

but possibly covering a substantial distance before it needs to gain height again. Vultures, some crows and gulls, and various other birds keep airborne for long periods in this way.

Albatrosses also remain airborne for hours, hardly ever flapping their wings, but their technique is quite different. It depends on the wind speed being lower near the surface of the sea than it is higher up. They glide down gaining speed, and then face into the wind and rise. A glider can always gain height at the expense of speed, until it is moving too slowly for its wings to produce enough lift. As the albatross rises it loses speed, relative to the ground. However, its speed relative to the air (on which lift depends) may actually increase, because the wind it is facing is faster at higher levels.

It can be shown by applying aerodynamic theory that a vulture can rise in small weak thermals only if it glides slowly, but that an albatross can only rise in a weak wind gradient without losing air speed if it glides fast. It has been shown that the ideal wing area is greater at low speeds than at high ones. Vultures appropriately have wing areas about double those of albatrosses of the same weight.

Flying birds use oxygen extremely rapidly. Fig. 11-11 shows that budgerigars use oxygen at up to 36 $cm^3 g^{-1} h^{-1}$ in ascending flight, and rates up to 55 $cm^3 g^{-1} h^{-1}$ have been measured in turbulent air. The oxygen consumption of the Black duck (*Anas rubripes*) has been measured in much shorter flights, not in a wind tunnel, and found to be 14 $cm^3 g^{-1} h^{-1}$. These are very much faster than the maximum metabolic rates of reptiles. At high body temperatures lizards similar in weight to the black duck could use no more than 0.5 $cm^3 g^{-1} h^{-1}$ (Fig. 9-5). Even mammals (except perhaps bats, see p. 416) cannot use oxygen as fast as birds of similar weight. Kangaroo rats (*Dipodomys merriami*) similar in weight to budgerigars use only 12–14 $cm^3 g^{-1} h^{-1}$ when running at top speed. Birds achieve their very high rates of oxygen uptake by means of peculiar lungs, which are described in the next section of this chapter.

Even in horizontal flight at the minimum power speed, budgerigars were found to use oxygen 10–15 times as fast as they would probably have done at rest. Hovering hummingbirds used oxygen seven times as fast as when resting at the same temperature. Since the resting rate is enough to maintain the body temperature, flying and hovering birds might be expected to overheat. However, the movements of air past the body which are involved must help heat loss by convection, and spreading the wings exposes the flanks which are rather sparsely covered with feathers. Heat loss by evaporation is important in mammals but relatively unimportant in birds, which do not sweat. Evaporation is mainly from the respiratory system. It accounted for only 15% of heat loss from budgerigars flying in Tucker's wind tunnel in air at 20 °C and 50% relative humidity. If birds were more dependent on heat loss by evaporation they would have to drink more often on long flights, and migrations across the Sahara or broad seas might not be practicable. The budgerigars in the experiment just mentioned were losing water by evaporation at a rate of 6% of body weight per 100 km, but some of this must have been replaced by water produced by metabolism. Metabolism of fat produces an approximately equal weight of water so if fat were being used at the rate of 3% of body weight per 100 km the net rate of loss of water would be only 3% of body weight per 100 km (p. 347).

RESPIRATION

Though birds take up oxygen so rapidly their lungs, as distinct from the air sacs connected to them, are relatively small. The lungs proper contain the respiratory surfaces, but change volume relatively little in breathing. The air sacs are much larger and are compressed and expanded greatly as the bird breathes, but their thin walls have a poor blood supply and it can be presumed that little of the oxygen uptake occurs in them. They fill a large part of the body cavity and are continuous with the air-filled cavities in bones which have already been referred to (p. 343).

The lungs contain a very complicated system of air passages (Fig. 11-15). The trachea divides in the usual way into two bronchi, one to each lung. Each bronchus gives rise to two groups of branches, the dorsobronchi and ventrobronchi, which are joined by very large numbers of fine parallel tubes, the parabronchi. There are two groups of air sacs which are connected to the anterior and posterior ends of the lungs.

When the bird breathes it moves its sternum up and down, mainly at the posterior end. This compresses and expands the body cavity, pumping air in and out of the air sacs. It is not obvious from anatomy what path the air takes through the lungs, as it travels to and from the air sacs. Various methods have been used in attempts to find out.

Some of the most revealing experiments have used the device shown in Fig. 11-15c. This is a short tube containing a pair of thermocouples and a hot-wire anemometer. The anemometer is simply an electrically heated filament which is cooled by air flow through the tube. Cooling alters its resistance, so it can be used to give a record of air speed in the tube. It responds in the same way to flow in either direction but, because it heats the air slightly, the air downstream of it is a little warmer than the air upstream of it. The thermocouples therefore indicate the direction of flow. In operations under local anaesthetic, devices like this have been fitted into the dorsobronchi of ducks. Records obtained from them while the ducks

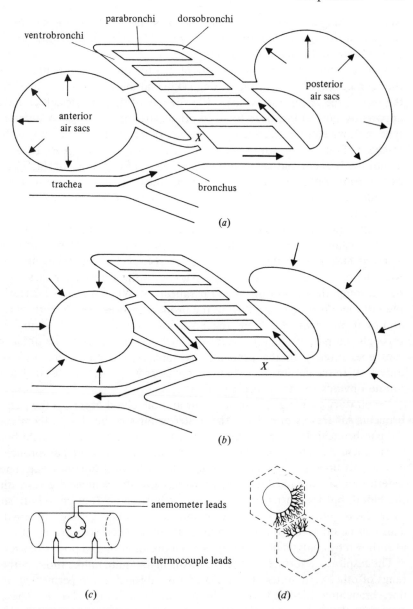

Fig. 11-15. (*a*), (*b*) Diagrams showing the connections between the parts of a bird lung and its air sacs, and the directions of flow determined as described in the text. (*a*) represents inspiration and (*b*) expiration. *X*, indicates that there is little or no flow of air.

(*c*) A flowmeter used in some of the experiments.

(*d*) A diagrammatic section through two parabronchi, showing the air capillaries radiating from them.

were breathing apparently normally, showed that air flows in the dorso-bronchi in the direction indicated in Fig. 11-15*a*, *b*, both while the bird is breathing in and while it is breathing out.

The dorsobronchi are conveniently placed for these experiments where they can be reached by cutting through skin and muscle without damaging the rest of the respiratory system. Flowmeters of the type described have apparently not, so far, been fitted in other parts of the lungs. A different type of flowmeter has been fixed to the end of a rod and pushed down the tracheae of ducks from an incision in the neck. It was found possible to get it into the bronchi and ventrobronchi, as well as the dorsobronchi. These experiments confirmed the information obtained by the other method and provided the additional information on flow which is presented in Fig. 11-15*a*, *b*.

These experiments show that when ducks (and presumably other birds) breathe in, air is drawn down the bronchi directly to the posterior air sacs. There is little flow at this stage in the ventrobronchi, near their junction with the bronchus, so most of the air going to the anterior air sacs must pass through the dorsobronchi and parabronchi. When the duck is breathing out there is little or no flow in the bronchi, between the dorsobronchi and ventrobronchi. Air from the posterior air sacs must leave mainly through the parabronchi and ventrobronchi while air from the anterior sacs goes directly to the bronchus. It appears that air normally flows anteriorly in the parabronchi, but it is not clear why it should do so. There are no obvious valves to direct its flow.

Respiratory exchange between the air and the blood occurs in the para-bronchi. Air travels directly to the posterior air sacs but leaves them via the parabronchi. It travels to the anterior air sacs via the parabronchi but leaves them directly. Thus half of the air passes through the parabronchi as the bird breathes in, and the other half as the bird breathes out. The anterior air sacs are filled with air which has already exchanged gases with the blood, but the posterior sacs with largely fresh air (not entirely fresh since the respiratory system is not completely emptied between breaths). This can be inferred from the results of the flowmeter experiments but is also shown by analysis of gas samples from the anterior and posterior sacs.

The respiratory surfaces in the parabronchi are quite unlike those in the lungs of other vertebrates. The walls of a parabronchus are permeated by fine, branching, air-filled tubes, the air capillaries (Fig. 11-15*d*). These open into the main channel of the parabronchus, and intertwine with the blood capillaries. Air is pumped through the main channel, but movement of gases along the air capillaries presumably depends on diffusion.

A rough calculation will indicate whether this diffusion path is likely to be important in limiting the rate of respiratory exchange. It is based on a bird such as a crow or pigeon, weighing 0.4 kg. In such a bird the parabronchi would be spaced about 0.1 cm apart, hexagonally packed, and the volume of the lungs would be about 10 cm³. Hence it can be estimated that the total length of the parabronchi would be about 13 m (i.e. that if the parabronchi were set end to end they would extend over this length). The diameter of the parabronchi would be about 0.05 cm, so their total surface area would be $0.05\pi \times 1300 = 200$ cm². The total cross-sectional area of the air capillaries radiating from them can be estimated as half of this, 100 cm². Since the centres of the parabronchi are 0.1 cm apart gases need never diffuse more than 0.05 cm, and 0.03 cm can be taken as an average diffusion path. When the bird was flying it would probably use about 14 cm³ oxygen $g^{-1} h^{-1}$ (this is the rate measured for the Black duck, see p. 353), so a total of about 5600 cm³ oxygen h^{-1} would have to diffuse along the air capillaries. The permeability constant for oxygen diffusing in air is 660 cm² $atm^{-1} h^{-1}$, so by equation 3. 1 the partial pressure difference required to drive this diffusion is $5600 \times 0.03/(100 \times 660) = 0.003$ atm. Carbon dioxide has a slightly lower permeability constant but the volume produced would probably be less than that of the oxygen being used and the required partial pressure difference would be about the same. The estimates are extremely rough; the careful reader will have noticed a succession of approximations and simplifying assumptions. They are, however, good enough to show that the partial pressure differences along the air capillaries must be small. The diffusion path through tissue between the air capillaries and the blood capillaries must be a much more serious barrier than the diffusion path through air along the capillaries. A similar conclusion was reached on p. 212 regarding the (very different) lungs of lungfishes.

Flight requires not only fast uptake of oxygen, but also fast transport of oxygen from lungs to flight muscles. Flying ducks use oxygen about thirty times as fast as can lizards of the same weight (p. 353). This is made possible partly by having blood which contains a lot of haemoglobin (per unit volume) and so can take up a lot of oxygen, and partly by pumping the blood rapidly round the body. The blood of the lizard in question (*Iguana*), has been found to have a capacity of 8.4 cm³ oxygen per 100 cm³ blood, which is a typical value for reptiles (and also lies in the usual range for fishes). Typical birds (probably including the duck) have capacities of 20–25 cm³ per 100 cm³. The heart of the lizard seems never to beat faster than about 2.5 times per second but that of the duck, immediately after a flight, was found to be making nine beats per second. The duck heart probably also pumps more blood at each beat.

It was explained on p. 288 that reptile hearts are so constructed that the left systemic artery may carry either oxygenated or deoxygenated blood, or a mixture of the two. Adult birds have no left systemic artery, and they have separate left and right ventricles. There are two completely separate pathways through the heart: all the blood from the right atrium and ventricle goes to the lungs and that from the left to the rest of the body. Since the left and right systemic arteries of reptiles are largely alternative pathways to the same parts of the body, one of them can be lost without major rearrangement of other arteries.

While in the egg birds, like reptiles, use the allantois rather than the lungs for respiration. The artery to the allantois branches from the aorta, not from the pulmonary artery. If the blood streams were separate in the embryo, as in the adult, all the blood which returned to the right side of the heart would be pumped through the lungs, and energy would be wasted. In fact the wall between the two atria is incomplete in the embryo, and there are also connections between the systemic and pulmonary arteries. Little blood flows through the lungs until hatching, when they become functional. Thereafter the pathways through the left and right sides of the heart quickly become separate.

LEGS

With their forelimbs so highly modified as wings, birds are necessarily bipedal. Bipedal lizards depend for balance on long tails (p. 304) and the same was apparently true of bipedal dinosaurs. Modern birds may have long tail feathers, too light to be important in balance, but they have nothing comparable to the long flesh-and-bone tails of lizards. If they stood with their feet directly below their hip joints, they would be liable to fall on their faces. They actually stand with their femurs nearly horizontal (much as in Fig. 11-7) so that their knee joints are on either side of the centre of gravity and their feet are centred more or less vertically below the knee joints. A vertical line through the centre of gravity would pass between the feet, as it must if the bird is not to fall. The femur is generally relatively short and the metatarsals are fused together to from a single bone which is often about as long as the femur. A few separate tarsal bones appear in embryos but they later fuse with other bones, with the tibia to form the bone known as the tibiotarsus and with the fused metatarsals to form the tarsometatarsus.

Many mammals have small feet (i.e. feet which make contact with only a small, compact area of ground). Small feet can be light, an adaptation for fast running (p. 385). However, small feet are not feasible for bipedal animals. A quadruped (or a stool) can stand on three small feet and be stable, but if an animal is to stand on one or two feet without having to perform a skilled balancing act, those feet must be reasonably large. Most birds have four rather long toes, with one pointing backwards. This is convenient for grasping and particularly for perching. There is a simple mechanism first described by Borelli in 1680, which helps to keep the foot firmly clamped round a branch. The muscles that bend the toes lie beside the tibiotarsus, with their origins near the knee joint. Their tendons run round the back of the ankle joint so that bending of the ankle under the weight of the body tends to tighten the tendons and bend the toes.

Newts and lizards use their back muscles to bend their bodies from side to side as they run, so lengthening their stride (p. 249). The use which mammals make of their back muscles when they gallop will be described in Chapter 12. There is no obvious way in which birds could use back movements in running, and if the back were flexible it would have to be kept stiff by muscle tension. Quite a long section of the vertebral column has its vertebrae fused to each other and fixed rigidly to the pelvic girdle, and needs no muscle to stiffen it. There is consequently very little muscle between the ilia and the skin. The shape of the pelvic girdle is very peculiar, with hollows on either side of the vertebral column ventral to the ilia, which house the kidneys. It will be seen when the fossil *Archaeopteryx* is described, that this peculiar girdle has evolved from one quite like the girdles of ornithischian dinosaurs.

The legs of birds show clear adaptations to various ways of life. Birds such as the heron (*Ardea*) which wade to find their food tend to have long legs. At the opposite extreme the swift (*Apus*) spends virtually all day on the wing and has very short legs. Birds of prey have strong legs with large muscles and sharp claws which they use to seize their prey (Fig. 11-7). Birds which swim usually have webbed feet, or feet with flaps on either side of the toes.

Ducks have webbed feet, and use them for swimming. Mallard ducks (*Anas platyrhynchos*) have been trained to swim in a channel (built in a laboratory) against a current of variable speed, so as to remain stationary. This was achieved by providing food at a particular point, much as in the experiments with pigeons flying in a wind tunnel (p. 345). Arrangements were made for measuring the rate at which they used oxygen which was found to increase sharply as the water speed was increased towards the maximum swimming speed of about 0.7 m s^{-1}. At this speed they use 4 cm^3 oxygen $\text{g}^{-1} \text{ h}^{-1}$, which is low compared to the $14 \text{ cm}^3 \text{ g}^{-1} \text{ h}^{-1}$ used in flight by *Anas rubripes* of similar size (p. 353. The oxygen consumption of flying mallard seems not to have been measured.) This difference in maximum oxygen consumption can be related to the difference in size of the muscles: the flight muscles of the mallard are 2.8 times as heavy as the leg muscles. One might suppose that a mallard with much larger leg muscles would be able to swim much faster, but this is probably not the case. The drag on a ship (and presumably on a duck) increases sharply at high speeds as the wavelength of the bow wave approaches the hull length. The maximum practicable speed fixed in this way is proportional to the square root of the hull length. The maximum swimming speed of the mallard seems to be close to the limit for a craft of its hull length, and larger muscles would be unlikely to improve it much.

BEAK AND SKULL

Like chelonians and some dinosaurs, modern birds have toothless, keratin-covered jaws. These jaws are not broad as in most reptiles, but form a generally narrow beak. Toothlessness may make the beak more suitable for preening feathers, but some fish-eating ducks such as the merganser, *Mergus*, have serrations on their beaks which serve like teeth to get a firm grip on prey. The narrowness of the beak has made it possible for the eyes of some birds to be arranged so as to have a forward view, with the fields of vision of the left and right eyes overlapping. This makes possible stereoscopic judgement of distance. Falcons, swifts and hummingbirds have two foveae (areas of detailed vision) on each retina: one fovea looks laterally and the other (which is presumably used for stereoscopy) faces forward.

Bird beaks are generally kinetic like the jaws of lizards (p. 301), but the mechanism is little more elaborate. The quadrate has movable joints with the cranium and with the palate and jugal bar (Fig. 11-16). There are flexible regions at the base of the bill, in the palate and in the jugal bar lateral to it, which act as a third joint. There is no joint between the frontal and parietal bones as in lizards but instead there is flexible bone further forward, between the cranium and the top of the bill. The hinges or flexible regions and the way they work are indicated in Fig. 11-16*b*. If the palate and jugal bar are pushed forward, the upper jaw is tilted up. There is a ligament running posterior to the eye, from the postorbital process to the lower jaw. When the lower jaw is depressed the ligament is tightened and the posterior end of the lower jaw swings up, pushing the quadrate forward and raising the upper jaw. Depressing the lower jaw raises the upper one (but raising the upper jaw does not force the lower one to fall). If the upper jaw is heavy enough or the joint between it and the cranium stiff enough, the weight of the lower jaw may not be enough to raise the upper one, and no muscle tension is needed to keep the mouth shut. The mouths of crows (*Corvus*) and probably many other birds are kept closed in this way without muscular effort.

Not all birds have exactly the same jaw mechanism: the description above is of the commonest type. One variant is found in birds such as the

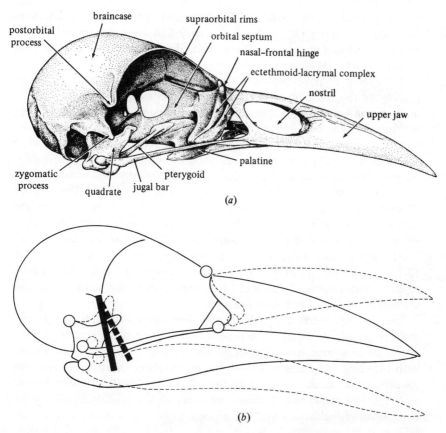

Fig. 11-16. (*a*) Skull of a crow, *Corvus*. From W. J. Bock (1964). *J. Morph.* **114**, 1–42.
(*b*) A diagram of the same skull, showing the position of the hinges and ligament referred to in the text, and movements which occur when the lower jaw is depressed.

woodcock (*Scolopax*) and kiwi (*Apteryx*), which use long bills to probe soft ground for earthworms. These birds do not have to force the whole bill open in the ground (which would require considerable strength) but because the flexible parts of the upper jaw are set well forward they can slide the palate forward and open the bill at the tip alone, to seize a worm.

Other birds have specialized bills for other feeding habits. Hawfinches (*Coccothraustes*) and other birds which crack open seeds to get the kernels have short bills which are very deep (and strong) at the base. Though hawfinches are small, weighing about 55 g, they can crack olive stones. This requires forces of around 60 kg wt. Birds of prey also have short, strong bills, but these are hooked and sharp. They are used to tear the prey into pieces small enough to swallow. By contrast birds which feed on insects generally have small bills. Warblers and others which search leaves and branches and pick up the insects they find have slender bills, while swallows (*Hirundo*) and other birds which intercept insects in flight tend to have short wide ones.

Birds seem to have evolved from archosaurs with diapsid skulls (see p. 294) but they have no postorbital bone and consequently have generally no partition between the upper and lower temporal fenestrae, or between the fenestrae and the orbit. (The postorbital process shown in Fig. 11-16*a* is a projection of the frontal bone.) Though there are sutures between the skull bones of embryos, many of them close in the course of development so that most adult bird skulls show very few sutures.

The large size of bird brains is obvious from their skulls. For animals of similar body weight, there is a very great difference in brain weight between birds and mammals on the one hand, and lower vertebrates on the other. The relationship between brain size and body size was discussed on p. 319. If the brain weights of fishes and amphibians were plotted in Fig. 10-16 they would lie in and immediately around the area occupied by reptile brain weights. If those of birds were added they would lie in and immediately around the mammal area.

The complex behaviour of birds and their fine co-ordination of movement (especially in flight) have probably only been made possible by the evolution of a large brain. Much of the complex behaviour is associated with breeding and helps to ensure the cooperation required between the parents, to maintain the temperature of the eggs without either parent starving. The vagueness of these comments on the relationships between co-ordination, behaviour and brain size suggest it as a field for future research.

The skulls of many sea birds have marked depressions above the eyes. These house salt glands, which function in the same sort of way as the salt glands of reptiles (p. 278). Reasons why salt glands are likely to be particularly valuable to marine animals which cannot produce concentrated urine were given on p. 279. Birds and mammals, unlike reptiles, can produce urine of higher osmotic concentration than the blood, but birds cannot achieve really high concentrations. The mechanism of concentration is apparently the same in mammals and birds, and is explained on p. 394.

SONG

Birds do not have vocal cords at the top of the trachea, like frogs (p. 258) and mammals. Instead they have a structure known as the syrinx at the junction of the trachea with the bronchi (Fig. 11-17). The walls of the

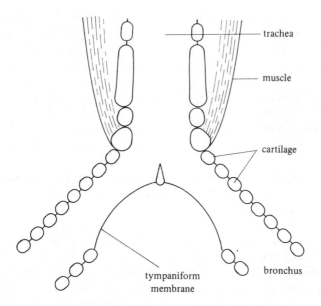

Fig. 11-17. A diagram showing the position and main parts of the syrinx of a typical bird.

trachea are stiffened by rings of cartilage, and those of the bronchi by incomplete rings. The first few rings of each bronchus have bigger gaps than the rest, so that there is quite a large patch of unstiffened membrane in the wall of each bronchus. This is the tympaniform membrane. It is sometimes exceedingly thin.

The sound-producing mechanism has been demonstrated with excised syringes.* The syrinx was enclosed in a chamber so that the pressure around it could be varied, and at the same time air could be blown through from the bronchi. So long as the chamber was kept above atmospheric pressure (making the tympaniform membranes cave in) blowing made the membranes vibrate and produced a sound. If the pressure was allowed to fall to atmospheric, blowing made the membranes bulge outwards and produced no sound. In intact birds one of the air sacs rests against the membranes, and apparently pressure is applied through it.

It seems that the membrane, caving in under pressure, partially blocks the bronchus. Air blown through the bronchus sets it vibrating at its resonant frequency, so that the air passage is alternately widened and con- stricted. Consequently the flow of air is not steady but pulsating, with the same frequency as the membrane. Thus the sound is produced.

The fundamental resonant frequency of a uniform circular membrane

* Syringes and syrinxes are alternative plurals of syrinx.

of mass m, under tension T, is $0.68 \ (T/m)^{1/2}$. Tympaniform membranes are small and thin, so m is low and high fundamental frequencies are possible. If there were water on one or both sides of the membrane it would vibrate with the membrane, increasing its effective mass and decreasing the frequency. The muscles of the syrinx can presumably alter the tension and so the frequency of the membrane. The syrinx shown in Fig. 11-17 has only a single pair of muscles running from the trachea to the bronchi, which presumably tighten the membranes by splaying the bronchi apart. Many birds, including most Passeriformes, have numerous syringeal muscles.

Pressures up to $35 \ cmH_2O$ have been recorded in the lungs of cocks while they were crowing. The pressures involved in normal respiration are far less, about $1 \ cmH_2O$.

The variety of bird song is well known. Some birds produce complexly patterned phrases, and have a remarkable range of pitch and sound quality. The subtleties of sound quality are not easily appreciated by man simply by listening, and Fig. 11-18 shows an example of what birds can achieve. The unprocessed waveform (top left) looks irregular, but this is because the bird is simultaneously producing different notes with its two tympaniform membranes: it is singing a duet with itself. The two notes can be separated from a recording by electronic filtering, and the separated waveforms (right) are much more regular than the original. The higher component (at 4–5 kHz) is not simply a harmonic of the lower one (at 2–3.5 kHz), which is why they are believed to be produced separately, one in each bronchus. Neither component is steady either in amplitude or, as the graph shows, in frequency. The amplitude and frequency of the higher component are modulated (i.e. they are fluctuating) in a rather complex way, the modulating frequency being about 140 Hz. Only one pulse of the lower component is shown but there were twelve of these pulses, each less loud than its predecessor, in a note lasting 0.4 s. The same species sings trills, producing short pulses of sound alternately from the two tympaniform membranes, tuned to different notes.

Sound quality can be modified by harmonics as well as by modulation. Any harmonics which may be present are very weak in the song fragment illustrated in Fig. 11-18, and in many other bird songs. However many bird songs, particularly of the type described as calls rather than songs, are rich in harmonics.

There seems so far to be no firmly based knowledge of the mechanisms used by birds to control the quality of their sound output.

Birds use numerous calls to co-ordinate the behaviour of members of the same species. Some calls are used to keep flocks together as they move. Others give warning of predators, or are used in courtship. About fifteen

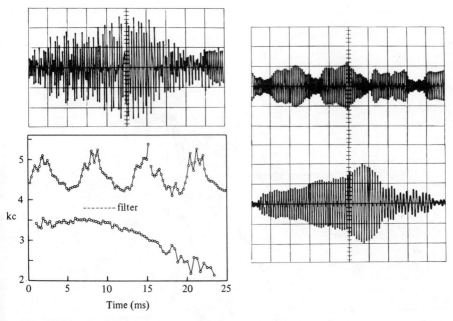

Fig. 11-18. An excerpt lasting 0.025 s from the song of a Wood thrush (*Hylocichla mustelina*). *Top left*, an oscillogram of the excerpt. *Right*, oscillograms of two components of the excerpt, separated by electronic filtering. *Bottom left*, graphs of frequency against time, for the two components shown on the right. From C. H. Greenewalt (1968). *Bird song, acoustics and physiology.* By permission of the Smithsonian Institution Press, Washington.

distinct calls have been recognized for each of several species of Passeriformes which have been studied carefully. Song is used in establishing and defending territories and often also as part of the sexual display that co-ordinates the breeding behaviour of pairs of birds.

THE EARLIEST BIRD
Class Aves, subclass Archaeornithes

The earliest known fossil birds are three reasonably complete specimens and a single feather from the Upper Jurassic (Fig. 11-19). All belong to the same species, *Archaeopteryx lithographica*. They are so different from other birds that they are given a subclass to themselves, with all other birds in one other subclass. The stone they are preserved in is a limestone, so fine grained as to be suitable for making lithographic stones. This exceptionally suitable stone retains quite detailed impressions of the feathers, which seem to be identical with modern feathers. The main wing feathers are arranged in exactly the same way as in modern birds.

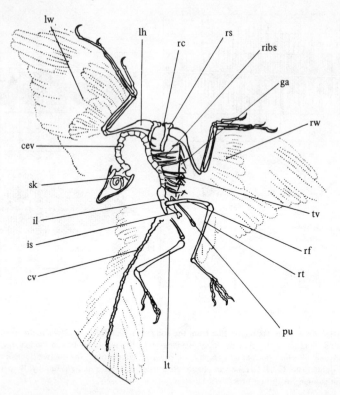

Fig. 11-19. Fossil of *Archaeopteryx lithographica* (the Berlin specimen) cv, Caudal vertebrae; cev, cervical vertebrae; ga, gastralia; il, ilium; is, ischium; lh, left humerus; lt, left tibia; lw, left wing; pu, pubis; rc, right coracoid; rf, right femur; rs, right scapula; rt, right tibia; rw, right wing; sk, skull; tv, trunk vertebrae. From G. R. de Beer (1954). *Archaeopteryx lithographica*. British Museum, London.

If impressions of feathers had not been preserved, *Archaeopteryx* might well have been identified as a reptile. The hand has only three fingers and though they are longer than one would expect in a reptile the metacarpals are not fused together as in modern birds. The metatarsals were long, as they are in birds and in many small dinosaurs, but do not seem to have been fused together as in modern birds. The first toe pointed backwards, but was shorter than is usual in modern perching birds. The pelvic girdle of *Archaeopteryx* is more like those of bird embryos and ornithischian dinosaurs (Fig. 10-12*b*) than those of most modern birds. The tail is long, with about twenty vertebrae, and has a row of feathers on either side.

The skull of *Archaeopteryx* is very like those of archosaurs. It has not got a bird-like bill but has teeth which are housed in sockets, and has the typical fenestrae between eye and nostril and in lower jaw. No specimen has the back of the skull well enough preserved to be clear whether it was diapsid. However, the size of the cavity for the brain is apparent from a natural internal cast, and it seems clear that the brain itself was intermediate in size between those of modern reptiles and birds of similar weight.

The sternum was small, apparently without a keel, and it has been questioned whether *Archaeopteryx* could fly. However, if estimates that its weight was 200 g and its wing area 480 cm^2 are correct, its wing loading was about average for modern birds of similar size. Lack of a keel need not necessarily mean small wing muscles.

MODERN BIRDS
Subclass Neornithes

All birds except *Archaeopteryx* are included in the subclass Neornithes. They are in most respects remarkably uniform in structure though they differ considerably in ways of life, and particularly in feeding habits. Peculiarities of bill and foot structure which are obviously adapted to particular ways of life have already been described. Because these peculiarities are easily observed, and are found in birds of otherwise not too different construction, the birds provide an excellent example of adaptive radiation. One can for instance show an elementary class how hawks, waders, ducks, finches and warblers are variations on a common structural theme, adapted to different ways of life.

A group of modern birds which in some respects seem to stand apart from the rest consists of the ostrich, emus, rheas, cassowaries and kiwis. These birds are sometimes referred to as ratites. None of them can fly (some zoologists include the tinamous, which can fly, among the ratites). Though their wings are tiny the bones of the hand are fused together in precisely the same peculiar way as in flying birds (except in cassowaries and kiwis, which have only one digit left). It can be inferred that the ratites have evolved from flying birds with normal wings. Their feathers are fluffy because the barbs do not interlock. Their pelvic girdles are rather like that of *Archaeopteryx*. Far more sutures persist in the adult skull than in other modern birds, and the mechanism of upper jaw kinesis is peculiar.

Many of these peculiarities can be interpreted as neoteny, as the persistence in adults of embryonic or juvenile characters. Neoteny in urochordates and urodeles was referred to on pp. 33 and 263. The chicks of many ground-nesting birds (including domestic fowl) have strong legs and run well while they still have rudimentary wings and cannot fly. Chicks have down feathers, without interlocking barbs, and they have sutures between their skull bones. Pelvic girdles like that of *Archaeopteryx* are found in

bird embryos. It has been argued that because the main points of similarity
of ratites are so largely attributable to neoteny, there is no good reason to
believe that all evolved from a single ratite ancestor. Rather, different
groups of ratites may have evolved by neoteny of different flying ancestors.
On the other hand, it has been claimed that the jaw mechanism peculiar
to ratites and tinamous (which is not easily explained as neotenic) is unlikely to have evolved more than once.

The largest flying birds weigh about 20 kg (p. 350) and larger birds even
with well developed wings would be likely to have difficulty in getting
airborne. Flightless running birds are not limited in size in this way.
Ostriches (*Struthio camelus*) may weigh 150 kg and stand 2.7 m high and
some moas (a group of extinct ratites) were larger still.

Among the many orders of modern birds, the Passeriformes is much the
the largest. It consists mainly of small and medium-sized birds, including
the crows, tits, warblers, thrushes, finches and many others. Their high
metabolic rates have already been noted (p. 333).

FURTHER READING

General

Farner, D. S. & J. R. King (eds.) (1971–2). *Avian biology*. Academic Press, New
York.

George, J. C. & A. J. Berger (1966). *Avian myology*. Academic Press, New York.

Marshall, A. J. (ed.) (1960–1). *Biology and comparative physiology of birds*.
Academic Press, New York.

Yapp, W. B. (1970). *Life and organization of birds*. Arnold, London.

Feathers and body temperature

Crawford, E. C. & G. Kampe (1971). Resonant panting in pigeons. *Comp.
Biochem. Physiol.* **40A**, 549–52.

Dawson, W. R. & J. W. Hudson (1970). Birds. In Whittow, G. C. (ed.) *Comparative physiology of thermoregulation*, vol. 1, pp. 223–310.

Jones, R. E. (1971). The incubation patch of birds. *Biol. Rev.* **46**, 315–40.

Porter, W. P. & D. M. Gates (1969). Thermodynamic equilibria of animals with
environment. *Ecol. Monogr.* **39**, 227–44.

Rijke, A. M. (1970). Wettability and phylogenetic development of feather
structure in water birds. *J. exp. Biol.* **52**, 469–79.

Schmidt-Nielsen, K., F. R. Hainsworth & D. E. Murrish (1970). Counter-
current heat exchange in the respiratory passages: effect on water and heat
balance. *Resp. Physiol.* **9**, 263–76.

Wings and flight

Berger, M., J. S. Hart & O. Z. Roy (1970). Respiration, oxygen consumption
and heart rate in some birds during rest and flight. *Z. vergl. Physiol.* **66**, 201–
214.

Brown, R. H. J. (1963). The flight of birds. *Biol. Rev.* **38**, 460–89.

Nachtigall, W. & B. Kempf (1971). Vergleichende Untersuchungen zur flugbio-logischen Funktion des Daumenfittichs (Alula spuria) bei Vogeln. I. *Z. vergl. Physiol.* **71**, 326–41.
Pennycuick, C. J. (1969). The mechanics of bird migration. *Ibis*, **111**, 525–56.
Pennycuick C. J. (1972). *Animal flight.* Arnold, London.
Tucker, V. A. (1968). Respiratory exchange and evaporative water loss in the flying budgerigar. *J. exp. Biol.* **48**, 67–87.
Tucker, V. A. (1970). Energetic cost of locomotion in animals. *Comp. Biochem. Physiol.* **34**, 841–6.
Tucker, V. A. & G. C. Parrott (1970). Aerodynamics of gliding flight in a falcon and other birds. *J. exp. Biol.* **52**, 345–67.
Weis-Fogh, T. (1972). Energetics of hovering flight in hummingbirds and in *Drosophila. J. exp. Biol.* **56**, 79–104.

Respiration

Bretz, W. L. & K. Schmidt-Nielsen (1972). The movement of gas in the respira-tory system of the duck. *J. exp. Biol.* **56**, 57–65.
Scheid, P. & J. Piiper (1971). Direct measurement of the pathway of respired gas in duck lungs. *Resp. Physiol.* **11**, 308–14.

Legs

Prange, H. D. & K. Schmidt-Nielsen (1970). The metabolic cost of swimming in ducks. *J. exp. Biol.* **53**, 763–78.

Beak and skull

Bock, W. J. (1964). Kinetics of the avian skull. *J. Morph.* **114**, 1–42.

Song

Greenewalt, C. H. (1968). *Bird song: acoustics and physiology.* Smithsonian Institution Press, Washington.
Thorpe, W. H. (1961). *Bird-song.* Cambridge University Press, London.

The earliest bird

de Beer, G. R. (1954). *Archaeopteryx lithographica.* British Museum (Natural History), London.
Jerison, H. J. (1968). Brain evolution and *Archaeopteryx. Nature, Lond.* **219**, 1381–2.
Yalden, D. W. (1971). The flying ability of *Archaeopteryx. Ibis*, **113**, 349–56.

Modern birds

Bock, W. J. (1963). The cranial evidence for ratite affinities. *Proc. XIII Int. ornithol. Congr.* 39–54.
de Beer, G. R. (1956). The evolution of ratites. *Bull. Brit. Mus. (Nat. Hist.) Zool.* **4**, 59–70.

12

Mammals and their ancestors

This chapter is about mammals in general, and about how they evolved from reptiles. The two chapters which follow deal with the adaptations of mammals to particular ways of life.

Mammals are homoiothermic, maintaining more or less constant high body temperatures generally a little below those of birds. Temperatures of 36–40 °C are usual in eutherian mammals (that is, in the great majority of mammals), 34–37 °C in marsupials and 30–33 °C in monotremes. The resting metabolic rates of placental mammals are close to those of non-passerine birds of similar size (Fig. 11-4).

Though mammals and birds both maintain high body temperatures, there are some differences between them in the mechanisms of temperature maintenance and control. The most obvious is that mammals do not have a heat-insulating covering of feathers. Instead most have fur, which traps a layer of air next to the body just as feathers do, and provides about the same degree of heat insulation as plumage of the same thickness. Like feathers, hairs consist of keratin (though in the α- rather than the β-form) and are formed from the epidermis in follicles in the skin. They are of course simple strands, but hairs lying parallel to one another have water-repellent properties like the parallel barbs of feathers. The waterproofing effect when the animal swims seems generally less good, because the hairs are not held at fixed spacing in the way barbs are held by barbules, so they tend to clump together and allow water to penetrate to the skin.

The water-repellent properties of plumage depend in part on the secretion of the oil gland. Hairs are coated with an oily secretion from sebaceous glands, which discharge into the hair follicles and so directly onto the hairs. Most hairs do not stand vertically, and the lie of the hair on different parts of the body is generally arranged so that when the animal is standing, water runs off easily. Ordinary hairs have sensory nerve endings around the follicle, but most mammals have a few stiff hairs (vibrissae) on the face, which have a particularly rich nerve supply and are very sensitive to touch and air movements.

Small strands of smooth muscle in the dermis attach to the hair follicles, and their contraction erects the hairs. The thickness of the fur is adjusted by means of these muscles just as the thickness of plumage is adjusted by fluffing out the feathers or flattening them against the body. This is one of the mechanisms of temperature regulation. Others are shivering (which increases heat production), panting (which increases evaporative heat loss) and adjustment of blood flow to uninsulated parts of the skin. In man, flow to the general body surface is adjusted and in the dog flow to the tongue, which hangs out while the dog is panting. Essentially the same mechanisms are used by birds but mammals have an additional one, sweating. Sweat is produced by tubular glands which develop from the epidermis but grow down into the dermis. It is discharged through pores in the surface of the skin. It is a solution of variable concentration of salts in water.

Men working in hot surroundings may produce two litres or more of sweat per hour. If the air is humid, much of the sweat may drip off the body before evaporating. If, however, the water evaporates from it first, most of its latent heat of vaporization is taken from the body. If two litres evaporates each hour the latent heat amounts to $5 \times 10^6 \, \mathrm{J \, h^{-1}}$. This quantity of heat would be produced by metabolism using $2.5 \times 10^5 \, \mathrm{cm^3}$ oxygen $\mathrm{h^{-1}}$, which for a 70 kg man amounts to 60 $\mathrm{cm^3}$ oxygen $\mathrm{kg^{-1} \, min^{-1}}$. This is about sixteen times the basal metabolic rate, and about equal to the maximum oxygen consumption which an athlete can achieve in severe exercise. The cooling potential of sweating, either in exercise or in hot sunny conditions, is tremendous. The cost is also high. A man losing 2 l sweat $\mathrm{h^{-1}}$ would soon be thirsty and after about $3\frac{1}{2}$ h without a drink he would have lost 10% of his body weight, and would be helpless. High sweating rates cannot be maintained unless plenty of drinking water is available. Sweating also involves salt loss at rates which can be serious. Miners and stokers used to suffer from heat cramps, until it was realized that this complaint was due to salt depletion by sweating and could be prevented by giving salt in the drinking water.

Sweating is much less important to some other mammals than it is to man. Dogs have sweat glands but apparently sweat little: panting with the tongue hanging out seems to be, for them, a far more important means of promoting heat loss by evaporation.

Mammals resting in cold environments can increase the heat output of their muscles, either by shivering or by another process; the metabolic rate of the muscle can be increased without any movement occurring. Brown fat may also be an important source of extra heat. When rats are exposed to cold they increase their heat production and the temperature of their brown fat rises above that of surrounding tissues. Heat production cannot be increased so much after most of the brown fat has been removed surgically. Not all mammals have brown fat, which is particularly common in new-born mammals and in ones which hibernate. Body temperature may fall very low in hibernation and a great deal of heat is needed to warm the body on arousal. It has been estimated from the weight of fat lost during arousal of ground squirrels (*Citellus*) from hibernation that most of the heat required is supplied by the brown fat.

Homoiothermic animals necessarily pass through a stage when they are too small to maintain their own body temperature. Birds keep their embryos at an appropriate temperature by incubating their eggs, and in some cases also brood their young (p. 337). Mammals (except monotremes, p. 404) are viviparous, so the embryo is maintained within the mother at her body temperature. Nevertheless, many new-born mammals are unable by themselves to maintain their body temperature in cool conditions. Marsupials are particularly small and immature at birth but are kept close to the mother's body temperature in the pouch until they have grown considerably.

Birds and mammals are in general active animals, capable of using oxygen very much faster than other vertebrates of similar size. Birds achieve fast oxygen uptake by means of peculiar lungs (p. 354). Mammals' lungs are much less peculiar, though their surface area is large. The alveoli are small, with an average radius in rat lungs of 20 μm (compare this with the *minimum* value of 50 μm in the lungfish *Protopterus*, p. 212). They do not simply form a spongy layer near the outer surface of the lung as in *Protopterus* and many amphibians and reptiles, but the whole lung is spongy. The total surface area of the lungs has been estimated as 5 m^2 kg body weight)$^{-1}$ for mice, but only 0.25 m^2 kg^{-1} for frogs.

Breathing depends partly on rib movements like those of lizards but partly, and in some mammals predominantly, on the diaphragm. This is a muscular partition in the body cavity separating the thorax (containing heart and lungs) from the abdomen (containing liver, stomach and intestines). When the lungs are fairly empty the diaphragm is domed (Fig. 12-1). Contraction of its muscles flattens it, enlarging the thorax and so drawing air into the lungs. When it relaxes the process is reversed, largely by the elasticity of the lungs.

More forceful breathing out involves contraction of the abdominal muscles, which draw in the belly and so force the diaphragm up.

There is no immediately obvious advantage of the diaphragm over moving ribs, as a mechanism of breathing. The abdominal wall must move with the diaphragm (since the contents of the abdomen are more or less incompressible) so though the diaphragm itself is light the total mass of tissue to be moved at each breath is not greatly altered. There may, however, be a little saving of mass. A reptile breathing by rib movements needs ribs or other stiffening structures all along the wall of the body cavity: if the abdominal wall were flabby and caved in when the thorax expanded, the rib movements would be ineffective. A mammal with a diaphragm, however, can have a lighter abdominal wall without ribs. Loss of posterior ribs was possibly a prerequisite for the evolution of galloping; this characteristically mammalian gait involves up-and-down bending of the abdominal region (see p. 381) which might not be possible if there were ribs in the abdominal wall.

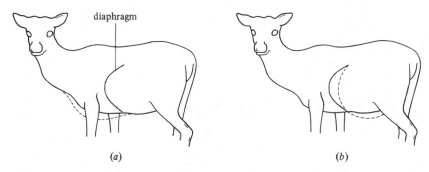

Fig. 12-1. Diagrams of a mammal showing how (*a*) rib movements and (*b*) contraction of the diaphragm contribute to the volume changes of the thorax involved in breathing.

In mammals as in birds, oxygenated and deoxygenated blood are kept entirely separate in a completely divided heart. A systemic artery has also been lost but while birds have lost the left one mammals have lost the right one.

REPRODUCTION

The peculiar monotremes lay eggs which are quite large, though smaller than those of reptiles of similar size (p. 404). The other mammals are viviparous and their ova are tiny, smaller even than those of teleost fish. They range in diameter from 0.1 to 0.25 mm. Like other small chordate ova these divide completely into cells at the beginning of development (only a small part of the large ovum divides in selachians, reptiles, birds and monotremes). Later, however, an amnion, chorion and allantois develop, and a yolk sac which is empty of yolk. These membranes are arranged in essentially the same way as in reptiles and birds, though they sometimes develop quite differently. The reptile ancestors of mammals presumably laid large eggs, in which these membranes served their usual reptilian functions. They presumably had the usual rich blood supply to the yolk sac, responsible

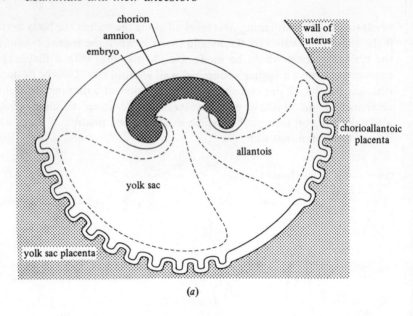

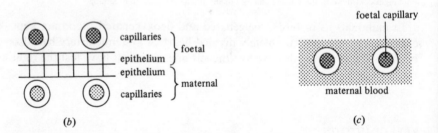

Fig. 12-2. (*a*) Diagrammatic section through a mammal embryo in its mother's uterus, showing how placentae are formed by embryonic membranes.

(*b*), (*c*) Diagrammatic sections through two types of placenta.

for collection of food materials from the yolk, and to the allantochorion, responsible for gaseous exchange through the shell. In mammals (except monotremes) one or both of these membranes become firmly attached to the wall of the uterus, forming a placenta (Fig. 12-2*a*). In most marsupials only the yolk sac forms a placenta, but *Perameles* (the bandicoot) and a few others have chorioallantoic placentae as well. In eutherian ('placental') mammals the chorioallantois forms the main placenta. Yolk-sac placentae also develop in some viviparous selachians (p. 120) and both yolk-sac and chorioallantoic placentae in various viviparous lizards and snakes.

The blood of mother and foetus come close together in the placenta. In the most primitive arrangement in which the uterus and embryonic membranes simply interlock (Fig. 12-2*b*), the two bloodstreams are separated by the foetal capillary wall, connective tissue and epithelium and by the maternal epithelium, connective tissue and capillary wall. This is the situation over most of the area of the placenta in pigs and sheep. In the most advanced arrangement, found for instance in the rabbit (*Oryctolagus*), bare foetal capillaries run through spaces filled with maternal blood (Fig. 12-2*c*) so that only a single layer of cells separates the bloodstreams. In all cases diffusion between the bloodstreams supplies the foetus with oxygen and foodstuffs, and removes carbon dioxide. Complex interlocking between maternal and foetal tissues ensures that the area available for diffusion is large: it has been estimated as 12 m² in the human placenta. The oxygen concentrations in maternal blood arriving at and leaving the placenta have been measured in various species. The results seem to show that the thinner the barrier between maternal and foetal blood, the higher the percentage of the oxygen in the blood that diffuses to the embryo: 70% of the oxygen is given up in the placenta of the rabbit, but only 30% in the placenta of sheep.

The foetal blood leaving the placenta necessarily has a lower partial pressure of oxygen than the maternal arterial blood. If its haemoglobin had the same properties as maternal haemoglobin it could never become saturated with oxygen. It does not have the same properties, but becomes saturated at lower partial pressures: the haemoglobin of a late human foetus becomes 50% saturated at about 0.025 atm oxygen, and that of its mother at about 0.035 atm.

In mammal embryos, as in bird ones (p. 358), the wall between the left and right sides of the heart is incomplete, and there is a connection between the systemic and pulmonary arteries. Little blood flows through the lungs until birth, and it is only after birth that the pathways through the left and right sides of the heart become completely separated. The arteries to the placenta of the mammal embryo, like the arteries to the allantois of the chick, branch from the aorta rather than from the pulmonary arteries.

After birth the young are fed at first on milk, secreted by the mammary glands of the mother. These are among the most characteristic features of the mammals, and indeed it is from the Latin word for breast that the name of the class is derived. Mammary glands are functional only in females, but are present as rudiments in males. They vary in number from two to more than twenty, and are most numerous in species which give birth to large litters. They may be anterior in position as in man or posterior as in the cow, or if numerous may be spread all along the ventral surface of the trunk. Each gland consists of branching tubules opening at a nipple (except in monotremes, which have no nipples). The tubules develop as invaginations of the ectoderm, like sweat glands and sebaceous glands. It is widely believed that mammary glands have evolved by modification of apocrine sweat glands, but the evidence is not entirely conclusive.

Milk is an aqueous emulsion of fat globules, ranging in concentration from about 1.5% by weight of fat in the horse to about 20% in the reindeer (*Rangifer*) and in the spiny anteater (*Tachyglossus*, a monotreme) and even more in whales and some seals. The aqueous phase of the emulsion contains sugars, salts and proteins, in solution or in colloidal suspension. The sugar in the milk of eutherian mammals is almost entirely lactose, but in marsupial milk other sugars are plentiful as well. The proteins are known as caseins. They are acid proteins containing phosphorus, and are present as calcium salts. Calcium and phosphorus are also present as colloidal calcium phosphate and citrate. These elements are needed by the young animal for building bone, and though they are present in milk in quantities similar to those of sodium and potassium they contribute little to its osmotic pressure, because they are present so largely in protein and colloidal salts. The osmotic pressure of the milk is about equal to that of the animal's blood.

Feeding the young on milk might be expected to be wasteful of energy. Energy from the food used to produce milk is inevitably lost in the processes of digestion and synthesis in the mother. Some of the energy from the milk must in turn be lost in digestion in the young. If the young could feed directly on the food taken by the mother, instead of receiving the energy indirectly as milk, an energy-wasting stage would (seemingly) be eliminated. Would energy really be saved?

The economic importance of cattle has stimulated a great deal of research on their growth and milk production, which makes possible an attempt to answer the question. Experiments have shown that the processes of milk synthesis by cows and of milk utilization by calves, are both remarkably efficient. Cattle have been given measured quantities of food of known heat of combustion, so that their total energy intake is known. The heats of combustion of their faeces and urine and of the methane produced by bacterial fermentation in the stomach (p. 401) have been measured. In experiments both with young steers and with lactating cows it has been found that of every 100 J of energy taken in as food, about 30 J is lost in the faeces, 3–4 J in the urine and 7–9 J as methane and as heat released in the fermentation process. This leaves about 60 J available for use in metabolism, growth or milk production, and this 60 J is referred to as the metabolizable energy of the food. The efficiency with which it could be used for growth was determined in experiments with young steers, which were fed varying daily rations of the same food. Suppose that an animal given R J daily grows at such a rate as to incorporate G J daily in its body, and that when fed $(R + \Delta R)$ it incorporates $(G + \Delta G)$. Then, if energy is being used at a constant rate for maintenance (as distinct from growth), the efficiency of the growth processes must be $\Delta G/\Delta R$. It was found for steers of a wide range of ages, that (within limits) every 100 J of food energy (60 J of metabolizable energy) in excess of the requirements for maintenance could be used to add 30 J to the body by growth. Experiments with adult cows showed similarly that 100 J food energy could be used to produce 42 J as milk, so milk production is a more efficient process than growth.

Milk is also highly digestible. Calves lose in their faeces and urine only about 5% of the energy fed to them as milk. Bacterial fermentation is not involved in milk digestion, so the losses associated with it are not incurred. Thus 42 J fed as milk provides 40 J metabolizable energy.

Even so, milk feeding compares poorly with direct feeding as a source of energy. The experiments described above show that 100 J supplied as food to a cow can yield 60 J metabolizable energy to the cow itself, but only 40 J metabolizable energy to a calf drinking the cow's milk.

The comparison is much more favourable to milk feeding if growth is considered. Growth rates of calves fed different daily rations of milk have been compared. In this way it was shown that of every 100 J given as milk in excess of the requirements for maintenance about 75 J was incorporated in the body by growth. Considering the processes of milk production and utilization together, 100 J fed to a cow can yield about 42 J as milk that can be fed to a calf to produce about $0.75 \times 42 = 32$ J growth. Thus 100 J as grass, etc. can produce about the same amount of growth, whether fed directly to growing cattle or to the mothers of unweaned calves.

Very large quantities of energy may be used in milk production. The energy output as milk of a moderate milking cow is about 0.9 times the basal (non-lactating) metabolic rate. This implies that the cow must eat more than twice as much as if it were not producing milk. Domestic cattle have, of course, been bred selectively for milk production, but the quantities of energy involved are also large in wild mammals. The quantity of food eaten and assimilated daily by female Bank voles (*Clethrionomys*) kept in a laboratory rose during lactation to more than twice the value for non-reproducing females. Pregnancy requires much less additional energy than lactation, both in cows and in Bank voles.

The mechanism of taking milk from the teat has been investigated by X-ray cinematography of goats. The milk in the mother's udder was replaced by a barium sulphate suspension which is opaque to X-rays and so showed up in the films. It appears that sucking involves up and down movements of the tongue. The tip of the tongue rises first, pressing the base of the teat against the palate, but eventually the whole teat is squeezed and emptied. It refills again as the tongue is lowered, and the films showed some barium sulphate flowing from teat to mouth while the tongue was still falling, presumably sucked out by reduced pressure in the mouth. Obtaining milk depends on stimulating the mother appropriately as well as on this pumping effect: little milk can be withdrawn through a cannula fitted to a cow's teat until an appropriate stimulus such as a milking action is applied.

LOCOMOTION

The American opossum *Didelphis* is a primitive marsupial which walks and stands in the manner which seems to be typical of small mammals in general. Its movements have been studied by X-ray cinematography (Figs. 12-3 to 12-6). Similar studies of tree shrews (*Tupaia*), rats (*Rattus*) and ferrets (*Mustela*) have shown that all stand and walk in very much the same way.

The opossum generally stands with its forelimbs in a position (Fig. 12-3, 4) with the foot vertically below the shoulder joint. When it walks it sets the foot down in this position or a little more anteriorly, and in the course of the step the body moves laterally so that the foot is more nearly below the vertebral column (position *c*). Thus the shoulders sway from side to side as the animal walks, bringing the chest over whichever forefoot is on the ground. As the limb is brought forward again it swings laterally (position *d*) so that it is well out of the way of the other forelimb which is on the ground. The movement which shifts the body over the foot and the reverse movement that shifts the foot to the side as it is brought forward involve rotation of the humerus about its own long axis. This is obvious in Fig. 12-3, from the changing views of the ridge known as the deltopectoral crest (dp) and of the distal end of the humerus.

These forelimb movements are very different from those of typical amphibians and reptiles, which stand and walk with their forefeet far lateral to the shoulder joints (Fig. 8-11). To support their weight as they stand and walk in this way, amphibians and reptiles need strong muscles running ventrally from the humerus across the chest. The main muscles involved are the pectoralis and supracoracoideus muscles, which have their origins on the sternum and on the ventral parts of the pectoral girdle. The coracoids are substantial bones running between the shoulder and the sternum, ensuring that the pectoralis muscles have the required effect and do not merely pull the shoulders towards the mid-line (Fig. 8-15*b*, *d*).

The typical mammalian manner of standing and walking does not require strong muscles across the chest. Monotremes have pectoral girdles

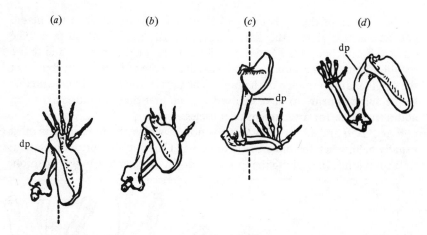

Fig. 12-3. Dorsal views drawn from X-ray cinematograph films, of the left forelimb skeleton and scapula of the opossum *Didelphis*. Four stages of a walking step are shown. The broken lines are parallel to the long axis of the body. dp, Deltopectoral crest of the humerus. From F. A. Jenkins (1971). *J. Zool., Lond.* **165**, 303–15.

very like those of lizards but in all other mammals the coracoid is reduced to a vestige (usually fused to the scapula). There is no bone connecting shoulder to sternum except the clavicle (collarbone), which is itself missing in many mammals. Even when the clavicle is present the joints at its ends are movable so the scapulae (shoulder blades) are not fastened rigidly together. The scapulae move relative to each other and to the underlying ribs, as a mammal walks. In fact in the stride represented in Fig. 12-4 there is very little movement at the shoulder joint while the scapula and humerus swing as a unit through about 45 °. Rotation about a high pivot (between scapula and ribs) rather than a low one (the shoulder joint) lengthens the stride a little.

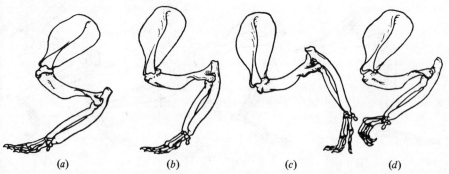

Fig. 12-4. As Fig. 12-3, but showing the same four positions in lateral view. From F. A. Jenkins (1971). *J. Zool., Lond.* **165**, 303–15.

The scapula of the opossum and of other mammals is not a flat blade, but has a tall ridge (the scapular spine) on its lateral surface. The clavicle is attached to the anterior edge of the scapula in reptiles and to the scapular spine in mammals, so the scapular spine is believed to represent the original anterior edge of the bone. The part of the blade that is anterior to the spine appears to have evolved for the first time in mammals as an additional area for attachment of muscles.

Figs. 12-5*a* and 12-6*a* show the position in which the hind limb is most usually held, when the opossum is standing. The foot is anterior and lateral to the hip joint. In walking the foot is set down more or less in this position,

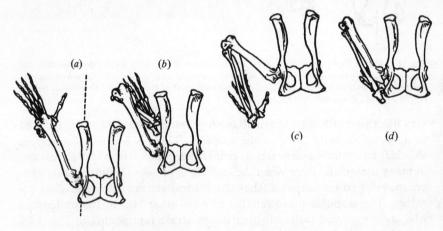

Fig. 12-5. Dorsal views drawn from X-ray cinematograph films of the left hind limb skeleton and pelvic girdle of the opossum *Didelphis*. Four stages of a walking step are shown. The broken line is parallel to the long axis of the body. From F. A. Jenkins (1971). *J. Zool., Lond.* **165**, 303–15.

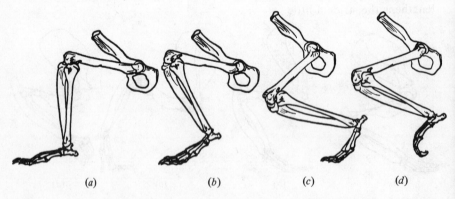

Fig. 12-6. As Fig. 12-5, but showing the same four positions in lateral view. From F. A. Jenkins (1971). *J. Zool., Lond.* **165**, 303–15.

and the body shifts to a position more nearly over it as the step proceeds (Fig. 12-5*b*): the hips sway as well as the shoulders.

Figs. 12-5, 12-6 also show the shape of the pelvic girdle, and the presence of a heel. These are distinctive features of mammals.

An advantage of the mammalian manner of moving the limbs, over the typical amphibian and reptilian manner, is that it makes a longer stride possible with limbs of given length and mass (see p. 317). The effective length of the limbs can be increased further by holding them straighter, as the dog does (Fig. 12-7). Notice the difference between the plantigrade stance of the opossum and the digitigrade stance of the dog: the opossum stands on the soles of its feet (Fig. 12-4*a*, 12-6*a*) but the dog stands on its toes, with metacarpals and metatarsals nearly vertical and with wrists and ankles well above ground level (Fig. 14-6).

Fig. 12-7 shows the two most important gaits used by dogs, the trot and the gallop. In the trot, pairs of diagonally opposite limbs move more or less simultaneously: frame 9 of the trotting sequence shows the left fore and right hind limbs moving while frame 25 shows the right fore and left hind limbs moving. The pattern is also shown in Fig. 12-8. Notice how similar this is to the lizard pattern (Fig. 8-13*c*). The limbs move in the same sequence in both cases but the lizard, like the newt (Fig. 8-11), uses lateral bends of the body while the dog does not. Lateral bends may extend the stride considerably in an animal which keeps its feet well out on either side of its body, but not in one which walks with them well under the body.

The pattern of the gallop is entirely different. The two forefeet move more or less together and the two hind feet more or less together. The animal travels in a series of leaps. In frame 1 of the galloping sequence it is landing on its forefeet and in frame 5 it has kicked off with them again and has no feet on the ground. In frame 9 it is landing on its hind feet and in frame 13 it is kicking off with them. Frame 17 shows the later part of this leap and frame 21 shows the same position as frame 1. Notice how galloping involves bending the back, not from side to side as in newts and lizards but vertically. The back muscles reinforce the leg action by bending as the dog kicks off with its forelegs and extending as it kicks off with its hind ones. The back is fully bent in frame 5 and fully extended in frame 13. It is only because the forelimbs move together, and the hind limbs together, that back movements can be used in this way.

In the galloping sequence the dog is moving four times as fast as in the trotting sequence. It is achieving this by making 1.5 times as many cycles of leg movements per second and travelling 2.7 times as far in each cycle ($1.5 \times 2.7 = 4$). The remarkable increase in distance per cycle is partly due to each leg swinging through a larger angle, but mainly to reduction of the percentage of the cycle for which each foot is on the ground (Fig.

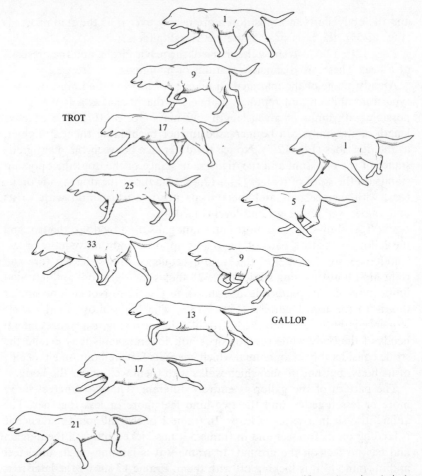

Fig. 12-7. Outlines drawn from cinematograph films of the same dog walking and running. A single cycle of limb movements is shown in each case. The serial numbers in each sequence of the frames reproduced are indicated (both sequences were taken at 64 frames per second).

12-8). In the trot each foot is on the ground for just over 50% of the time. In the gallop each is on the ground for only 25% of the time, and for 20% of the time there are no feet at all on the ground. These leaps are quite modest: films of a greyhound and of a cheetah (*Acinonyx*) galloping show all feet off the ground for 40–50% of the time.

Mammals change from trot to gallop as they increase speed, suggesting that the trot is advantageous at low speeds and the gallop at high ones. Why should this be? Consider an animal with hip and shoulders joints at a height h from the ground, and with limbs of moment of inertia I about these joints. It is running with velocity U. Every time a limb swings back

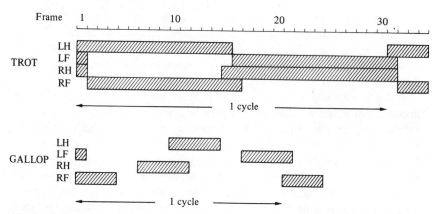

Fig. 12-8. Diagrams based on the same film sequences as Fig. 12-7, showing which feet are on the ground in each frame. Hatching indicates that the foot is on the ground. LF, left forefoot; RH, right hindfoot, etc.

the foot must be accelerated to velocity $-U$ relative to the trunk, so the limb must be given angular velocity $-U/h$ and kinetic energy (relative to the trunk) $IU^2/2h^2$. This energy is lost when the limb is halted in preparation for the forward swing. The same amount of energy could be used to raise the centre of gravity in a leap: if the mass of the body were m it could be raised $IU^2/2h^2m$ g. The duration of the leap would be the time taken to rise to this height and fall back again under the influence of gravity. During the leap the animal would continue to travel at velocity U. It can easily be shown by elementary dynamics, that the distance covered in the leap would be $(2U^2/g\,h)(I/m)^{1/2}$. Notice that this distance is proportional to U^2. The distance covered in a step with the foot on the ground is, by contrast, more or less fixed. As U increases there must come a time when a leap takes the animal further than a step using the same amount of energy. At high speeds the animal can travel faster, for given power output, by interpolating leaps between its steps.

That argument is too simple. In particular it ignores the fluctuations of the velocity of the trunk which occur in running and may use up much more energy than the fluctuations in the velocity of the limbs. The importance of these fluctuations of trunk velocity, and the reasons for them, are emerging from research which is in progress while this is being written. If they had been considered the argument would have been more complicated but the conclusion would have been strengthened.

A great deal of energy is saved in locomotion by tendon elasticity. Kinetic and potential energy lost at one stage during a step may be stored as elastic energy in a stretched tendon and restored to the body in an elastic recoil. The likely importance of tendon elasticity is illustrated by experiments in which the oxygen consumption of men was measured while they repeatedly bent their knees and stood straight again. The exercise needed 30–40% less oxygen if the subjects bounced up immediately from the bent position, than if they paused with knees bent to allow muscle tension to fall. The saving is attributed to elastic recoil of the tendons.

When a man walks he always has at least one foot in contact with the ground, but when he runs he has periods with both feet off the ground. In this respect the difference between human walking and running resembles the difference between the trot and gallop of four-footed mammals. Fig. 12-9 shows measurements of the oxygen consumption of men walking and running at various speeds. The measurements were made by the Douglas bag technique which is described in textbooks of medical physiology: it involves collecting and analysing the expired air. The change from walking to running was made, as is done naturally, at about 2.5 m s^{-1}. Extrapolation of the graphs indicates that they would cross at about this speed, so that above 2.5 m s^{-1} walking would demand more energy than

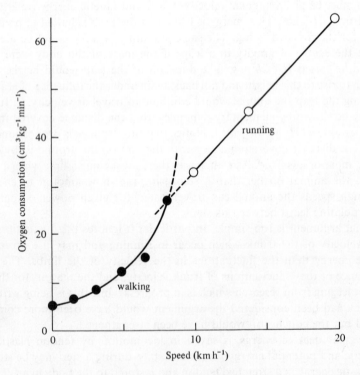

Fig. 12-9. Graphs of oxygen consumption against speed, for men walking and running. Data from R. Passmore & J. V. G. Durnin (1955). *Physiol. Rev.* **35**, 801–40.

running, and below it running would demand more energy than walking. Walking is apparently more economical below the critical speed, and running above it. The same sort of thing could presumably be shown for the trotting and galloping gaits of four-footed mammals, by similar experiments.

The trot and gallop are not of course the only gaits used by mammals. A slow walk is sometimes used, with the legs moving one at a time in the order which allows constant stability (see p. 248). Giraffes and camels use the amble, in which both left legs move together and both right legs together. They have very long legs, relative to the distance from hip to shoulder, which may make the amble more convenient than the trot: if the fore and hind legs of the same side moved out of phase with each other, they might get in each other's way.

It was shown above that at given speed, the kinetic energy given to the limbs at each step is proportional to I/h^2. The distance which could be covered at each step would be proportional to h so the total kinetic energy given to the limbs in covering a given distance would be proportional to I/h^3. Hence locomotion should be most economical if the limbs are long with low moments of inertia. These requirements are largely in conflict: long limbs in general will have large moments of inertia, but for given length the moment of inertia will be least if most of the mass of the limb is near its upper end. Most of the muscle of mammal limbs and indeed of vertebrate limbs in general is found in the upper part of the limb. The change from the plantigrade to the digitigrade stance, referred to above, increased limb length without too much increase in moment of inertia, and the process was taken a stage further in the evolution of hoofed mammals (p. 451).

The mammal stance and gaits make different demands on the vertebral column, than the stance and gaits of newts and lizards. The latter stand with their feet well to either side of the body, so the column probably has to withstand considerable twisting moments when one limb of a pair is off the ground. Mammals stand with their feet under or nearly under the body, and twisting moments can be expected to be relatively small. Lizards and newts bend from side to side as they run but galloping mammals bend vertically. It was explained on p. 252 how the articulating processes of amphibians and reptiles are arranged so as to restrict twisting but allow lateral bending. In mammals the vertebrae of the posterior part of the trunk have articulating processes with faces more or less parallel to the median plane. These still resist twisting, but allow very free vertical bending. The vertebrae of the more anterior part of the trunk (i.e. most of the thoracic vertebrae) have roughly horizontal articulating processes, which offer no resistance to twisting.

TEETH AND CHEWING

Mammals have differently shaped teeth in different parts of their jaws, and some of the shapes are very complicated. Only in characinoid fishes (p. 178) is comparable complexity found. The details of tooth structure vary greatly among mammals, showing striking adaptations to various diets. Some of these adaptations will be described in later chapters, but the dentitions of all marsupial and eutherian mammals seem to have evolved from a common pattern, found in some early fossil mammals and also some modern ones. Only this pattern is considered in this chapter.

The opossum *Didelphis* will be taken again as an example of a primitive mammal. Its teeth are very like those of some early mammal fossils, and its chewing movements have been carefully studied. It eats a wide variety of foods including fruit, insects and small vertebrates. The general arrangement of its teeth is shown in Fig. 12-10. The most anterior are the incisors which are small, peg-like teeth probably used mainly for picking up and manipulating food. The lower incisors are not vertical, but tilt forwards, and so can the more easily be slid under objects which are to be picked up. Posterior to the incisors are the much larger canines, used by other mammals and presumably by the opossum for tearing at prey. They are appropriately sharp and curved. Next come the premolars, which are perhaps most useful for piercing food in the early stages of chewing. Finally there are the molars, which are much the most complicated teeth in the mouth and much the most interesting. They are used for chewing the food before it is swallowed. By breaking the food up into small pieces they speed the action of digestive enzymes on it. Fast digestion is important to animals with high metabolic rates. Cyprinid fishes (p. 189) and lizards (p. 303) chew their food to some extent, but mammals break their food up much more thoroughly.

Fig. 12-11*A*, *B* show the form of opossum molar teeth. Note that the upper and lower molars are quite different. The upper ones are triangular in plan with an inner relatively low part (stippled) and an outer generally higher part. Three main cusps (labelled *a*, *b* and *c*) rise above these levels, and sharp ridges run between the cusps and along the edges of the tooth. Each lower molar consists of a relatively high triangle and a generally lower squarish heel. Again there are cusps (labelled with Greek letters) and ridges. All the cusps have accepted scientific names, but as the names are long and not particularly memorable, letters seem more appropriate in this book.

The chewing movements in which the opossum uses these teeth have been observed and recorded by X-ray cinematography. Lateral views show that chewing a mouthful of food occurs in two stages. First come a series of bites

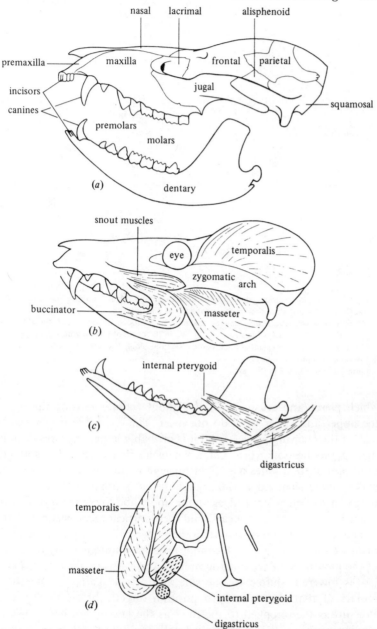

Fig. 12-10. (*a*) Lateral view of the skull of the opossum *Didelphis*. (*b*) The same, showing the positions of muscles. The arrow is explained in the text. (*c*) Median view of the lower jaw with muscles in position. (*d*) Diagrammatic transverse section through the head of *Didelphis*, posterior to the eye.

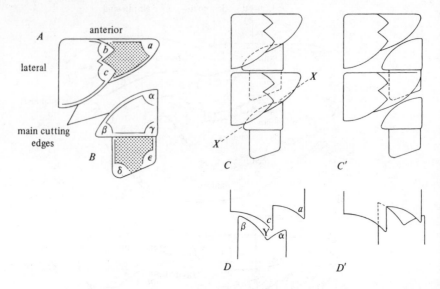

Fig. 12-11. (A), (B) Diagrams showing the main cusps and ridges on upper and lower molar teeth, respectively, of a primitive therian mammal such as *Didelphis*.

(C) (C') Diagrammatic plans showing the relative positions of upper and lower molars at the beginning (C) and end (C') of the effective stroke in the cycle of chewing movements.

(D) (D') Diagrams representing thick vertical sections along XX, showing upper and lower molars in the same two positions as (C), (C').

Lettering is explained in the text.

which pierce and crush the food without actually cutting through it, so the upper and lower teeth do not meet. The cusps are important in this part of the chewing process, and it is probably largely as a result of it that their points become worn. Later, when this first process has softened the mouthful of food, bites pass right through it, and upper and lower molars make contact. The ridges rather than the cusps are important in this chopping phase of chewing. Ridges on the upper and lower teeth slice past each other like the blades of shears, and their vertical faces become worn.

The lower jaw is less wide than the upper one, so it has to be moved to one side for chewing, which cannot proceed simultaneously on both sides of the mouth. X-ray cinematographs in vertical view show this. The lower jaw is lowered, shifted to one side and raised again, so that the lower molars of that side take up a position directly below the upper ones. Pressure is then applied to the food as the jaw moves back towards the median position. It is first applied with the lower molars in the position shown in Fig. 12-11C and is continued as they move to the position shown in Fig. 12-11C'. In the latter position the high triangles of the lower molars occupy the triangular gaps between the upper molars, and the low heels of the lower molars (stippled in Fig. 12-11*B*) rest against the low inner corners of the upper molars (stippled in Fig. 12-11*A*).

The main cutting edges are indicated in Fig. 12-11*A*, *B* and shown in the sectional view in Fig. 12-11*D*. This shows the position when they are beginning to slice past each other. Notice that because the cutting edges are concave, food trapped in the space *Y* cannot easily slip out of position and escape being cut as the lower molar moves to the position shown in Fig. 12-11*D'*. The same principle is applied in the design of garden shears with concave cutting edges, which will cut branches that would slip out of straight-bladed shears. The edges indicated as the main ones are probably the most important, but every edge apparently slices past another at some stage in the movement from the position of Fig. 12-11*C* to that of Fig. 12-11*C'*.

Each side of the lower jaw in mammals consists of a single bone (the dentary), not several as in other vertebrates. There is a very large temporal fenestra (filled by the temporalis muscle in Fig. 12-10*b*), usually with no bone dividing it from the orbit which accommodates the eye (see also Fig. 12-17). The evolution of the lower jaw and of the fenestra are considered later in this chapter: here they are simply noted in relation to the jaw muscles. The bar formed by the jugal and squamosal bones along the ventral edge of the fenestra is known as the zygomatic arch. The coronoid process of the dentary lies between it and the braincase.

In reptiles the main jaw muscles are the posterior adductor and the pterygoideus (Fig. 9-10). In mammals the pterygoideus persists as the internal pterygoid, running forward from the inner face of the lower jaw to the (small) pterygoid bone. (The adjacent external pterygoid, which is very small in *Didelphis*, seems to have evolved from the posterior adductor.) The posterior adductor has divided into two main parts. One of these is the temporalis, a pinnate muscle that inserts on either side of the coronoid process and on a tendinous extension of it (Fig. 12-9*D*, and compare Fig. 9-12). It slopes backwards to its origins, which are on the lateral wall of the braincase and on the tough sheet of connective tissue which covers the temporal fenestra. The other part is the masseter, which inserts on the ventral edge of the outer face of the lower jaw and runs forward to its origin on the zygomatic arch. The masseter has distinct superficial and deep portions. The deep masseter is itself generally distinct from the adjacent temporalis muscle but in the opossum they merge into one another with a gradual change in the inclination of the fibres.

The masseter and pterygoid muscles tend to pull the jaw anteriorly when they contract, and the temporalis tends to pull it posteriorly. The temporalis is set at a particularly favourable angle for use when the animal is tugging at prey with its canine teeth. The force exerted by the food on the lower jaw is then a forward and somewhat downwardly-directed one which is likely to be more or less in line with the force exerted by the temporalis. If the masseter muscles alone were used to hold the mouth shut while the animal tugged, the whole force of the pull would have to be taken by the ligaments of the jaw articulation and there might be a danger of dislocating the jaw. It will be shown later in this chapter, when the ancestors of mammals are discussed, how having the jaw muscles pulling at various angles can reduce forces at the jaw articulation when biting with the molars. Electromyographic observations on a bat, *Myotis*, which seems to chew in the same way as *Didelphis*, indicate that the internal pterygoid plays an important part in producing the side-to-side component of the chewing action.

Some other muscles of the head are shown in Fig. 12-10*b*, *c*. The digastricus opens the mouth. Note its peculiar structure, with two bellies joined in tandem by a tendon. Note also that the muscle which opens the mouth in reptiles is quite differently placed (Fig. 10-9, muscle 1). The snout muscles move the sensitive snout which is typical of mammals. The buccinator is the muscle of the cheek, which is another special feature of mammals. Cheeks prevent partly masticated food from leaking from the side of the mouth and are used, with the tongue, to adjust the position of food during chewing.

Breathing can continue while the mouth is full of food, because mammals have secondary palates like those of crocodilians (p. 325). The nasal passages run above the secondary palate, by-passing the mouth. Thus the duration of chewing is not limited by the need to breathe. The manner in which the secondary palate evolved is discussed later (p. 401).

The teeth of mammals, like those of archosaurs, have roots housed in sockets in the jaw and held in place by collagen fibres.

In the life of a reptile, the teeth are shed and replaced many times. Waves of replacement travel backwards along the jaw in such a way that alternate teeth are always very different in age. A young tooth, or an old one which is being resorbed at the base prior to being shed, is flanked on either side by mature, firmly fixed teeth. Thus at least half of the teeth in any given part of the jaw are always fully functional. In mammals, however, very little replacement occurs. In eutherians the original ('milk') incisors canines and premolars are replaced only once, and the molars not at all. In marsupials only one tooth in each jaw (the last premolar) is replaced.

The teeth of a lizard can perform their functions of holding and piercing (see p. 303) provided only that there is always a reasonable proportion of mature teeth. The chewing action of mammal molars such as those of the opossum depends on a precise fit being maintained between each upper tooth and a pair of adjacent lower teeth. This fit could not be maintained while alternate teeth were being replaced, especially if the animal was growing and the new teeth were bigger than their predecessors. Continual replacement in the reptilian manner would make the molars much less effective. 'Milk' premolars are generally much more like molars than adult premolars, and are used in chewing by young animals. They are not shed and replaced until there are molars to take over their chewing function.

Reptiles feed in the adult manner using their teeth, as soon as they emerge from the egg. If mammals took solid food and needed teeth while still so small a fraction of the adult size, their teeth would surely have to be replaced many times during growth to maintain a reasonable proportion between tooth size and head size. Young mammals are born quite large and subsequently suckled, so that only quite a modest amount of head growth occurs while teeth are in use. Suckling may have made possible the mammalian pattern of (very limited) tooth replacement.

EARS

The ears of mammals are very different from those of other vertebrates. Their most obvious peculiarity is the pinna (external ear) which acts in part like the horn of an old phonograph recorder. In such a machine the energy required for indenting the record was provided entirely by the sound, without electrical amplification. The energy was collected from the large area of the mouth of the horn and concentrated on the small diaphragm at its apex which operated the indenting needle. The horn also had an impedance matching effect, making the amplitude of pressure changes greater and of velocity changes less, at its apex than at its mouth (impedance matching and its importance in the ear were explained on p. 237). The pinna seems to act as an energy collector and impedance matcher, collecting sound from a large area and providing the first stage of impedance matching between the air and the oval window. The eardrum is relatively small and occupies a deep, protected position. It has been shown that the threshold of hearing of cats is raised by about 10 dB when the pinna is removed.

If these were the only functions of the pinna the most appropriate shape for it would be an exponential horn. In fact, very much more complex shapes are found. They seem to play an important part in the localization of sound sources. Though accurate judgement of direction depends partly on comparisons between two ears (the ear nearer the source receives louder sound, and receives it sooner) men can still locate sound sources quite well with one ear blocked. The importance of the shape of the pinnae can be demonstrated by bending them, which makes judgement of direction less good. Reflection from the pinna affects sound arriving at the eardrum in complex ways that depend on the direction of the source of sound. Different frequencies are delayed by different amounts. It is believed that this provides information from which the three-dimensional location of the sound source can be judged.

Another peculiarity of mammal ears is that each has three auditory ossicles, not just a stapes. The manner in which the additional ossicles were added to the ear is considered later in this chapter (p. 402). The chain of ossicles functions as a complex lever system, so the amplitude of vibration of the oval window can differ from that of the eardrum. Observations on dissected cats showed that at most frequencies the oval window had about half the amplitude of the eardrum. The ossicles thus doubled the forces and halved the velocity changes, so augmenting the impedance matching effects of the pinna and of the difference in area between eardrum and oval window.

In amphibians, reptiles and birds impedance matching between air and oval window seems generally to depend solely on the difference in area between eardrum and oval window (though the stapes of the lizard *Crotaphytus* acts as a lever, halving the amplitude of vibration). The ratio of areas is about twenty both in man and in most of a selection of birds which have been investigated, but man has two impedance-matching devices (the pinna and the three-ossicle system) which the birds lack. Presumably the impedance per unit area at the oval window is greater in man than in birds. This would not be surprising, in view of the great enlargement of the sound-sensitive part of the ear to form the cochlea in mammals.

In amphibian ears (Fig. 8-3) sound is detected by the amphibian and basilar papillae, which are both small. Higher vertebrates have no amphibian papilla, and depend on the basilar papilla. Notice that the route from oval window to round window via the basilar papilla passes through perilymph, the endolymph of the papilla, and perilymph again. In mammals the basilar papilla, with the portions of the perilymph on either side, has become drawn out into a long cochlea, coiled as a helix of several turns (Fig. 12-12a). Birds and crocodilians have similar but much shorter cochleae, which are curved but do not form even one complete turn.

In mammals, as in amphibians, the round and oval windows vibrate together. When one caves in the other must bulge out, and vice versa. The fluids of the inner ear must move with them. These movements could be restricted to the perilymph since there is a pathway, entirely through perilymph, between the windows. However, this pathway is a long one around the tip of the cochlea, and the partitions within the cochlea are flexible. Movements of the windows deflect the partitions, so that endolymph moves as well as perilymph.

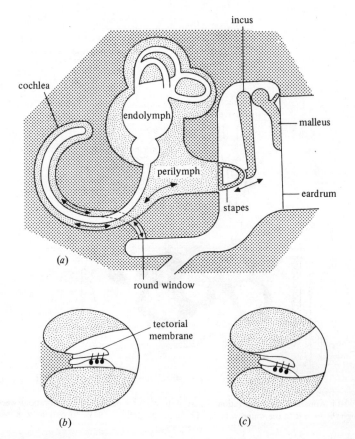

Fig. 12-12. (*a*) A diagram showing the structure of the ear of a mammal. (*b*) A diagrammatic section through the cochlea.

Fig. 12-12*b, c* show how these deflections are detected. A long cupula (known as the tectorial membrane) runs like a shelf along the cochlea, close to one of the partitions. Hair cells in the partition have their sensory 'hairs' embedded in the cupula, and are stimulated by the shearing movement of the cupula relative to the partition, as both are deflected up and down.

Dr G. von Békésy ground small openings in cochleas from cadavers so that part of the partition could be seen. Silver filings sprinkled on it made it easier to see. He observed it by stroboscopic illumination while the round window was vibrated by apparatus fitted to it (Fig. 12-13). For any frequency of vibration, different parts of the cochlear partitions vibrated up and down with different amplitudes. The greatest amplitudes were near the base when the frequency was high, and further along the cochlea if the

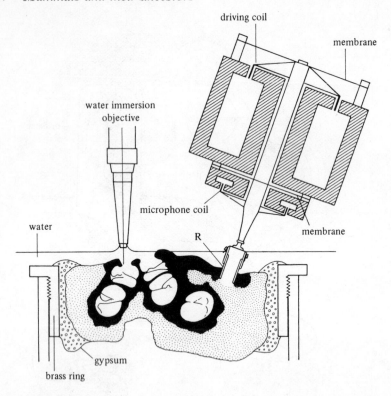

Fig. 12-13. Apparatus for observation of part of the cochlear partition while vibrations were applied through the tube R to the round window. From G. von Békésy (1943). *Akustische Z.* **8**, 66–76.

frequency was lower. The vibrations were not in phase all along the cochlea, but waves seemed to travel along the partitions from the base.

Sounds of different frequencies are very probably distinguished by the different positions in the cochlea of the peak amplitude of vibration. However, the observed peaks are broad ones, so it is hard to believe that small differences could be perceived in this way. A recent theoretical discussion of the problem raises the possibility that there may be sharp resonances in the intact cochlea which were lost in von Békésy's preparations.

KIDNEYS

Reptiles apparently cannot produce urine of higher osmotic concentration than the blood (p. 276) but birds and mammals can. The mechanism seems to be the same in both but the facility is so much more highly developed in mammals that description of it has been deferred to this chapter. Most birds which have been tested, even when deprived of drinking water, do not produce urine more concentrated than about 0.7 osmol l^{-1} (about twice the concentration of the blood). Even a desert bird (*Geococcyx*, the road-runner) and a sea bird (*Pelecanus*, the pelican) failed to exceed this concentration. An exception was *Passerculus*, the Savannah sparrow, which inhabits salt marshes in California and can produce urine at 2 osmol l^{-1}. Some mammals apparently have no more ability than typical birds to produce concentrated urine, but the dog, cat and rat can all produce urine of more than 2.5 osmol l^{-1}, and some desert rodents can produce very concentrated urine. The Australian hopping mouse, *Notomys*, can achieve 9 osmol l^{-1} (including 5 mol l^{-1} urea). The ability to produce concentrated urine is particularly important to mammals since they excrete their nitrogenous waste mainly as urea. A reptile or bird can conserve water by passing almost dry urate precipitates but a mammal can only excrete urea in solution.

The shape of typical mammal kidneys is well known. The ureter joins the kidney on its concave inner side. There are two distinct layers of tissue, the medulla on the concave side and the cortex on the outer convex side. The capsules are in the cortex (Fig. 12-14). The structure which enables mammal and bird kidneys to produce concentrated urine is a section known as the loop of Henle, interpolated in each kidney tubule. Most of its wall is flattened epithelium, thinner than the rest of the tubule wall. The tubules do not run straight from the cortex through the medulla to the ureter but down into the medulla and back before joining one of the collecting ducts which run through the medulla again to the ureter. The section which runs into the medulla and back is the loop of Henle.

The sequence of concentration changes which occur as fluid passes along a tubule has been investigated in experiments with various rodents, which were anaesthetized and opened to expose the kidneys. Extremely fine pipettes (outside diameter 5 μm) were used to puncture individual tubules where they touched the kidney surface, and withdraw minute quantities of fluid. The osmotic concentrations of the samples were determined by finding their freezing points, and the concentrations of samples of blood plasma and urine were determined in the same way. The animal was killed and the kidney treated chemically to separate it into its component tubules. The punctured tubule was found and the position of the puncture noted.

It was only possible to collect from tubules where they touched either the outer surface of the cortex or the inner surface of the medulla. In Fig. 12-14, which is based on the results of these experiments, only the concentrations in the superficial parts of the tubules (at the top and bottom

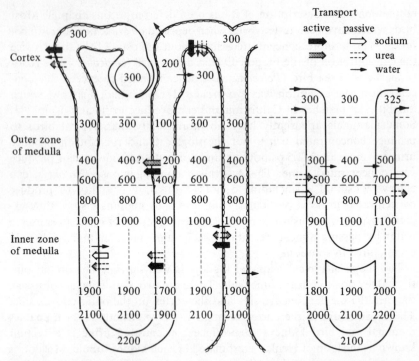

Fig. 12-14. A diagram illustrating the supposed mechanism of urine concentration in mammal and bird kidneys. A tubule with a long loop of Henle is shown on the left, and a blood vessel loop on the right. The numbers are concentrations in $mosmol\,l^{-1}$. From C. W. Gottschalk & M. Mylle (1959). *Am. J. Physiol.* **196**, 927–36.

of the figure) are based on actual measurements. The other concentrations are plausible interpolations. The figure represents an animal with blood plasma of about $0.3\,osmol\,l^{-1}$ producing urine at $2.1\,osmol\,l^{-1}$. The proximal part of the tubule, close to the capsule, contains fluid which has almost the same osmotic concentration as the blood plasma. By the time it reaches the turn in the loop of Henle the concentration is about the same as that of the urine but as it returns to the cortex its concentration falls again, below that of the blood. The concentration is high again by the time the urine leaves the collecting duct. Not all loops of Henle extend right through the thickness of the medulla, and it is probably only in those that do that concentrations equal to that of the urine are reached.

Blood vessels also run in loops, from the cortex into the medulla and back again. Blood samples taken from the medullary ends of the loops while concentrated urine was being produced, have been found to have as high osmotic concentrations as the fluid in the adjacent collecting ducts.

The remarkable changes in osmotic concentration in the tubule fluid and the blood, seem to be due to the loop of Henle functioning as a counter-current multiplier. It produces high osmotic concentrations in essentially the same way as the retia mirabilia of the swimbladder produce high partial pressures of gases (p. 150). Just as the retia build up a high partial pressure without there being a large difference at any point between adjacent parts of their arterial and venous capillaries, so the loop of Henle builds up large differences in osmotic concentration without there being at any level large differences between its ascending and descending limbs.

The details of the process are not known with as much certainty as could be wished. The essential active process seems to be active transport of sodium ions out of the ascending limb of the loop of Henle. (There is also evidence that active transport of sodium occurs elsewhere, as indicated in Fig. 12-14.) Chloride and other anions follow the sodium, preserving electrochemical equilibrium, but water apparently cannot easily follow to maintain osmotic equilibrium: the wall of the ascending limb seems to be relatively impermeable to water. The interstitial fluid between the tubules thus becomes a little more concentrated than the fluid in the ascending limb. The difference has been shown as 200 mosmol l^{-1} in Fig. 12-14. The interstitial fluid also surrounds the descending limbs, which have walls that are highly permeable to water and probably to salts. Water diffuses out of the descending limbs and salts diffuse in, so keeping the fluid inside more or less in osmotic equilibrium with the interstitial fluid. The tubular fluid is thus concentrated as it travels down the descending limb and it is this concentrated fluid that arrives at the base of the ascending limb. Transport of sodium from it in the ascending limb makes the interstitial fluid even more concentrated, and so the concentration builds up.

The walls of the collecting ducts seem to be effectively impermeable to ions. Their permeability to water varies, but is high in the presence of the hormone which stimulates production of concentrated urine. As the tubular fluid descends the collecting duct it passes through regions of more and more concentrated interstitial fluid. Ions cannot diffuse into it to maintain osmotic balance, but if the hormone is present water diffuses out, and the urine eventually reaches the same concentration as the tubular fluid at the bottom of the loop of Henle.

The water which diffuses out of the descending limbs and collecting ducts is removed in the blood, but blood flow would make the mechanism ineffective if it resulted in substantial dilution of the interstitial fluid. This is avoided because the blood vessel loops act as counter-current exchangers in essentially the same way as the retia in the blood supply of tuna swimming muscles (p. 199). Operation of the counter-current principle enables the blood to flow through the medulla where the osmotic concentration is high, without removing substantial quantities of salts, just as it enables tuna blood to pass through warm muscle without removing much heat. Diffusion of salts and water between blood and interstitial fluid concentrates the blood as it travels into the medulla and dilutes it again as it returns.

ANCESTORS OF MAMMALS

Class Reptilia, subclass Synapsida, orders Pelycosauria and Therapsida

The mammals apparently evolved from reptiles of the subclass Synapsida. The earliest known synapsid fossils are from the upper Carboniferous, and so are among the earliest known reptiles, but it was in the Permian and Triassic that the subclass flourished. The more primitive synapsids are found mainly in North America and are grouped together in the order Pelycosauria. The advanced ones have been found most plentifully in South Africa, and are placed in the order Therapsida. Both orders show considerable adaptive radiation. Pelycosaurs must have looked rather like lizards, but though some were quite small others reached a length of about 3.5 m. Some such as *Varanosaurus* (Fig. 12-15) had pointed teeth and were

Fig. 12-15. The skeleton of the pelycosaur *Varanosaurus* (length about 1.5 m). From A. S. Romer (1956). *Osteology of the Reptiles*. University of Chicago Press, Chicago. Copyright © 1956.

presumably carnivorous (or piscivorous). Some others had blunt teeth which look suitable for crushing molluscs or soft plants, and as no fossil molluscs have been found in the same habitat they are believed to have eaten plants. The therapsids were the dominant terrestrial animals of the late Permian and early Triassic, but were later supplanted from that position by the archosaurs (see p. 314). Some seem to have been carnivores and some herbivores. *Thrinaxodon* (Fig. 12-16) was one of the smaller carnivores: others grew to the size range of bears. The herbivores included species as large as a hippopotamus.

Fig. 12-15 shows that the skeleton of *Varanosaurus* followed the primitive reptile pattern in most respects. Apart from a small temporal fenestra in the skull the skeleton is very like those of cotylosaurs and even of primitive amphibians (Fig. 8-1). The tail is long. The limbs were held in typical reptile fashion, with humerus and femur more or less horizontal (this is apparent from the way the articulating surfaces of the hip and shoulder

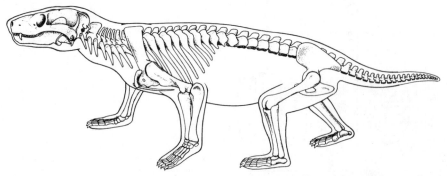

Fig. 12-16. The skeleton and supposed body outline of the therapsid *Thrinaxodon* (length about 40 cm). From F. A. Jenkins (1970). *Evolution*, **24**, 230–52. © Society for the Study of Evolution.

joints fit together). The limb girdles have the usual reptilian form and vertebrae right back to the sacrum bear ribs.

Fig 12-16 shows that *Thrinaxodon* was far more like a mammal. The tail was apparently short. The glenoid cavity in the pectoral girdle, where the humerus articulates, faces somewhat posteriorly and ventrally, and the head of the femur is turned inwards rather as in mammals (Fig. 10-12*d*). These features suggest a stance approaching that of the opossum (Figs. 12-3 to 12-6). The anterior edge of the scapula is turned a little outwards, anticipating the evolution of the scapular spine of mammals. There is a round opening in the pelvic girdle between pubis and ischium, as in mammals. Five sacral vertebrae attach to the ilium, not two or three as in pelycosaurs, but these sacral vertebrae are separate instead of being fused as in mammals. The heel has appeared but the tarsal bones are still not arranged quite as in mammals. The ribs stop short well anterior to the pelvic girdle, suggesting that *Thrinaxodon* may have had a diaphragm. The first two vertebrae (the atlas and axis) are much more mammal-like than in pelycosaurs.

So far the skull has been ignored in this comparison. An increase in the size of the temporal fenestra can be seen by comparing Fig. 12-16 with Fig. 12-15, but these figures are too small to show other details of the skull at all well. A primitive mammal skull has already been illustrated (Fig. 12-10). The diagrams in Fig. 12-17 shows some of the major differences in skull structure between typical pelycosaurs, therapsids and mammals.

The temporal fenestra is small in pelycosaurs and occupies the synapsid position with the squamosal and postorbital meeting above it (see Fig. 9-11). It is so much larger in therapsids that these bones often fail to meet it, and indeed there are therapsids in which they have come to meet below the fenestra: in such cases it could be said that the skull has evolved from the synapsid to the parapsid condition. The fenestra is larger still in

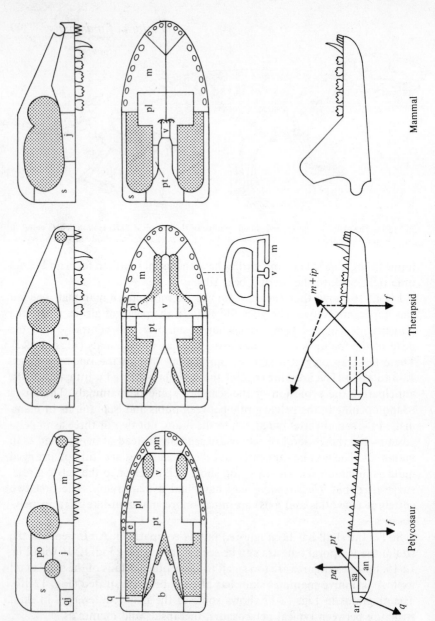

Fig. 12-17. Diagrams of the skulls of a typical pelycosaur (based on *Varanosaurus* and *Dimetrodon*), therapsid (based on *Thrinaxodon* and *Procynosuchus*) and mammal (based on *Didelphis*).

an, angular; ar, articular; b, braincase (basioccipital and basisphenoid); j, jugal; m, maxilla; pl, palatine; pm, premaxilla; po, postorbital; pt, pterygoid; qj, quadratojugal; s, squamosal; sa, surangular; v, vomer.

Lower case italic letters indicate forces exerted on the lower jaw by *f*, the food, *m* + *ip*, the masseter and internal pterygoid muscles; *pa*, the posterior adductor muscle; *pt*, the pterygoideus muscle; *q*, the quadrate articulation; *t*, the temporalis muscle.

mammals, and since the postorbital has disappeared it is confluent with the orbit. Some other dermal bones disappeared as the mammals evolved, including the quadratojugal which was already small in therapsids. The premaxillae separate the external nostrils in pelycosaurs and therapsids, but not in mammals.

Fig. 12-17 also shows progressive changes in the palate. Pelycosaurs have no trace of a secondary palate: their palates are essentially like those of amphibians and most other reptiles. In therapsids shelves grew inwards from the maxillae and palatines under the primary palate. The vomers fused together as a single bone with a flange growing down from it between the left and right nasal passages. The pterygoids no longer articulate with basipterygoid processes, but are firmly sutured to the underside of the braincase: the same thing happened in the evolution of turtles and crocodilians, but the kinesis of lizard skulls depends on the basipterygoid articulations being movable. In many therapsids and in mammals the left and right maxillae and palatines meet, completing the secondary palate. The pterygoid is small in mammals and the quadrate has become an ear ossicle.

Finally, Fig. 12-17 shows changes in the lower·jaw. Pelycosaurs-have jaws of normal reptilian type, incorporating several substantial bones in addition to the dentary. Many therapsids have a very large dentary which, with the large temporal fenestra, seems to indicate large, powerful jaw muscles. The other bones in the lower jaw are rather small. It is these posterior jaw bones that articulate with the skull, so we are faced with a paradox: the jaw articulation apparently got weaker as the jaw muscles got stronger. The paradox can be resolved if account is taken of the lines of action of the jaw muscles. The most primitive pelycosaurs probably had jaw muscles arranged very much as in *Iguana* (Fig. 9-10). The posterior adductor and pterygoideus muscles probably inserted on different parts of the jaw pulled in different directions, as indicated by the arrows pa and pt in Fig. 12-17. If the animal were biting with its back teeth the food would exert a force f on the jaw, more or less as shown (this is of course equal and opposite to the force exerted by the jaw on the food). f is well out of line with the resultant of pa and pt, so the jaw could not have been in equilibrium under the action of these forces alone. There must have been in addition a force q, acting at the jaw articulation. In many therapsids the dentary has a tall coronoid process, apparently for insertion of mammal-like temporalis muscles which would exert a force t. The other jaw muscles were also apparently arranged as in mammals; a masseter and an internal pterygoid muscle probably inserted close together (though on opposite sides of the jaw) and pulled in similar directions, exerting a force $m + ip$. Note in the diagram that t and $m + ip$ are shown crossing on f.

Their resultant could have been in line with f, so that there was no force on the jaw articulation. If as is likely the resultant was actually a little out of line, a small force at the articulation would suffice. Thus while larger jaw muscles were evolving they became rearranged, so that forces at the jaw joint were probably actually reduced and the posterior jaw bones could become weaker. As evolution proceeded the dentary became even larger and eventually made contact with the squamosal, lateral to the original articulation between the articular and quadrate. This made the articular/ quadrate joint redundant (though fossils are known with both joints) and all bones except the dentary were lost from the lower jaw. The articular became an ear ossicle (the malleus) and the quadrate became detached from adjacent bones to form another (the incus). The angular, from the lower jaw, became the tympanic bone which encircles the eardrum (sometimes incompletely) and often contributes to a bony wall enclosing the middle ear cavity.

This sequence of changes has not been fully elucidated. Some of the uncertainty is due to the difficulty of deciding exactly where the eardrum was in some fossil synapsids. In early synapsids it presumably lay in the usual reptile position, posterior to the quadrate. The angular bone soon evolved a notch which could have housed the anterior corner of the eardrum, but it is not clear at what stage, or why, the eardrum moved to the angular. The stapes in synapsid reptiles has a process which rests against the quadrate so that stapes, quadrate and articular had the same relative positions as the stapes, incus and malleus of mammals, even before the quadrate and articular ceased to constitute the jaw articulation. However, it is not clear at what stage, or why, the eardrum became attached to the articular (malleus) and detached from the stapes.

Pelycosaurs have teeth of different sizes in different parts of their jaws but some therapsids such as *Thrinaxodon* show much greater differentiation (Fig. 12-17). They have small conical anterior teeth like mammalian incisors, large conical teeth like canines and multicusped teeth like premolars or molars. Though their dentition looks strikingly mammalian, tooth replacement seems to have proceeded in an essentially reptilian manner.

The brains of most mammals are ten or more times as heavy as those of reptiles of the same body weight (Fig. 10-16). The necessary expansion of the braincase was achieved largely by enlargement of the frontal and parietal bones (Fig. 12-10). These are dermal bones evolved originally as ossifications of the skin (p. 133). In primitive reptiles they still lie immediately under the skin, and though they form the roof of the braincase they do not contribute to its side walls. As the synapsids evolved and the brain got larger they extended down the sides of the braincase, as in amphisbaenians and snakes (p. 307). Thus large parts of these dermal bones came to lie quite deep in the head. The squamosal, which is a dermal bone not originally involved in the braincase at all, came to form part of its side wall. The alisphenoid, which in typical reptiles lies lateral to the membranous part of the side wall of the braincase (Fig. 10-5), became an integral part of the braincase. These changes are quite well advanced in some therapsid reptiles.

Many of the characteristic features of mammals are not preserved in fossils, and it is by no means certain that *Thrinaxodon* and similar therapsids would be classed as reptiles if they were alive today. We cannot tell whether they were homoiothermic and hairy, nor whether they suckled their young, so the convention has been adopted that a fossil is not regarded as a mammal unless the dentary is the only bone in the lower jaw.

Some therapsid skulls have foramina in the maxilla and adjacent bones, which have been supposed to be for nerves and blood vessels to a sensitive, mammal-like snout. Some also have pits in the maxilla which suggest the presence of vibrissae. If therapsids had vibrissae they very probably had fur (since vibrissae are modified hairs) and if they had fur they were probably homoiothermic. Unfortunately for a fascinating chain of reasoning, similar foramina are present in some modern lizards that do not have mammal-like snouts, and the pits are present in some therapsids that lack the foramina. It seems unlikely that the pits could have housed vibrissae where there is no indication of rich innervation or blood supply.

EARLY MAMMALS

Class Mammalia, subclasses Theria and one or more others

The earliest mammal fossils known are from the late Triassic, when the therapsids were declining and the archosaurs were taking over as dominant terrestrial vertebrates. In the Jurassic and Cretaceous mammals may have been quite numerous but they were small, apparently with much the same size range as modern rodents. Large mammals seem not to have appeared until the Tertiary, after the dinosaurs became extinct.

It is conventional to divide mammals into about four subclasses, but it has been suggested recently that two may be enough. The subclass Theria includes the marsupials and eutherians, which still flourish and which seem to have originated in the Cretaceous. Primitive members of both have molars like those of the opossum (Fig. 12-11). Also included in the Theria are some primitive groups in which can be traced the gradual evolution of such teeth.

Other early mammals have teeth much less like those of the opossum.

Some seem to have been carnivorous but a large group (the multituberculates) were superficially rodent-like, and presumably herbivorous. There are some interesting similarities between the multituberculates and the only living non-therian mammals, the monotremes, (described in the next section of this chapter). In Theria the side wall of the braincase is formed largely by the alisphenoid and squamosal bones (Fig. 12-10). In multituberculates and monotremes it is formed mainly by the periotic (the bone of sclerotome origin which encloses the inner ear). Theria have a jugal bone but multituberculates and monotremes have none: an extension of the maxilla takes the place of the jugal in the zygomatic arch. Theria have no ectopterygoid but multituberculates have one, and monotremes have a bone of uncertain homology which seems likely to be one. There are thus substantial reasons for believing that multituberculates and monotremes are closely related. The other (extinct, apparently carnivorous) non-therian mammals have rather similar braincases. It seems quite likely that future classifications will include all non-therian mammals in a single subclass.

MONOTREMES
Class Mammalia, subclass uncertain, order Monotremata

There are only three genera of monotremes: the duck-billed platypus (*Ornithorhynchus*) of Australia and Tasmania, and the echidnas or spiny anteaters (*Tachyglossus* and *Zaglossus*) of Australia, Tasmania and New Guinea. Adults range in weight from about 2 kg (*Ornithorhynchus*) to 15 kg (*Zaglossus*). They lay eggs like reptiles, but suckle their young like other mammals. Their eggs are small (less than 2 cm long). *Ornithorhynchus* lays its eggs in a burrow in the bank of a stream, and keeps them and subsequently the young warm by curling round them. *Tachyglossus* lays a single egg and carries it in a temporary pouch which forms on the female's belly at the beginning of the breeding season. The cloaca can reach the pouch if the animal curls up, and it is suspected that the egg is laid directly into the pouch.

Since they come from small eggs, newborn *Tachyglossus* are necessarily small. They weigh about 0.4 g, and are entirely dependent on the mother, whose mammary glands open at two areolae (not raised into nipples) within the pouch. The young can suck milk without leaving the pouch, where it remains until it weighs 170–400 g. By the time it leaves the pouch it is developing the spines, which are a formidable defence for an adult animal but could hardly be tolerated in her pouch by a mother. The temperature in the pouch is about equal to the mother's body temperature but by the time it leaves the pouch the young can maintain its own temperature. The mother leaves it in a burrow and returns periodically to suckle it.

Monotremes resemble reptiles rather than other mammals in some other respects, as well as laying eggs. Urine, faeces and genital products are all voided through a single opening, the cloaca. The testes do not descend into a scrotum but remain in the body cavity. Though the penis is tubular as in other mammals it lies within the cloaca, from which it emerges only when erected. The pectoral girdle is constructed on the reptilian pattern though the pelvic girdle is just like those of marsupials. The brain is reptile-like in some respects. Though they are homoiothermic monotremes have low body temperatures, of only 30–33 °C. The basal metabolic rate is also lower than for other mammals of comparable size.

The spiny anteaters are strange looking animals with slender snouts, strong short limbs and a formidable covering of spines (Fig. 12-18). These

Fig. 12-18. The platypus (*Ornithorhynchus*, left) and a spiny anteater (*Tachyglossus*, right). From J. Z. Young (1950). *The Life of Vertebrates*. Clarendon Press, Oxford.

consist of α-keratin, like porcupine quills, and like them have presumably evolved by modification of hairs. The strong limbs are used both for constructing burrows and for digging into ant hills. The forelimbs are particularly strong and the hands are large with broad claws. Ants and termites are the main foods. They are picked up with the long tongue (which is coated with sticky saliva) and ground up between horny surfaces on the palate and the base of the tongue. There are no teeth. Many other ant- and termite-eating mammals also have long snouts and tongues, large salivary glands producing sticky saliva, peculiar teeth or no teeth at all, and digging forelimbs: they include the anteater (*Myrmecophaga*) and its immediate relatives, some armadilloes such as *Cabassous*, the pangolins (*Manis*), and aardvark (*Orycteropus*) and the marsupial anteater (*Myrmecobius*). This group of adaptations has apparently evolved independently in monotremes, marsupials and several groups of eutherians. This is a remarkable example of convergent evolution.

The platypus is furry, with a broad bill shaped like that of a duck but covered with soft leathery skin (Fig. 12-18). Teeth appear transitorily in the course of development but are replaced by horny plates. The platypus swims well, and feeds largely on invertebrates from the bottoms of streams.

MARSUPIALS
Class Mammalia, subclass Theria, infraclass Metatheria

Marsupials are at the present time confined to America and Australasia. The American ones are nearly all in Central and South America: only the opossum *Didelphis* extends into North America. The Australasian ones occur in Australia, Tasmania, New Guinea and Celebes, and show a remarkable adaptive radiation that parallels to a large extent the radiation of eutherian mammals in other parts of the world.

The females of nearly all marsupials have a pouch. This is permanent, not transitory like the pouches of spiny anteaters (p. 404). The newborn young are very much smaller than those of eutherians, and though they are in most respects at a very early stage of development they have strong clawed forelimbs, which they use in making their own way to the pouch. The mother adopts a standard posture when giving birth, sitting with the tail extended forward between the legs. She does not seem to assist the young to make its way to the pouch except by licking the fur over which it must travel. The young of the red kangaroo (*Macropus rufus*) reaches the pouch within a few minutes of birth. It quickly finds and attaches itself to one of the nipples, which lie inside the pouch. The nipple swells inside its mouth, so that it is for the time being firmly attached. The American mouse opossums (*Marmosa*) have no pouch, and the young simply hang on to the nipples on the mother's belly.

The difference in life history between marsupials and eutherians can be illustrated by comparing two fairly large grazing mammals, the red kangaroo (adult females weigh about 30 kg) and the sheep (70 kg). The young kangaroo is born about one month after copulation, when it weighs about 0.7 g. A sheep embryo (excluding embryonic membranes) has about the same weight at this age but is only born after five months, when its weight is around 6 kg. (Twins are a little lighter than single lambs.) The kangaroo remains in the pouch for about seven months (i.e. to an age of eight months from conception) and at the end of that time weighs about 2 kg. There is a short period when it returns to the pouch occasionally, but thereafter it keeps out of the pouch while continuing to feed partly on milk. The nipple may dangle over the rim of the pouch or the young may put its head into the pouch to suck. Milk is still taken to an age of well over a year (weight 10–15 kg) but grass is also eaten. The sheep also changes gradually from a milk to a grass diet. The half-way point when equal amounts of metabolizable energy are obtained from milk and grass occurs at about three months from birth (eight months from conception) and a weight of over 30 kg.

There are other more or less consistent differences between marsupials and eutherians. Most marsupials have only a yolk-sac placenta (see p. 374). Marsupials have peculiar female reproductive tracts. Their brains lack the corpus callosum which in placentals is a major pathway for axons crossing from one cerebral hemisphere to the other. There is a small bone in each side of the ventral body wall, anterior to the pubis, both in marsupials and in monotremes. (It is not shown in Figs. 12-5, 12-6.) There are small and rather inconstant differences between the skulls of marsupials and eutherians. Marsupial body temperatures are rather low, generally 34–37 °C.

EUTHERIANS
Subclass Theria, infraclass Eutheria

The eutherians are often referred to as the placental mammals, although marsupials also have placentae (see p. 374). They include about 94% of the species of mammals. The main differences between them and the marsupials have already been described. The adaptive radiation of the eutherians is the main subject of the next two chapters, though marsupials are also mentioned.

FURTHER READING
General
Matthews, L. H. (1960). *British mammals*, 2nd edit. Collins, London.
Southern, H. N. (1964). *The handbook of British mammals*. Blackwell, Oxford.
Walker, E. P. (1968). *Mammals of the world*, 2nd edit. (3 vols.). Johns Hopkins Press, Baltimore.
Young, J. Z. (1957). *The life of mammals*. Clarendon Press, Oxford.

Temperature and Respiration
Bligh, J. (1973). *Temperature regulation in mammals and other vertebrates.* North-Holland, Amsterdam.
Schmidt-Nielsen, K. (1964). *Desert animals*. Clarendon Press, Oxford.

Reproduction
Ardran, G. M., A. T. Cowie & F. H. Kemp (1957). A cineradiographic study of the teat sinus during suckling in the goat *Vet. Rec.* **69**, 1100–1.
Blaxter, K. L. (1962). *The energy metabolism of ruminants*. Hutchinson, London.

Kon, S. K. & A. T. Cowie (1961). *Milk: the mammary gland and its secretion.* Academic Press, New York.

Metcalfe, J., H. Bartels & W. Moll (1967). Gas exchange in the pregnant uterus. *Physiol. Rev.* **47**, 782–838.

Sadler, R. M. F. S. (1969). *The ecology of reproduction in wild and domestic mammals.* Metheun, London.

Locomotion

Cavagna, G. A., F. P. Saibene & R. Margaria (1964). Mechanical work in running. *J. appl. Physiol.* **19**, 249–56.

Hildebrand, M. (1959). Motions of the running cheetah and horse. *J. Mammalol.* **40**, 481–95.

Jenkins, F. A. (1971). Limb posture and locomotion in the Virginia opossum (*Didelphis marsupialis*) and in other non-cursorial mammals. *J. Zool., Lond.* **165**, 303–15.

Alexander, R. McN. (1974). Mechanics of jumping by a dog. *J. Zool., Lond.* **173**, 549–73.

Smith, J. M. & R. J. G. Savage (1956). Some locomotory adaptations in mammals. *J. Linn. Soc. (Zool.),* **42**, 603–22.

Teeth and chewing

Crompton, A. W. & K. Hiiemae (1970). Molar occlusion and mandibular movements during occlusion in the American opossum, *Didelphis marsupialis* L. *Zool. J. Linn. Soc.* **49**, 21–47.

Hiiemae, K. & F. A. Jenkins (1969). The anatomy and internal architecture of the muscles of mastication in *Didelphis marsupialis. Postilla,* **140**, 1–49.

Kallen, F. C. & C. Gans (1972). Mastication in the Little Brown Bat, *Myotis lucifugus. J. Morph.* **136**, 385–420.

Smith J. M. & R. J. G. Savage (1959). The mechanics of mammalian jaws. *School Science Rev.* **40**, 289–301.

Ears

von Békésy, G. (1960). *Experiments in hearing.* McGraw-Hill, New York.

Huxley, A. F. (1969). Is resonance possible in the cochlea after all? *Nature, Lond.* **221**, 935–40.

Kidneys

Gottschalk, C. W. & M. Mylle (1959). Micropuncture study of the mammalian urinary concentrating mechanism: evidence for the countercurrent hypothesis. *Am. J. Physiol.* **196**, 927–36.

MacMillen, R. E. & A. K. Lee (1969). Water metabolism of Australian hopping mice. *Comp. Biochem. Physiol.* **28**, 493–514.

Ancestors of mammals

Crompton, A. W. (1963). On the lower jaw of *Diarthrognathus* and the evolution of the mammalian lower jaw. *Proc. zool. Soc. Lond.* **140**, 697–753.

Hopson, J. A. (1966). The origin of the mammalian middle ear. *Am. Zool.* **6**, 437–50.

Jenkins, F. A. (1970). Cynodont postcranial anatomy and the 'prototherian' level of mammalian organization. *Evolution,* **24**, 230–52.

Early mammals

Kermack, D. M. & K. A. Kermack (eds.) (1971). Early mammals. *Zool. J. Linn. Soc.* **50**, suppl. 1, 1–203.

Monotremes

Griffiths, M. (1968). *Echidnas*. Pergamon, Oxford.

Marsupials

Sharman, G. B. & P. E. Pilton (1964). The life history and reproduction of the red kangaroo (*Megaleia rufa*). *Proc. zool. Soc. Lond.* **142**, 29–48.

Tyndale-Biscoe, H. (1972). *Life of marsupials*. Arnold, London.

Waring, H., R. J. Moir & C. H. Tyndale-Biscoe (1966). Comparative physiology of marsupials. *Adv. comp. Physiol. & Biochem.* **2**, 237–376.

13

Mainly carnivorous mammals

The mammals are so diverse and have been so extensively studied that even a brief review of the major groups cannot easily be fitted into a chapter of normal length. The review which follows is divided between two chapters. This chapter is about mainly carnivorous groups including the insectivores, bats, carnivores, seals and whales. The chapter which follows it (and ends the book) is about mainly herbivorous groups.

INSECTIVORES
Infraclass Eutheria, order Insectivora

The insectivora are rather primitive Eutheria which are not by any means confined (as their name suggests) to a diet of insects. Their primitive features include limbs and molar teeth more or less like those of *Didelphis* (Chapter 12). The most numerous of the Insectivora are the shrews (*Sorex* and related genera), which are all very small and include the smallest of mammals. Though common, shrews are seldom seen because they spend most of their time in tunnels in the soil (made by themselves or by other small mammals), in runways through dense vegetation and in leaf litter. They are intermittently active both by day and by night. Their eyes are small but they have long mobile snouts with vibrissae pointing forward and to the side. They seem to depend mainly on these vibrissae for finding food such as earthworms and beetles.

Moles such as *Talpa* are also members of the Insectivora. Their skulls and particularly their teeth are very primitive, like those of *Didelphis*. Their eyes are extremely small but they have mobile sensitive snouts like shrews, well provided with vibrissae. Their forelimbs are extraordinarily specialized for burrowing (Fig. 13-1). The hand is very broad. The shoulder is set well forward, where one might expect a neck, so the huge muscles of the forelimb lie mainly anterior to the rib cage rather than lateral to it. If they were lateral to the ribs the animal would be broader and would have to make a broader burrow, and it would presumably use more energy in digging a given distance. The humerus is held in a peculiar position, almost vertical, with the elbow dorsal to the shoulder (X-radiographs show this clearly). Moles have been put in glass-sided boxes of soil so that the digging action could be watched when they burrowed near the glass. Digging

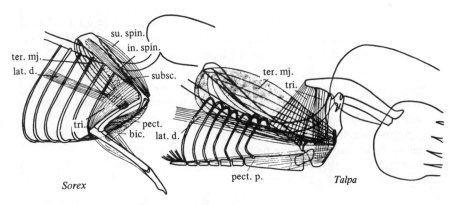

Fig. 13-1. Diagrams showing the principal bones and muscles of the forequarters of a shrew (*Sorex*) and a mole (*Talpa*). bic., biceps; in. spin., infraspinatus; lat. d., latissimus dorsi; pect., pectoralis; pect. p., pectoralis posticus; su. spin., supraspinatus; subsc., subscapularis; ter. mj., teres major; tri., triceps. From D. W. Yalden (1966). *J. Zool., Lond.* **149**, 55–64.

involves alternate movements of the two forelimbs. The effective stroke starts with the limb in the position shown in Fig. 13-1 (right). The humerus is then rotated about its own long axis so that the hand swings laterally and posteriorly. The muscles which are apparently responsible insert on the edge of the humerus, which is very broad. The largest is the teres major: Fig. 13-1 shows how much larger it is, relative to body size, in moles than in shrews. The pectoralis posticus and latissimus dorsi (also illustrated) presumably co-operate with it. When runs are built close to the surface the soil is simply pushed up into a ridge over the tunnel, but soil from deeper burrows is pushed back along the burrow and to the surface, where it forms a mole-hill.

It has been shown by examining the stomach contents of *Talpa* that earthworms are by far its most important food. These are presumably caught mainly underground, but *Talpa* is occasionally seen on the surface. Indeed, moles spend enough time on the surface to be caught frequently by surface-hunting predators. The Tawny owl (*Strix aluco*), like other predaceous birds, ejects pellets containing the bones of its prey. From analyses of a large collection of pellets it has been estimated that its diet in summer includes 45% by weight of moles.

Radioactive tags have been attached to moles to enable their movements underground to be followed. A mole was captured and a metal ring incorporating a small radioactive source was clipped round the base of its tail. It was then released and the observer patrolled the area, carrying a Geiger counter on a long handle. By sweeping the counter over the ground, he could locate the mole below. It was found that moles are active day and night, but that periods of about four hours' activity alternate with periods of about four hours' rest in a nest of grass or dry leaves in a chamber in the tunnel system. Fig. 13-2 is a plan of a tunnel system, showing the movements of a mole during an unusually long ($6\frac{1}{2}$ h) active period. Much of this time was spent digging, extending the system, but a lot of time was also spent patrolling the system, presumably finding and eating worms and other soil animals which had wandered into the tunnels. During this six and a half hours the mole travelled about 750 m.

Apart from the shrews and moles the only Insectivora likely to be familiar to most readers are the hedgehogs (*Erinaceus*, etc.)

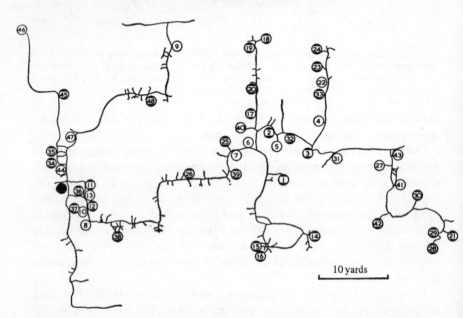

Fig. 13-2. A plan of the tunnel system of a mole in an arable field, showing successive positions of the mole during a period of 8 h. The filled circle shows the nest and the partially filled ones show where digging occurred. From G. Godfrey & P. Crowcroft (1960). *The life of the mole*. Museum Press, London.

BATS

Infraclass Eutheria, order Chiroptera

The bats are the only mammals which can fly, but they are very numerous both in individuals and in species. About 1000 species of bat are recognized, out of a total of about 4200 species of mammals. The smallest bats weigh only about 3 g when adult, and most bats are fairly small. The largest are flying foxes (*Pteropus*) weighing up to about 1 kg.

Most bats (including all British species) feed on insects. Most of them catch their prey in the air, but the Long-eared bat (*Plecotus*) picks insects off foliage: it hovers to catch its prey without alighting. Bats bite the wings off large insects before swallowing them. This is often done in the air, but some bats return to a perch to deal with prey. Large quantities of moth wings, beetle wings and elytra, etc. are found under such perches. Insect-eating bats have teeth very like those of the opossum (Figs. 12-10, 12-11) but their lower incisors are not tilted forward.

There is also a considerable number of tropical bats (including *Pteropus*) which feed on fruit, and have much simpler molar teeth with flatter crowns. Such teeth would be ineffective for chopping insects, but are presumably effective for crushing fruit. Other species have long tongues which they use for feeding on nectar and pollen. They often hover in front of a flower like a hummingbird, and feed without alighting. There are large flowers which open at night and are usually pollinated by bats just as some large day-opening flowers are pollinated by hummingbirds. A few bats feed on fish which they seize with long claws as they skim over water. The vampire bats (*Desmodus*, etc.) drink the blood of sleeping mammals. They have very sharp, blade-like incisors and canines which they use to draw blood.

Bats depend for flight on a wing membrane stretched between limbs and body (Fig. 13-3). The second to fifth digits are enormously long, with the

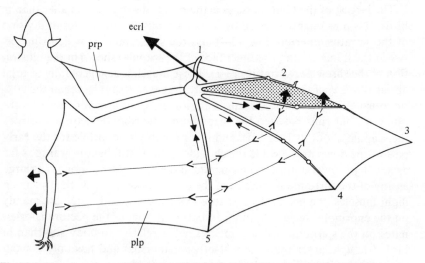

Fig. 13-3. The wing of a bat showing how it is kept spread. The extensor carpi radialis longus pulls on the second digit as indicated by the large arrow marked ecrl, and this pull is transmitted by a ligament to the third digit. Other muscles pull the hind limb medially. These opposing pulls set up tension in the directions indicated in the wing membrane and compressive stresses in the phalanges. plp, plagiopatagium; prp, propatagium. From C. J. Pennycuick (1971). *J. exp. Biol.* **55**, 833–45.

membrane running between them. Both elbow and wrist joints are hinges that can only bend in the plane of the wing. The ulna is rudimentary and it is the proximal end of the radius that forms the hinge joint with the humerus: in consequence, the forearm cannot be twisted. The advantages of such limited mobility in a wing have been mentioned (p. 338).

In flight the wing membrane must be kept taut. Fig. 13-3 shows how this is apparently done. The extensor carpi radialis longus is a muscle lying in the forearm which inserts on the metacarpal of the second digit, swinging it forward. This tends to keep the third (longest) digit extended, because there is a ligament from the tip of the second digit to the anterior edge of the terminal phalanx of the third. Other muscles pull the hind limb medially, towards the tail. Thus opposing forces on the third digit and the hind limb set up tensions in the intervening wing membrane, as indicated in the figure by lines with open arrowheads. Note that the direction of the tension changes at each digit, so that the bones of the digits are in compression. There are elastic fibres in the wing, most of which run more or less in the directions of tension. There are also some small muscles running within the wing membrane which must affect its tension: they include the coraco-cutaneus and the tensor plagiopatagii (Fig. 13-4). Part of the wing membrane (the propatagium) is anterior to the arm. Its free edge is kept taut by the occipito-pollicalis muscle (Fig. 13-4).

The largest of the flight muscles is the pectoralis, but it is smaller than in birds. It is presumably assisted by two other muscles, the posterior division of the serratus anterior (Fig. 13-4) and the subscapularis, which runs between the humerus and the inner face of the scapula (shown in the illustration of the shrew *Sorex* in Fig. 13-1). These muscles seem generally to total about 10–12% of the weight of the body, but even this is less than the 20% or so usual in birds. The serratus anterior does not attach directly to the humerus, but runs between the scapula and the ribs. Its action is to rock the scapula about its long axis, and it is probably responsible for the early part of the downstroke while the shoulder joint is still bent upwards as far as it will go. Several muscles co-operate in the upstroke but none is as large as any of the three main muscles of the downstroke. In birds the two main flight muscles, the pectoralis and supracoracoideus, both originate mainly on the enormous keeled sternum. In bats only part of the pectoralis originates on the sternum, which is much smaller relative to body size than in birds (though larger than in most other mammals) and has only a small keel. All the other flight muscles originate elsewhere.

Photographs of bats flying show that they move their wings in much the same way as birds. Aerodynamic forces presumably act in the manner indicated for birds in Fig. 11-9. The wing membrane balloons upwards between the digits during the downstroke, showing that substantial lift

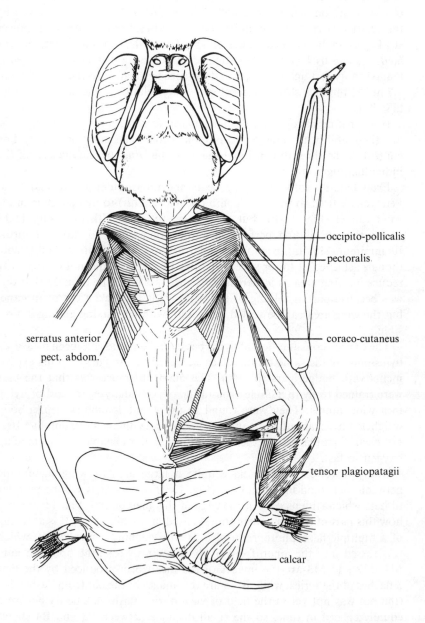

Fig. 13-4. A dissection of a bat, showing various muscles. From T. A. Vaughan (1970). In W. A. Wimsatt (ed.), *Biology of bats*, vol. 1, pp. 139–94. Academic Press, New York.

is being produced, but its shape during the upstroke indicates that the forces acting on it then are small. This confirms the impression given by the relative sizes of the muscles that the downstroke is the main power stroke, as in birds. The range of speeds at which a small bat, *Myotis lucifugus* can fly has been investigated by releasing it in a wind tunnel. It tended to fly fast against headwinds at speeds (relative to the air) of up to 7.7 m s^{-1} but flew slowly, almost stationary relative to the air, in a strong tailwind.

Insect-eating bats do not fly particularly fast in comparison with birds. but their ability to manoeuvre is remarkable. This may be due to fine control of the three-dimensional shape of the wing, by the muscles of the individual digits.

Fruit bats fly with slow wing beats and intermittent gliding. Dr C. J. Pennycuick tried to train six Tomb bats (*Rousettus*) to fly and glide in his wind tunnel (Fig. 11-10) but was successful only with one. Fig. 11-13 shows that its gliding performance was comparable to those of birds. Its minimum sinking speed was similar to those of the vulture and falcon, though achieved at lower speeds. Unlike the birds, it did not consistently reduce its wing area as it increased its speed. It might be thought that this was because partial folding of the wing tends to slacken the membrane, but the wing area can be varied quite a lot without making gliding impossible.

The oxygen consumption of flying bats (*Phyllostomus*) has been measured by essentially the same technique as was used by Dr Tucker in his experiments with budgerigars (p. 346). The main difference was that the bats were trained to fly in a circle around the oxygen analyser instead of flying in a wind tunnel. They were found to use about 25 cm^3 oxygen (g body weight)$^{-1}$ h^{-1}: about the same as budgerigars flying horizontally. It is not clear what features of their respiratory systems enable bats to take up oxygen so fast.

Many insect-eating bats have quite a large area of wing membrane between their hind limbs and tail (Fig. 13-4: the calcar is a bone peculiar to bats which stiffens part of the edge of this membrane). Fig. 13-5 shows how this part of the wing membrane is used in catching prey. It is a tracing of a multiple-flash photograph of a *Myotis* catching a mealworm, which was tossed into the air in front of the camera by a spring-operated gun. M1, M2 and M3 are the images of the mealworm produced by the first 3 flashes of the series, while B2–5 are the images of the bat from flashes 2–5 (the bat was not yet in the field of view during flash 1). The flashes were equally spaced in time, so the small distance between B3 and B4 shows that the bat slowed down considerably as it took the mealworm. In B3 the membrane between the legs is spread with the mealworm M3 just in

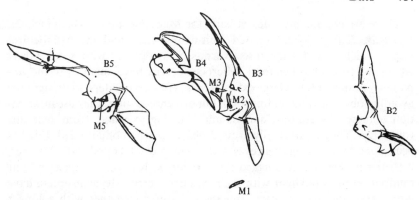

Fig. 13-5. Tracing from a multiple-flash photograph of a bat (*Myotis lucifugus*) catching a mealworm which was tossed into the air. Flash rate 12 s⁻¹. Further explanation is given in the text. From F. A. Webster & D. R. Griffin (1962). *Anim. Behav.* **10**, 332–40.

front of it. In B4 the bat has bent its body to bring this membrane, with the mealworm enclosed, over its face. In B5 the bat is holding the meal-worm in its mouth. Other photographs show *Myotis* catching thrown meal-worms in its wings and Fig. 13-6 shows the Horseshoe bat, *Rhinolophus*, using its wing to catch a moth which had been released near it (this bat does not seem to use the membrane between its legs for catching insects). Use of some part of the wing membrane must make the capture of insects much easier, than if they were caught directly in the mouth: the area of the open mouth of *Myotis* is only about 1 cm², but the area of its spread wings is about 80 cm².

Fig. 13-6. Tracing of two images from a multiple-flash photograph of a Greater horseshoe bat (*Rhinolophus ferrum-equinum*) catching a moth. From F. A. Webster & D. R. Griffin (1962). *Anim. Behav.* **10**, 332–40.

There are two main groups of bats: the Megachiroptera, which includes *Pteropus, Rousettus* and related fruit-eating bats, and the Microchiroptera, which includes the insectivorous bats together with some others which eat fruit or other food. Microchiroptera have small eyes and apparently poor sight, but large external ears. They do not depend on sight for avoiding obstacles and finding food, but on echolocation: they use information obtained by emitting sounds and listening to echoes from their surroundings. They fly normally around obstacles when blindfolded (this has been done by covering their eyes with opaque black collodion). However, if their ears are tightly plugged they collide with obstacles apparently at random, even in daylight with their eyes uncovered. Megachiroptera have larger eyes and smaller ears, and these features (together with a longer snout than is usual in bats) give the head of *Pteropus* the rather dog-like appearance which presumably inspired the name 'Flying fox'. Most Megachiroptera seem to depend on eyesight for orientation and do not use echolocation; specimens forced to fly blindfold have been observed to fly slowly backwards, reaching out with their feet for a perch. An exception is *Rousettus*, which uses echolocation when flying in darkness. It can fly blindfold, but can also fly well with its ears covered and eyes uncovered if there is sufficient light.

The sounds which most bats use for echolocation are inaudible to human ears, because of very high frequency. They can, of course, be detected by suitable microphones and displayed as wave forms, for instance on a cathode-ray oscilloscope. They vary considerably between groups of bats. Bats of the family Vespertilionidae, which includes *Myotis* and most of the other bats of Britain and the northern United States, make very brief but intense chirps. (A sound inaudible to man on account of high frequency may nevertheless be intense in terms of energy flux.)

Bats chirp more frequently as they approach an obstacle or insect prey, and this gives an indication of the distance at which objects are detected. Many observations have been made of bats flying along a room, steering between wires stretched from floor to ceiling. *Myotis* started increasing their chirping rate while still 2 m from wires 1 mm in diameter. They must have been able to detect the wires at *at least* that distance.

The chirps become shorter as a bat approaches an insect. *Myotis* have been filmed and their chirps recorded, as they chased mealworms thrown into the air. Each chirp lasted about 1.5 ms while the bat was 0.5 m from the mealworm (as shown in the film), but only 0.3 ms at 0.1 m. Sound travels at 330 m s^{-1} in air, so in 1.5 ms it travels 0.5 m and in 0.3 ms, 0.1 m. In each case the first sound waves must have been reaching the mealworm as the last were emitted by the bat. If chirps 1.5 ms long were still being emitted when the bat was only 0.1 m from the mealworm, the echo would return to the bat before the chirp was over. The relatively weak echo might not be easily heard, during the intense chirp.

There is rather surprising evidence that their echoes from large objects may actually sound louder to bats than the chirps themselves. Electrodes have been inserted into the ears of a Mexican bat (*Tadarida*) to record electrical potentials (cochlear microphonics) produced by sounds. These electrodes were connected to long, light leads so that the potentials could be recorded while the bats crawled and even flew around the laboratory. It was found that echoes of chirps from a wire window screen, over 1 m from the bat, produced microphonics almost as large as did the original chirps. This is probably partly due to the sound being directed forward from the mouth, not back towards the pinnae. However, there is another effect. Microphonics were recorded as before while a sound of constant amplitude was emitted by a loudspeaker. The size of the microphonics due to this sound diminished immediately before each chirp, and returned to normal immediately after it. The ear was apparently made less sensitive to sound while the chirp was being emitted, and allowed to return to normal before the echo returned. There is evidence that this is achieved by brief contraction of muscles attached to the ear ossicles.

High-frequency sounds are most appropriate for locating small objects. This is because the intensity of sound reflected or scattered by an object depends on the ratio between the dimensions of the object and the wavelength of the sound. Objects which are large relative to the wavelength reflect sound in particular directions, as a mirror reflects light. Those that are small scatter sound in all directions and the intensity of the scattered sound falls off very rapidly indeed as the diameter falls below about 0.1 wavelengths. The frequency falls in the course of a chirp but may be 10^5 s^{-1} or even more at the beginning. Sound of this frequency, in air, has a wavelength of only 3 mm, and so is scattered reasonably effectively by insects down to the size even of gnats.

How much information can a bat obtain from an echo? It seems obvious from the facility with which insects are caught and obstacles avoided, that it can distinguish the direction from which the echoes come. A vespertilionid bat with one ear blocked is nearly as helpless as with both blocked, so the sense of direction presumably depends largely on comparisons between the two ears (as in other mammals. See p. 391.) It also seems clear that distance can be judged. This has been confirmed by training *Eptesicus* to fly for food to whichever of two similar platforms was the nearer. It could make the correct choice even with one platform at a distance of 60 cm and the other at 58.5 cm. In other experiments it was shown that the bats could distinguish between triangular plates differing in area by 17% or more, and between triangles of the same area but different proportions.

Readers are reminded that the above account of echolocation is based mainly on Vespertilionidae. Bats of other families make rather different sounds and in some cases apparently use distinctly different mechanisms. For instance, horseshoe bats (Rhinolophidae, Fig. 13-6) emit rather long pulses of sound which overlap with the returning echo, and their ability to avoid obstacles is not seriously impaired by blocking one ear.

CARNIVORES

Infraclass Eutheria, order Fissipedia

The order Fissipedia includes the dogs, cats, bear, hyaenas and many smaller mammals. Some of them are dangerous predators but others feed largely on invertebrates or even plants. The differences in food habits are reflected by differences in structure, but there is some degree of uniformity in the structure of the jaws and teeth. Most Fissipedia have large canine teeth, specialized posterior teeth, large temporalis muscles and a hinge-like jaw articulation. Large canines are not a peculiarity of Fissipedia; even *Didelphis* (Fig. 12-10) has quite large canines. The value of large temporalis muscles, when canines are to be used for tearing food and tugging at prey, has already been explained (p. 389). Table 13-1 illustrates the general rule that the temporalis makes up a larger proportion of the jaw musculature in carnivores than in herbivores. The masseter is relatively small in carnivores, and the pterygoideus muscles are particularly small. The latter muscles lie median to the jaw and would restrict the gullet if they were too large. Fissipedia tend to bolt their food in large hunks, and so need a wide gullet. The articular processes of the lower jaw form relatively long cylinders (Fig. 13-7A). These fit into transverse grooves in the squamosals, where they are held by ligaments. The only possible jaw movements are opening and closing with the articulation acting as a hinge, and slight sliding of the whole jaw from side to side. The former movements involves

TABLE 13-1.   *The weights of the jaw-closing muscles of some mammals. Each weight is given as a percentage of the total. Data from Becht, G. (1953).* Proc. K. Ned. Acad. Wet. **56,** *508–27*

	Temporalis	Masseter	Pterygoideus
Dog (*Canis*)	67	23	10
Tiger (*Panthera*)	48	45	7
Zebra (*Equus*)	11	50	40
Bison (*Bison*)	10	60	30
Marmot (*Marmota*)	19	66	15

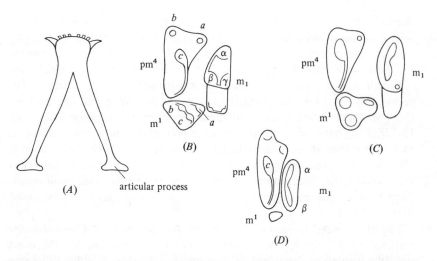

Fig. 13-7. (A) Dorsal view of the lower jaw of a cat (*Felis*) (B), (C), (D) Diagrammatic plans of the posterior teeth of a mongoose (*Herpestes*), a Hunting dog (*Lycaon*) and a cat, respectively. pm⁴, fourth upper premolar; m₁, first lower molar, etc. Cusps are labelled as in Fig. 12-11 A, B.

rotation of the cylindrical processes in the grooves and the latter sliding along the grooves.

Most Fissipedia have particular premolars and molars specialized for cutting and for crushing. In primitive mammals such as *Didelphis* all the molars are of the type shown in Fig. 12-11, and the premolars are simpler. In the various families of Fissipedia the molars and premolars are differentiated to different extents. The mongooses and genets (family Viverridae) are rather primitive Fissipedia, but even in them the last upper premolar and the first lower molar have disproportionately large main cutting edges (Fig. 13-7*B*). These teeth are referred to as the carnassial teeth. The edges in question, which cut against each other, involve the cusp labelled *c* in Fig. 12-11 in the upper carnassial and cusps α and β in the lower one.

The carnassial teeth are more sharply differentiated in dogs (family (Canidae). Cusps *a* and *b* in the upper carnassial, and γ in the lower one, are greatly reduced (Fig. 13-7*C*). However, the lower carnassial retains quite a substantial heel on which the first upper molar presses. This molar, in contrast to the carnassial immediately anterior to it, has lost all cutting edges but retains three blunt cusps: it has become specialized for crushing. There are also smaller, more posterior crushing teeth. Ferrets and their relatives (family Mustelidae) have the first upper molar even more highly modified as a crushing tooth.

When a dog gnaws at a bone with the side of its mouth it is using its carnassial teeth to slice or scrape off flesh. The cutting action of these teeth depends on the edges passing close together, just as the effectiveness of scissors depends on a tight rivet. The lower carnassials are closer together than the upper ones and the jaw has to be slid a little to the left to bring the left carnassials into the cutting position, or to the right for the right ones.

In cats (family Felidae) the teeth have lost their crushing function. The carnassials have very large cutting edges but the lower one has no heel, the first upper molar is tiny and there are no more posterior teeth (Fig. 13-7 *D*). Cats have also lost some anterior premolars, and so have been able to evolve rather short jaws and snouts. In contrast to cats, bears (family Ursidae) have their premolars and molars specialized for crushing. They are very like the corresponding teeth in pigs, rather flat with blunt cusps. They have no cutting edges and none of them could reasonably be called carnassial teeth.

The differences in dentition between the families of Fissipedia are matched to a large extent by differences in diet. Felidae are the most strictly carnivorous, and feed on mammals and birds. Ursidae kill some small mammals but also feed largely on insects (including grubs from bees' nests), honey, fruit and roots. Canidae and Mustelidae, with less uncompromisingly specialized teeth, are more varied in diet. For instance, among African Canidae the Hunting dog (*Lycaon pictus*) seems to eat only mammals while the Golden jackal (*Canis aureus*) eats beetles and other large insects and even fruit, as well as rodents and the young of larger mammals. Among British Mustelidae the stoat (*Mustela erminae*) feeds mainly on small mammals and birds while the badger (*Meles meles*) feeds largely on beetles and partly on bulbs and acorns, as well as eating small mammals. The teeth of badgers are more specialized for crushing and less for cutting, than those of stoats.

Fissipedia show differing degrees of specialization in their limbs, as well as in their teeth. The difference between the plantigrade stance of primitive mammals such as the opossum and the digitigrade stance of the dog has already been explained and illustrated (Figs. 12-3 to 12-7). It has been shown by X-ray cinematography that the ferret (*Mustela putorius*) stands and walks in very much the same way as the opossum. Other Mustelidae, Viverridae and Ursidae are also plantigrade. Canidae and Felidae are digitigrade, and some of them can run very fast. The best greyhounds (which have of course been bred selectively for speed) reach 65 km h^{-1} and the cheetah (*Acinonyx*) is even faster. Many records of cheetah speeds are unreliable estimates made in difficult field conditions, but a captive cheetah has been filmed galloping at 90 km h^{-1}.

Both dogs and cats use their speed in catching large prey, but their hunting methods are rather different. The Hunting dog lives in the grasslands of East and South Africa. It hunts in packs, often about a dozen strong, which travel in single file trotting at about 11 km h^{-1} on morning and evening hunting forays. They feed largely on gazelles (*Gazella* spp.) up to about twice their own weight. These are pursued at speed, often for several kilometres. The speed of an exceptionally long but unsuccessful chase of 5 km was estimated by observers following in a Land-Rover as nearly 50 km h^{-1}, and higher speeds have been estimated in shorter chases. The fastest dogs get ahead of the rest of the pack but if the prey changes direction the slower dogs may take the lead temporarily by cutting the corner. Of 91 chases followed by Land-Rover, 39 were successful. The leading dog usually brings the prey down by seizing a hind leg or flank in its jaws. The pack then tear open the belly of the prey (without taking further steps to kill it) and eat its flesh in large hunks. Wolves (*Canis lupus*) feed largely on reindeer or caribou (*Rangifer*) and moose (*Alces*). They tend to select young or sick animals to pursue and kill.

Cats seem generally to stalk their prey and then make a sudden fast dash at it: they do not pursue prey over long distances. Lions (*Panthera leo*) hunt singly or in groups. They feed on large herbivores including zebra (*Equus* spp.) and wildebeest (*Connochaetes*). Most of their prey weigh 100 to 300 kg, while lions themselves weigh around 150 kg. They usually knock or drag large prey down by grabbing the rump with their forepaws. Once down it is killed; the lion usually bites it in the throat and maintains the grip until it asphyxiates. Other cats are generally solitary hunters but all seem to kill their prey in the same way as the lion, before eating it.

The claws of Felidae (excluding the cheetah) can be withdrawn into bony sheaths when they are not in use. They are thus protected from wear in locomotion, and can be kept very sharp.

Carnivorous mammals, like owls and other predatory birds, tend to have their eyes facing forwards so that the fields of vision of left and right eyes overlap. This makes it possible for them to use the stereoscopic effect in judging distance, which may be important in pouncing on prey. Fig. 13-8 shows that the fields of vision of a cat's left and right eyes overlap by about 130°. Carefully designed tests have confirmed that cats make use of the stereoscopic effect (as people do) in judging distance. The field of vision of an eye cannot be much more than a hemisphere, unless the eye protrudes a long way from the surface of the head, so substantial overlap of fields of view implies loss of backward vision. A cat must turn its head to see something directly behind it. For a herbivorous animal likely to fall victim to predators, all-round vision may be more valuable than stereoscopic vision. Rabbits have their eyes mounted in the sides of their heads, not in front as in cats. Consequently, as Fig. 13-8 shows, they have all-round vision but little overlap between the fields of view. The stereoscopic vision of man and other primates is discussed in Chapter 14.

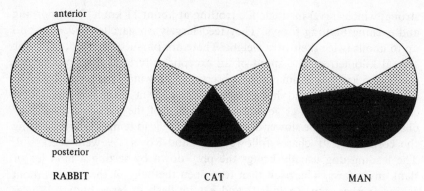

Fig. 13-8. Diagrams showing in plan view the fields of vision of a rabbit (*Oryctolagus*), a cat (*Felis*) and a man. The white segments are in the fields of vision of both eyes, the stippled ones of one eye and the black ones of neither eye. Based on data in K. Tansley (1965). *Vision in vertebrates*. Chapman & Hall, London.

MARSUPIAL CARNIVORES
Infraclass Metatheria, order Marsupialia, suborder Polyprotodonta

The common opossum, *Didelphis marsupialis*, has been referred to frequently as a typical primitive mammal. It is only one of a substantial number of very similar marsupial species, in America and Australasia, which range in length (excluding the tail) from about 9 to 50 cm. They feed on insects, lizards, rodents, fruit, etc. There are two larger, more specialized carnivores in Australasia: *Sarcophilus*, the Tasmanian devil, and *Thylacinus*, the Tasmanian wolf. The latter is almost, if not entirely, extinct. It preys on kangaroos, etc. It is superficially remarkably like a dog, and its molars are strikingly different from those of *Didelphis*. They have been modified in the same way as carnassial teeth, but instead of there being one upper and one lower carnassial on each side they are all modified. There are also marsupials which are strikingly like moles in structure and habits, and others which resemble anteaters. The variety of herbivorous marsupials is described in Chapter 14. The adaptive radiation of Australian marsupials is as remarkable as the radiation of South American characins (p. 178).

SEALS
Infraclass Eutheria, order Pinnipedia

Seals have probably evolved from terrestrial Fissipedia but have short tails and short limbs, with large, paddle-like hands and feet. Sealions (family Otariidae) swim mainly by movements of their forelimbs, like otters and dogs. Typical seals (family Phocidae), when swimming fast and straight, keep their fore-flippers against their sides and use their hind flippers. Sometimes these are held soles together behind the body and used like the tail of a fish: they are swept from side to side by lateral bending of the body. More usually they are used in turn, the left flipper being extended as the pelvis is swung to the right, and vice versa. The fore-flippers are used in turning, as hydrofoils or paddles, and sometimes for paddling the seal slowly along.

Seals come on to land (or in many cases ice) to breed. Sealions can turn their feet to what is in other mammals the normal position, with the toes pointing forward. Their manner of crawling is not too ungainly. Phocid seals cannot turn their toes forward, and on land can only hitch themselves forward in a series of jerks. The hindquarters are drawn forward, the head and thorax are raised on the forelimbs, and the body is pushed forward from behind. Fig. 13-9 shows how the forelimbs are used. Notice how the digits are bent, so that the claws grip the ground. Speeds of about 10 km h^{-1} are possible over short distances.

Most seals feed mainly on fish and cephalopods. Most have large canines and a hinge-like jaw articulation, like Fissipedia, but have simpler conical or triangular premolars and molars. Though modified as flippers, the forelimbs are still quite effective for manipulating food: fish are held between them while the flesh is eaten.

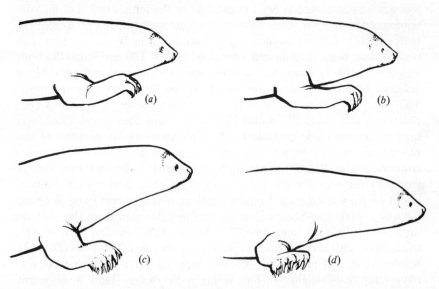

Fig. 13-9. Drawings from selected frames of a cinematograph film of a young Grey seal (*Halichoerus grypus*) crawling on land. From K. M. Backhouse (1961). *Symp. zool. Soc. Lond.* **5**, 59–75.

The natural diving habits of seals are awkward to study. It is not easy to follow a seal under water, and it may surface some distance from where it dived. This difficulty was overcome in a study of the Weddell seal, *Leptonychotes weddelli*, which spends the winter under the Antarctic ice sheet. It breathes at holes in the ice: these probably originate as natural breaks but it uses its teeth to keep them open. Captured Weddell seals were taken to a man-made hole in an ice-sheet which was apparently otherwise unbroken over a wide area. They were released and dived, but had to return to the same hole to breathe. Instruments were attached to them that made records of pressure changes (and so depth changes) and could be recovered when the seal returned. Records of 381 dives by 27 seals were obtained. Most dives were to less than 100 m and most lasted less than 5 min, but the deepest was to 600 m (this was about the depth of the bottom) and the longest lasted 43 min. Deep dives were generally fairly brief and seem to have been hunting forays. The longest dives were relatively shallow, and it is thought that the seals may have been searching for other holes in the ice. Information on the diving habits of other seals is sparse, but several species have been hooked on lines at 180 m or more.

How does a seal survive a 40 min dive without asphyxiating? The quantity of oxygen it can take with it is strictly limited. Typical seals have lungs which occupy about 10% of the volume of the body, when full. Seals generally breathe out before they dive (the probable advantage of this will be considered later), but even when the lungs are full of air containing 20% oxygen, they can only contain about 20 cm^3 oxygen (kg body weight)$^{-1}$. More oxygen can be carried in the blood. Various seals which have been investigated contained about 120 cm^3 blood (kg body weight)$^{-1}$. This is more than is usual in terrestrial mammals: the blood volumes of men and dogs are about 70 and 90 cm^3 kg^{-1}, respectively. The oxygen capacity of the blood of various species of seals has been measured, and values of around 300 cm^3 l^{-1} have been found. (Such very high values are made possible by the high haemoglobin contents of seal blood – about 20% by weight.) Hence when all the blood is saturated it contains about 40 cm^3 oxygen (kg body weight)$^{-1}$. Oxygen can also be stored in the muscles, combined with myoglobin. Seal muscle contains about 80 g myoglobin kg^{-1} (eight times as much as beef) and is consequently a dark blue-black colour. It can be calculated from this that the myoglobin of a seal containing 35% by weight of muscle, must be able to combine with about 40 cm^3 oxygen (kg body weight)$^{-1}$. Thus the lungs, haemoglobin and myoglobin between them can take up about 100 cm^3 oxygen (kg body weight)$^{-1}$. This is far more oxygen than a terrestrial mammal could take with it on a dive, but still not enough to support metabolism for long. The basal metabolic rates of various seals have been measured, and found to be considerably higher than Fig. 11-4 would suggest. As an example, consider *Phoca vitulina*, which is known as the Common seal in Britain and as the Harbor seal in the United States. A 30 kg specimen was found to use 7 cm^3 oxygen (kg body weight)$^{-1}$ min^{-1}, so the estimated oxygen store of 100 cm^3 oxygen kg^{-1} would only last for about 14 min if the seal was resting, or less if it was active. In fact this species has survived 28 min submergence in a laboratory experiment.

Long dives are apparently made possible by anaerobic metabolism. In one series of experiment seals were submerged in a bathtub, and samples were taken periodically from their back muscles. When the samples were analysed it was found that the oxygen (combined with myoglobin) in the muscle declined almost to zero in the first ten minutes or so of submergence, and that the lactic acid concentration rose thereafter.

Ultrasonic flowmeters have been used to investigate changes in blood flow during diving. A seal was anaesthetized and opened, flowmeters were fitted to selected blood vessels, and it was sewn up and allowed to recover. In some experiments the seals were immersed forcibly but slightly different results were obtained if diving was voluntary: seals were specially trained for this purpose, to submerge their heads in a bowl of water or even to dive in a pool on command. A record of a typical experiment is shown in Fig. 13-10. Each hump in the record represents a single heart beat. The height of the hump shows the velocity of the blood and the area under the hump shows the volume of blood pumped in that beat. The frequency of the heart beat falls considerably during the dive but the volume of blood

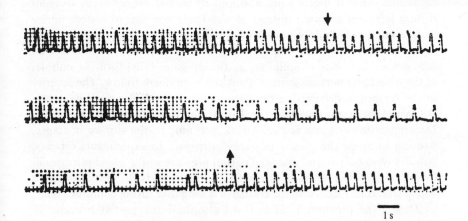

Fig. 13-10. A record of blood flow in the pulmonary artery of a Californian sealion (*Zalophus californianus*). The sealion immersed its head in water at the first arrow, and raised it again at the second. From R. Elsner, D. L. Franklin & R. L. Van Citters (1964). *Nature, Lond.* **202**, 809–10.

pumped at each beat stays the same. This particular experiment, with the flowmeter on the pulmonary artery, shows the total output of the heart. Other experiments with flowmeters on other arteries indicate that blood flow to the brain is reduced less, during a dive, than blood flow to the muscles and viscera. Brain function in man, and presumably in seals, is rather easily disturbed by lack of oxygen.

The extent to which the muscles are isolated from the blood during a dive is illustrated by further results of the experiments in which muscle samples were taken from submerged seals. Blood samples were taken as well. During submersion, the partial pressure of oxygen in the muscles fell well below the partial pressure in the blood, and the concentration of lactic acid in the muscles rose far above the concentration in the blood. It is only after surfacing that the concentration of lactic acid in the blood rises at all substantially.

Some of the veins of seals are enlarged to form big sinuses. Blood excluded from the muscles, during a dive, apparently accumulates in them.

The partial pressure of dissolved nitrogen in the blood of animals breathing air at the surface is 0.8 atm, the same as the partial pressure of nitrogen in the atmosphere. If the animal dives the air in its lungs is compressed, the partial pressure of nitrogen in it rises, and more nitrogen dissolves in the blood and tissues. Two of the hazards of diving, for man, result from this. The first is nitrogen narcosis: high partial pressures of dissolved nitrogen in the blood have an anaesthetic effect. Human divers, breathing compressed air, suffer only slight effects at 30 m but many become incapable of useful work at about 80 m. For deeper dives, oxygen/helium mixtures are used instead of air. The other hazard is decompression sickness, which does not occur during the dive but afterwards, when the diver returns to the surface. It is due to the additional dissolved nitrogen coming out of solution, as the pressure falls, forming bubbles in the blood and nervous system. Pain and even death follow. The severity of the symptoms depends on the duration of the dive as well as its depth, but symptoms are rare even if the dive is long if the depth is 13 m or less. Decompression sickness is avoided by returning to the surface in stages, pausing to allow the release of excess nitrogen. In experiments on cats, bubbles were only found after the partial pressure of dissolved nitrogen in the blood had been allowed to rise to 3.3 atm. This is the partial pressure which would be reached in the course of a long dive, using compressed air, to 31 m. (The pressure at 31 m is 4.1 atm, and the partial pressure of nitrogen in air compressed to 4.1 atm is $0.8 \times 4.1 = 3.3$ atm.)

Seals dive to depths far greater than 30 m, and so might be expected to suffer from both nitrogen narcosis and decompression sickness. How do they avoid these hazards? The problem has been investigated by Dr G. L. Kooyman and his colleagues. They fitted seals (*Mirounga* and *Phoca*) with a catheter in a vein or artery so that they could take samples of blood, and also with electrocardiograph electrodes so that they could ensure that the experiments were not harming them. They were then subjected to simulated dives. A seal was put in a pressure chamber which was completely filled with water, and the pressure was increased. Most of the experiments involved pressures of 4–15 atm, simulating dives of 30–140 m. Blood samples were taken periodically and analysed. It was found that the partial pressure of nitrogen in the blood rose in the first few minutes of the dive, but generally reached a steady value of about 2 atm. This is lower than the partial pressures which cause narcosis, or serious danger of decompression sickness, in man. How is it kept so low, even at pressures of 15 atm?

One important factor is that the seal dives with only a limited volume of air in its lungs. This volume was determined by measuring the volume change of the seal during compression, in essentially the same way as the volumes of fish swimbladders were determined in the apparatus shown in Fig. 5-5. It was found to be only about 20 cm^3 (kg body weight)$^{-1}$ (at atmospheric pressure): the seals dived with their lungs only partially filled. This means that they only took 16 cm^3 nitrogen kg^{-1} down on the dive. The solubility of nitrogen in water or blood is about 14 cm^3 (kg water)$^{-1}$ atm^{-1}, and about 0.12 of the weight of a seal is blood, so if all this nitrogen were dissolved in the blood it would increase the partial pressure of dissolved nitrogen by about $16/(14 \times 0.12) = 10$ atm. It could not exceed this, even at depths where the pressure is far greater. This is in contrast to the situation in human divers using compressed air, where nitrogen supply is virtually unlimited. Even this value of 10 atm would never be reached or even approached if much of the nitrogen in the lungs were prevented from dissolving, or if much nitrogen diffused from the blood to other tissues.

In fact, remarkably little of the nitrogen dissolves. This has been demonstrated by taking X-radiographs of seals in the compression apparatus at pressures up to 31 atm (equivalent to a depth of 300 m). Powdered tantalum was blown into the bronchi to coat their walls and make them show up better in the radiographs. It was found that the trachea was compressed considerably at the highest pressures, but that the bronchi and bronchioles were hardly compressed at all. The bronchi and bronchioles are reinforced by cartilage and are much more rigid than in land mammals. It seems that high pressures make the lung alveoli collapse and drive all the air into the more rigid air passages. From there, its nitrogen is much less easily absorbed than if it remained in the alveoli.

The fur seals (*Arctocephalus* and *Callorhinus*) have dense fur which provides heat insulation both on land and in water. It is apparently effective in retaining an insulating layer of air under water. This has been demonstrated by slipping a thermocouple immediately under the skin of a captive *Callorhinus*, which was then put in water at 10 °C. The thermocouple registered about 35 °C, little below the deep body temperature of about 37 °C. The fur does not cover the flipper, and temperatures in the flippers are apt to change rapidly. Body temperature is apparently controlled largely by varying the blood supply to the flippers: heat can be conserved by reducing their blood supply, or dissipated by increasing it.

Other seals have sparser hair which is easily wetted, and experiments with thermocouples show that it has hardly any insulating effect under water. When *Phoca* is immersed in water the temperature of its skin falls almost exactly to the temperature of the water. When immersed, such seals are almost entirely dependent for heat insulation on the layer of fat (blubber) under the skin. The blubber does not cover the flippers, so heat loss can be controlled as in fur seals, by adjusting blood flow to the flippers. Heat loss through the blubber is probably also adjusted, by adjusting blood flow through it.

The thermal conductivity of hunks of seal blubber has been measured, and found to be about four times the conductivity of fur. A 4 cm thickness of blubber would be needed, even if no blood were allowed to flow through it, to give the insulating effect of a 1 cm thickness of fur. The fur of *Callorhinus* is about 2 cm thick, and the blubber of *Phoca* about 4 cm. The insulating effect of fur must decrease as depth of submergence increases, as the trapped air is compressed. Blubber is not affected in this way.

WHALES
Infraclass Eutheria, order Cetacea

The whales include the largest living animals. The largest species of all is the Blue whale (*Balaenoptera musculus*) with an average adult length of about 25 m and a weight of about 100 tons. Weighing a large whale, is, of course, a laborious process, involving cutting it up carefully into pieces of manageable size. Blue whales up to 134 tons have been actually weighed, and others have been estimated to be even heavier. Large whales are very much heavier than any of the sauropod dinosaurs (p. 317), which included the heaviest land animals. The problem of support which probably sets an upper limit to the size of terrestrial animals does not arise for whales. which live in seawater of about the same density as themselves.

Whales are much more specialized aquatic animals than seals, and never leave the water; indeed, many whales die as a result of accidental stranding.

The young are born and suckled in the water. The forelimb is more highly modified as a flipper than in seals; the digits are not distinguishable externally, and little movement is possible at any of the joints except the shoulder. No hind limbs are visible but nearly all whales have rudiments of the pelvic girdle embedded in the ventral body wall (Fig. 13-11). In some species rudiments of a femur and even a tibia are attached to the girdle rudiment. These rudiments are well anterior to the tail flukes,

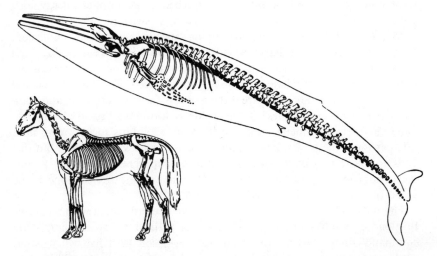

Fig. 13-11. Skeletons of a Blue whale (*Balaenoptera musculus*) and, for comparison, of a horse (*Equus caballus*). From E. J. Slijper (1962). *Whales.* Hutchinson, London.

which are not derived from limbs but have evolved as new structures, like the tail flukes of ichthyosaurs (Fig. 10-17*b*). However, they are not vertical like fish tails and ichthyosaur flukes, but horizontal. The swimming action is essentially like that of fish with hydrofoil tails (Fig. 4-2) except that the movement is up and down, not side to side. The up-and-down beat may be a legacy from galloping ancestors; galloping involves up-and-down bending of the back, and the vertebrae and muscles of whale ancestors were probably adapted for this rather than for lateral bending.

Most whales, like ichthyosaurs, have a dorsal 'fin'. Their bodies are well streamlined, tapering smoothly at the rear. The flukes, flippers and dorsal fin are also streamlined in section. A slender neck would be a departure from the ideal streamlined shape, and would increase drag by generating eddies. There is effectively no neck; the neck vertebrae are greatly shortened and in many species fused together, and the head merges without interruption into the trunk. There is no protruding external ear, and the penis and nipples are housed in slits in the body wall from which they emerge only when in use. The skin is very smooth and is hairless, apart from short vibrissae.

One would suppose from their more streamlined shape that whales would leave a less disturbed wake than seals. A chance observation seems to confirm this. Seals and dolphins were watched swimming in the evening, through plankton which emitted flashes of light when the water was disturbed. Both seals and dolphins left bright wakes, but those of the seals were much larger and brighter. A disturbed wake implies energy lost, giving kinetic energy to the water and setting it swirling.

Many estimates of whale swimming speeds have been published. Dolphins in particular have often been credited with remarkably high speeds, but not all records are reliable. Most depend on comparisons between the speed of a whale and the measured speed of a ship. Whale and ship speeds are hard to compare if the whale is far from the ship, and comparison can be meaningless if the whale is too close because of hydrodynamic interaction between it and the ship. Dolphins such as *Delphinus* often swim in the bow waves of ships, keeping pace with the ship, but this does not mean they could swim so fast if the ship were not there. By adopting an appropriate position in the bow wave they can be carried along at the speed of the ship, without having to swim.

The most reliable estimates of speed have been obtained with a captive Bottlenose dolphin (*Tursiops*) which was trained to swim rapidly on command along a 61 m course. It was rewarded with fish whenever it made a particularly fast run. Its best speed for the course was 30 km h^{-1} (8.3 m s^{-1}). Equation 4. 1 can be used to estimate the power required for swimming at this speed. The power needed to propel a body is (drag $\times$ velocity) or $\frac{1}{2}\rho A U^3 C_D$. The density ρ of seawater is just over 1000 kg m^{-3}. The frontal area A of the dolphin must have been about 0.11 m^2. The velocity U was 8.3 m s^{-1}. The drag coefficients of rigid streamlined bodies rise at high Reynolds numbers because the fluid in the boundary layer becomes turbulent. A coefficient of 0.055 would be expected for a rigid body of conventional streamlined shape at the Reynolds number appropriate to the dolphin. However, the dolphin has a special shape (also seen in tunnies) with the widest part well back, it is not rigid but undulates its tail, and its skin has peculiar mechanical properties. There are reasons for thinking that all these features may help to restrict turbulence and reduce drag. On the other hand the efficiency of swimming cannot be as much as 100%, so the mechanical power output of the muscles would have to be more than that required simply to overcome drag on the body. We will probably not be too far wrong if we estimate from the formula that the power required is $\frac{1}{2} \times 1000 \times 0.11 \times 8.3^3 \times 0.055 = 1700$ W.

Measurements have been made of the power outputs of men, pedalling against a resistance on a bicycle ergometer. The maximum power which athletes can develop in a burst of activity lasting a few seconds seems to be about 1400 W. The dolphin weighed 89 kg and a typical athlete would weigh about 70 kg, so their power outputs, per unit body weight, seem to be comparable. The propulsive muscles make up roughly 15% of the body weight in each case. It is often regarded as a great puzzle how dolphins can swim so fast, but the maximum speed seems in fact to be about what might be expected, on the basis of the likely power output of the muscles.

Whales breathe through their blow-holes which are nostrils which have shifted, in the course of evolution, to the top of the head. This shift has been largely responsible for extraordinary distortion of the skull (Fig. 13-12). As the nostrils moved posteriorly the nasal bones moved back with them and the premaxillae and maxillae extended back. Since there is no constriction at the neck large trunk muscles originate directly on the occipital bone, which is appropriately large. In some whales (Fig. 13-12, bottom) the maxilla and occipital actually meet on top of the skull. A likely advantage of having the blow-hole on top of the head is that the whale can remain horizontal when breathing.

In diving ability, whales seem generally comparable to seals. The large Sperm whale *Physeter* has been known to stay submerged for over an hour, but dolphins have not been observed to submerge for more than about 12 min. It seems to be a general rule that the large species make the longest dives. This is as might be expected. Basal metabolic rates of mammals tend to be proportional to (body weight)$^{0.75}$ (Fig. 11-4), but the oxygen store of a diving whale and the oxygen debt it can develop are presumably proportional to body weight. Hence the maximum duration of a dive, which is the time taken for metabolism to use up the oxygen store and develop the maximum oxygen debt, should be proportional to (body weight)$^{0.25}$. A fifty-ton Sperm whale should be able to submerge for $1000^{0.25} = 5.6$ times as long as a 50 kg dolphin. Weddell seals weigh only about 400 kg, so the 43 min dive by this species mentioned on p. 426 should perhaps be considered more remarkable than any known dive by a whale. Several Sperm whales have been found entangled in telephone cables at depths around 1000 m, and it is presumed that this must be within the normal depth range of the species. After long dives whales stay close to the surface for some minutes, taking frequent breaths.

Whales share various adaptations for diving with seals. Their bronchi and bronchioles are reinforced with cartilage rings. The main veins are capacious. The oxygen capacity of the blood is little higher than in typical terrestrial mammals but the concentration of myoglobin in the muscles is very high. Experiments with trained dolphins, like those with trained seals, have shown that the heart rate is much lower during a dive than at the

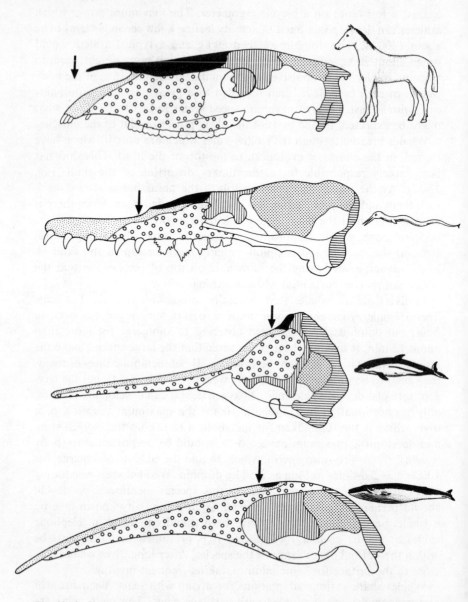

Fig. 13-12. Skulls of the horse (*Equus*), an early fossil whale (*Basilosaurus*), the Common dolphin (*Delphinus*) and the Common rorqual (*Balaenoptera*) showing progressive distortion of the skull. The horse is included as an example of a normal mammal. Fine stipple, pre-maxilla; black, nasal; coarse stipple, parietal; circles, maxilla; vertical hatching, frontal; horizontal hatching, occipital. Arrows point to the nostrils. After E. J. Slijper (1962). *Whales.* Hutchinson, London.

surface. Whales have no fur, and heat insulation is provided by a layer of blubber.

Dolphins are believed to use echolocation, much as bats do (p. 418). A captive *Tursiops* was persuaded by patient training to allow an experimenter to fit opaque plastic suction cups over her eyes. She swam without hesitation round the tank immediately after blindfolding. She could navigate blindfold between obstacles, find and pick up food without touching nearby obstacles, and perform tricks such as pressing an electric bell. Echolocation has also been demonstrated convincingly for porpoises (*Phocaena*) and Killer whales (*Orcinus*) but not, so far, for the larger whales. The sounds which are apparently used in echolocation are high-frequency clicks. It is not known how they are produced.

Echolocation will only work if the animal can emit sound in narrow beams, or judge the direction of returning echoes, or both. Echolocation by dolphins probably depends on directional emission, and partly on directional hearing. It has been shown by sound measurement in the water around captive dolphins that at any given distance the echolocation sounds are much more intense ahead of the animal than in other directions. Speculations as to how this is achieved involve the idea that the sound may be focused into a beam by the concave upper surfaces of the maxillae, which may serve as a concave mirror, and by the lens-shaped mass of fat which gives the bulbous shape to a dolphin's 'forehead'. Sound travels faster in fat than in water so a submerged convex mass of fat acts as a converging lens for sound. Possible mechanisms of directional hearing have also been suggested. There is no pinna to aid directional hearing, and indeed a pinna would not work in water in the way it does in air, because there is little contrast in acoustic properties between water and tissues. For the same reason the head would not shield the left ear from sounds coming from the right, were it not that the ears are surrounded (except on the lateral side) by foam. Sound approaching from the same side can enter freely but sound from other directions tends to be reflected by the gas in the foam. (The volume of gas is rather large so its natural frequency is low, and it cannot be expected to amplify vibrations appreciably, in the way swimbladders do (p. 172), except at very low frequencies.)

The modern Cetacea are divided into two suborders, the Odontoceti (toothed whales) and Mysticeti (whalebone whales). Only the toothed whales retain teeth after birth. The teeth are simple in shape, not differentiated into incisors, canines and molars (but there is more differentiation in some fossil whales, Fig. 13-12). Dolphins have numerous small teeth in their long jaws, and feed on fish such as the herring (*Clupea*) and cephalopods. The Killer whale (*Orcinus*) has fewer, larger teeth and feeds largely on seals and the smaller whales, which are swallowed whole. As it grows to a length of 9 m, its stomach can accommodate several of these animals at once. The Sperm whale (*Physeter*) and the Bottlenose whale (*Hyperoodon*) feed largely on squids. The Sperm whale has substantial teeth in the lower jaw but only rudiments hidden in the gum of the upper jaw. In the Bottlenose whale and some related species the only teeth which ever cut the gums are one or two pairs at the front of the lower jaw, and even they often fail to do so.

The Mysticeti have no teeth except as embryos, but they have in their mouths the whalebone or baleen that used to be used to stiffen corsets and the peaks of caps. This consists of keratin, not bone, and develops from the epidermis of the roof of the mouth as closely spaced vertical plates with fringed edges (Fig. 13-13). The fringes of adjacent plates become tangled together.

Whalebone whales feed largely on zooplankton, but also take larger prey including cephalopods and fishes. The plankton is strained from the water by the baleen. Its most important constituents are euphausiacean crustaceans about 3–4 cm long (excluding appendages).

There are two main groups of whalebone whales, known as the right whales and the rorquals. The right whales (*Balaena* and related genera) are right from the point of view of old-fashioned whalers: they swim rather slowly and they do not sink when killed. The rorquals (the Blue whale and other species of *Balaenoptera*, and the Humpback whale *Megaptera*) swim rather faster and sink after death, so that better equip-

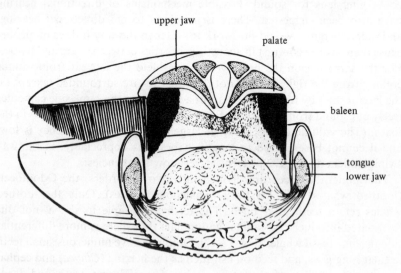

Fig. 13-13. A diagram of the chopped-off snout of a rorqual showing the positions of the baleen and tongue. From E. Hentschel (1937). In *Handbuch der Seefischerei Nordeuropas*, vol. 3 (1).

ment is needed for hunting them. The two groups feed differently. Right whales simply swim through plankton with their mouths open. The plankton is caught on the tangled fringes of the baleen as the water flows between the plates, and is periodically swallowed. The baleen is very long, averaging 3 m in one species. Rorquals have shorter baleen. They take a large mouthful of water, more or less close the mouth, so that water cannot escape except through the tangled fringes of the baleen, and then raise the floor of the mouth. The water is forced out, leaving the plankton behind on the baleen. Pleated ventral skin (Fig. 13-13) is a peculiarity of rorquals that may make the mouth more distensible.

FURTHER READING

General

See the list at the end of Chapter 12.

Insectivores

Crowcroft, P. (1957). *The life of the shrew*. Reinhardt, London.
Quilliam, T. A. (ed.) (1966). The mole: its adaptation to an underground environment. *J. Zool., Lond.* **149**, 31–114.

Bats

Cahlander, D. A., J. J. G. McCue & F. A. Webster (1964). The determination of distance by echolocating bats. *Nature, Lond.* **201**, 544–6.
Griffin, D. R. (1958). *Listening in the dark. The acoustic orientation of bats and men*. Yale University Press, New York.
Norberg, U. M. (1972). Bat wing structures important for aerodynamics and rigidity. *Z. Morph. Tiere*, **73**, 45–61.
Pennycuick, C. J. (1971). Gliding flight of the dog-faced bat *Rousettus aegyptiacus* observed in a wind tunnel. *J. exp. Biol.* **55**, 833–45.
Simmons, J. A. & J. A. Vernon (1971). Echolocation: discrimination of targets by the bat *Eptesicus fuscus*. *J. exp. Zool.* **176**, 315–28.
Thomas, S. P. & R. A. Suthers (1972). The physiology and energetics of bat flight. *J. exp. Biol.* **57**, 317–35.
Webster, F. A. & D. R. Griffin (1962). The role of the flight membrane in insect capture by bats. *Anim. Behav.* **10**, 332–40.
Wimsatt, W. A. (ed.) (1970). *Biology of bats*. Academic Press, New York.

Carnivores

Kruuk, H. (1972). *The spotted hyaena. A study of predation and social behaviour*. University of Chicago Press, Chicago.
Neal, E. (1969). *The badger*, 3rd edit. Collins, London.
Schaller, G. B. (1972). *The Serengeti lion. A study of predator–prey relations*. University of Chicago Press, Chicago.
van Lawick-Goodall, H. & J. (1970). *Innocent killers*. Collins, London.

Seals and whales

Andersen, H. T. (ed.) (1969). *Biology of marine mammals*. Academic Press, New York.

Harrison, R. J. (ed.) (1972). *Functional anatomy of marine mammals*. Academic Press, New York.

Harrison, R. J. & J. E. King (1965). *Marine mammals*. Hutchinson, London.

Kooyman, G. L., J. P. Schroeder, D. M. Denison, D. D. Hammond, J. J. Wright & W. P. Bergmen (1972). Blood nitrogen tensions of seals during simulated deep dives. *Am. J. Physiol.* **223**, 1016–20.

Lang, T. G. & K. S. Norris (1966). Swimming speed of a Pacific bottlenose porpoise. *Science* **151**, 588–90.

Slijper, E. J. (1962). *Whales*. Hutchinson, London.

14

Mainly herbivorous mammals

This chapter is the last in the book. It is about the major groups of mammals which feed mainly on plants, especially the rodents, the hoofed mammals and the primates.

FEEDING ON PLANTS

The main difference between plant and animal tissues, as food, is that plant cells are enclosed in cellulose cell walls. Cellulose is not digested by any known enzyme produced by vertebrates. It is a polymer of glucose, like glycogen, but glycogen has the glucose units linked together in a different way and is easily digested. Plant cell walls do not consist of cellulose alone but contain hemicelluloses (which are polymers of various sugars and sugar acids) and lignin (which is another organic polymer, of which the structure is not fully understood). These substances, like cellulose, are not digested by vertebrate enzymes.

The proportions of cell wall materials in different plant tissues vary enormously. For instance, the dry matter of wheat grains consists of about 2% salts, 12% cellulose, hemicellulose and lignin, and 86% fat, protein and easily digestible carbohydrate (mainly starch). The dry matter of mature ryegrass (*Lolium*) contains about 5% salts, 60% cellulose and hemicellulose, and 11% lignin, and less than 25% easily digestible material.

Not only are cell wall materials indigestible to animals lacking appropriate enzymes, but they may protect the contents of the cells from digestion. The digestible materials are not accessible until the cells have been broken open. Grain, fruit and other plant materials, containing a large proportion of digestible material enclosed in relatively thin cell walls, may only need crushing to make most of their food content available. Materials such as grass which contain relatively little digestible material, and that enclosed in thick cell walls, may require fine grinding. This can be very effective, in releasing cell contents, as the data on the Chinese grass carp shows (p. 190).

The grass carp can only digest the cell contents, however finely it rasps and grinds its food. Herbivorous mammals can make use of cell wall materials as well. They do not themselves produce enzymes that attack these materials but they have in their guts bacteria and protozoa which do. These organisms can digest cellulose and hemicellulose, but not lignin.

They are present in the stomachs of ruminant mammals such as cows and sheep. They are also present in other herbivorous mammals including pigs, horses and rodents, in the large intestine and in the caecum which branches off from its proximal end. In each case the part of the gut which contains them is large. In cattle and sheep the stomach accounts for some 70% of the capacity of the gut, and the large intestine and caecum for about 12% In the horse the stomach accounts for a mere 10% and the large intestine and caecum for about 60%. Carnivorous mammals also have bacteria inhabiting the large intestine but these are relatively unimportant: digestion is almost completed in the small intestine.

The stomachs of cattle and sheep consist of four connecting chambers. These are in order, from the oesophageal end, the reticulum, rumen, omasum and abomasum (Fig. 14-1). The rumen is much the largest. The active micro-organisms occupy the reticulum and rumen, where the pH is close to neutrality. The abomasum however has acid contents, like the whole stomach of most vertebrates. It is only into it that gastric enzymes and hydrochloric acid are secreted, and the pH of its contents is about 3. The omasum lies between the rumen and the abomasum. It is crossed by parallel sheets of tissue, like the leaves of a book, between which material must pass as it travels through the chamber. This arrangement must restrict mixing of the neutral rumen contents with the acid abomasum contents.

Whether the micro-organisms live in the stomach or the large intestine, their environment is anaerobic. The use they can make of foods eaten by

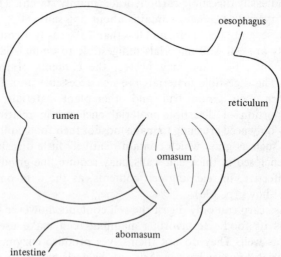

Fig. 14-1. A diagram showing the structure of the stomach of a cow.

their host is limited by lack of oxygen. Though polysaccharides can be hydrolysed and energy obtained from their constituent sugars by converting them to fatty acids, they cannot be oxidized. Acetic and propionic acids are produced, with other acids in smaller quantities. These can be absorbed and used by the host. Gases are also produced, mainly carbon dioxide and methane.

In horses and other mammals which have their micro-organisms in the posterior part of the gut the food is partly digested by the time it reaches the micro-organisms. The cell contents are largely digested and absorbed in more anterior parts of the gut, leaving little but cell wall materials for the micro-organisms. In ruminants, all constituents of the food are available to the micro-organisms. This could be regarded as a disadvantage of having the micro-organisms in the stomach, but there is an important advantage. The products of digestion leaving the rumen carry some of the micro-organisms with them. They go to the abomasum and intestine, where the micro-organisms themselves are digested. Thus foodstuffs which the micro-organisms use for growth are not permanently lost to the host. The rumen is never cleared of micro-organisms: there are always plenty left behind to maintain the population by division and growth.

Many experiments have been carried out with rumen contents removed from the animal, and kept *in vitro* in appropriate conditions. Oxygen is excluded, and the temperature is kept close to the natural body temperature. In such conditions, the micro-organisms can flourish. They can make use of foodstuffs which are provided, and grow. When glutamic and aspartic acids labelled with ^{14}C are added, labelled fatty acids are produced but little labelled protein. These amino acids are converted mainly to fatty acids rather than incorporated in the protein of the micro-organisms. When ammonia labelled with ^{15}N is added, ^{15}N gets incorporated in microbial protein. It is concluded that the micro-organisms break down protein in the hosts' food into fatty acids and ammonia, but can re-synthesize amino acids for their own use from these materials.

The fate of the energy content of food has been investigated in other experiments with rumen contents *in vitro*. In one set of experiments, cellulose was provided to sheep rumen micro-organisms, which broke it down. The fatty acids that were produced accounted for 76% of its energy content (heat of combustion), the methane for 7%, growth of the micro-organisms for 11% and heat which was released for 6%. If this had been happening in the sheep the methane and heat would have been lost, but the fatty acids would have been absorbed and the micro-organisms would eventually have been digested and absorbed. Thus 87% of the energy content of the cellulose which was digested would have been made available to the sheep.

The micro-organisms of the rumen typically consist of about equal weights of ciliate protozoa and bacteria. Though equal in weight, they are probably very unequal in importance. The bacteria are very much smaller than the protozoa, and consequently can be expected to have a much higher metabolic rate and to play the major part in digestion. It is a general rule that metabolic rate per unit weight tends to be higher for small organisms than for large ones: Fig. 11-4 illustrates the rule, but for immensely larger organisms than are being discussed here.

The micro-organisms are not present in the rumen at birth, but are gained subsequently. Enough to start the population are transferred from a cow to her calf when she licks and grooms it, and it swallows some of her saliva. The saliva is particularly likely to contain micro-organisms, because ruminants return rumen contents to the mouth when they chew the cud. Young ruminants may also acquire rumen micro-organisms in other ways. Rumen protozoa have been found on the grass in a sheep pasture, and rumen bacteria have been found in the air in a cowshed. It is more or less inevitable that the rumen will be infected by appropriate micro-organisms, unless the young animal is separated from other ruminants soon after birth.

If food is broken up before it is swallowed, more of it is immediately accessible to digestive enzymes and digestion can proceed faster. The premolar and molar teeth of herbivorous mammals play an important part in breaking up the food. They are of two main types. The more primitive is the bunodont type, with low cusps, which is found for instance in pigs, monkeys and man (Fig. 14-2a). Bears have similar teeth (p. 422). These are essentially crushing teeth, with no special provision for excessive wear. The more advanced is the hypsodont type, found for instance in horses and cattle (Fig. 14-2b). The cusps are high but their points soon wear away, and the fissures between the cusps are largely filled by cement. This is a form of bone which is present in small quantities only, around the roots, in bunodont teeth. Hypsodont teeth are adapted for grinding rather than mere crushing. They are particularly suitable for grinding grasses, which tend to be more abrasive than other foods. Most grasses have abrasive silica crystals in their leaves, and they are very apt to have grit on their surfaces because they grow so near the ground. An indication that the external grit may abrade the teeth more than the internal silica in the grass is provided by the observation that the teeth of horses kept on sandy pastures get particularly severely worn. The remarkably complete fossil record of the evolution of the horses shows that the early ancestors of modern horses had bunodont teeth but that horses with hypsodont teeth appeared in the late Tertiary, at about the time when grasses first became widespread and abundant.

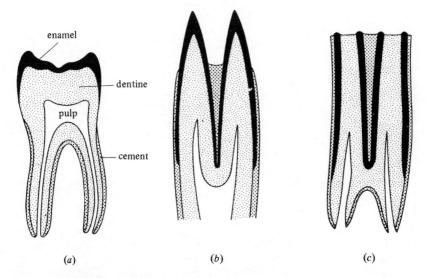

Fig. 14-2. Diagrammatic transverse sections of molar teeth of herbivorous mammals. (*a*) A bunodont molar; (*b*) an unworn hypsodont molar and (*c*) a worn hypsodont molar.

As a hypsodont tooth wears away its constituent layers of enamel, dentine and cement become exposed edge-on (Fig. 14-2*c*). Enamel has a much higher inorganic content than the other constituents and is accordingly more resistant to wear. As wear proceeds, the enamel comes to stand in ridges slightly above the worn surfaces of the dentine and cement. This leaves it rather more exposed to wear than the dentine and cement, so the latter do not wear into progressively deeper hollows. Once the difference in level has been established all the materials wear more or less evenly and the difference remains more or less constant. The rough file-like surface which is so effective for grinding is maintained.

The simplest type of herbivore molar is bunodont and square, with four cusps (Fig. 14-3*A*, *B*). The upper and lower molars are more or less identical. Monkeys and people have molars like this. Three of the cusps of the upper molars are apparently homologous with cusps *a*, *b* and *c* of primitive mammals (Fig. 12-11*A*) while the fourth is new. The cusps of the lower molars are apparently homologous with cusps β, γ, δ and ϵ of primitive mammals: α has been lost (Fig. 12-11*B*). Fig. 14-3*C* shows how the upper and lower molars fit together. Hypsodont, grinding teeth have elongated cusps, which when worn provide a more effective file-like surface than round cusps would. Ruminants grind their food by side-to-side movements of the jaw, and have longitudinally elongated cusps (Fig. 14-3*D*). Fig. 14-3*E* shows how the cusps of the lower teeth move transversely between the cusps of the upper ones, at right angles to the lines of

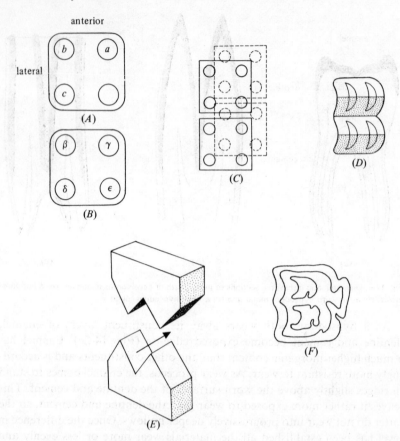

Fig. 14-3. Diagrams of molar teeth of herbivorous mammals. (*A, B*) Plan views of the crowns of simple bunodont molars, of the type possessed by monkeys and man. Upper and lower molars are shown. Cusps are labelled as in Fig. 12-11.

(*C*) A plan showing how the lower molars (broken outlines) fit against the upper molars (continuous outlines) in monkeys and man.

(*D*) A plan view of the crown of a molar of a ruminant mammal.

(*E*) A diagram showing how the lower molars of ruminant mammals move across the upper ones in chewing.

(*F*) A plan view of the crown of an upper molar of a horse.

exposed enamel. Notice that there are tall ridges running across each tooth from cusp to cusp, as well as the much lower ridges of enamel running along the tooth. It is the low ridges of enamel which have the file-like action. Horses have molars which wear much flatter, with nothing like the tall ridges of ruminants. Their cusps have complicated shapes so that the crown of the worn tooth has very long convoluted lines of enamel (Fig. 14-3*F*).

A human tooth erupts fully before it comes into use. Hypsodont teeth with high crowns that allow for wear erupt gradually, over a long period. For instance, a horse premolar or molar may have a crown 7 cm high but have less than 2 cm projecting above the jaw at any time. As it wears the tooth erupts further to keep the worn surface at this level. The tooth continues to grow in the young horse, and the root does not begin to form until the horse is about 5 years old.

The ruminant mammals owe their name to the habit of rumination, or chewing the cud. After a period of fermentation in the rumen food is returned to the mouth and chewed again, so that it is ground more finely than when it was first swallowed. Cattle fed on lush clover ruminate relatively little, but more time is spent ruminating more resistant foods such as hay. Up to eight hours per day may be spent ruminating. 30–85 cycles of jaw movements are used in chewing each bolus of food that is returned to the mouth.

Carnivorous mammals generally have large temporalis muscles and smaller masseter and pterygoid muscles. In herbivorous mammals the masseter and internal pterygoid tend to be large, and the temporalis small. The difference was illustrated by specific examples in Table 13-1, and an explanation was offered of the large temporalis muscles of carnivores. It was shown that a large temporalis muscle would counteract the force tending to dislocate the jaw, when the canine teeth were used to tear flesh. Herbivores do not generally have occasion to pull forcibly with their teeth, and most have no canines. They do not have the same need as carnivores for large temporalis muscles, and there seems to be an advantage in having large masseters and internal pterygoids instead.

Most herbivorous mammals grind food by a side-to-side jaw movement, essentially like the chewing action of the opossum (p. 386). When a cow, for instance, is chewing food at the left side of its mouth it opens its jaws and moves the lower jaw to the left. It then closes the left lower teeth against the upper ones and moves them, in the effective grinding stoke, towards the mid-line. The jaw muscles must press the left teeth together and at the same time pull the lower jaw towards the right. The pull to the right can only be provided effectively by the left internal pterygoid muscle (see Fig. 12-10d). Chewing on the right side would similarly demand a pull to the left, exerted by the right internal pterygoid. Thus the internal pterygoids apparently play a major role in chewing, and it is not surprising that they are large. Since the jaw is shaped in such a way as to make room for large internal pterygoids on its inner side, there is room also for large masseters on the outer sides. The masseters pull more or less vertically and can contribute to the vertical component of the force exerted in the effective grinding stroke, but not to the transverse component.

There is a striking difference in shape between the lower jaws of typical carnivorous mammals, and typical herbivorous ones (Fig. 14-4). In carnivores the large temporalis muscles have a tall coronoid process to attach to, but the angle of the jaw, where the masseter and pterygoid muscles insert, is shallow. In herbivores the angle of the jaw is deep, making room for the large masseter and internal pterygoid muscles, while the coronoid process is small. There is a corresponding difference in the skull. Carnivores generally have the jaw articulation approximately in the plane of the secondary palate, but herbivores have it higher. The two types of herbivore skull shown in Fig. 14-4 are compared later in the chapter.

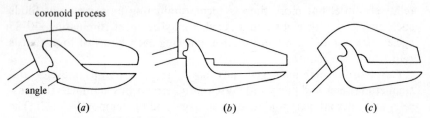

Fig. 14-4. Diagrams of the skulls of three mammals. (*a*) is a carnivore, and (*b*) and (*c*) are herbivores. Further explanation is given in the text.

The biomass of herbivores that can live in an area is limited by the rate at which the plants they feed on grow. Consider the herbivorous mammals that live in and around the Serengeti National Park, Tanzania. This is an area of 25000 km², which consists mainly of wooded grassland. It has been estimated from censuses of mammals that their total mass is about 108000 tonnes, or about 4 t km⁻². This total consists mainly of wildebeest (*Connochaetes*, 1.7 t km⁻²), zebra (*Equus*, 1.0 t km⁻²) and buffalo (*Syncerus*, 0.8 t km⁻²), weighing 80–500 kg each. The resting metabolic rates of mammals in this range of weights are around 2.5 cm³ oxygen kg⁻¹ min⁻¹ (Fig. 11-4). Since metabolism involving 1 cm³ oxygen releases about 20 J, this is equivalent to about 50 J kg⁻¹ min⁻¹ or 0.8 kW t⁻¹. A population of 4 t km⁻² would use about 3 kW km⁻² in resting metabolism alone. Since metabolizable energy is only about 60% of food intake (p. 376) food would be used at a rate of about 5 kW km⁻² for resting metabolism. The animals do not rest all the time, but feed and move about. The amount they eat has not been determined, but it seems likely, from investigations of sheep and cattle, to be two to three times the amount needed for resting metabolism, or 10–15 kW km⁻². This food energy must be derived from photosynthesis. It is only a tiny fraction of the average rate, over the year, at which solar energy falls on the Serengeti but it is probably a very large fraction of the average rate at which solar energy is incorporated by photosynthesis in grass. I have no data on grass production in the Serengeti but the average rate over the year of grass production in the tall-grass prairies of the United States of America, expressed in terms of energy, is about 15 kW km^{-2}.

SMALL HERBIVORES

Infraclass Eutheria, orders Rodentia and Lagomorpha

The Rodentia is by far the largest order of mammals. It includes some 1700 of the 4200 or so living species of eutherian mammals. The Lagomorpha is a much smaller order which consists only of the rabbits and hares, and the pikas (*Ochotona*). No member of either order is really large. The largest of all is the capybara (*Hydrochoerus*, a South American rodent) which attains weights of 45 kg and more. Most rodents and lagomorphs are much smaller than this, in the size range which extends from mice to hares.

The rodents are mainly herbivorous. The most familiar of them are the rats and mice, which have come, through association with man, to eat a wide variety of the foods which man's presence makes available. Many other rodents also take some animal food. Deer mice (*Peromyscus* spp.) include substantial proportions of beetles and caterpillars in their food, as well as seeds, etc. Mice and caterpillars are important foods of different species of ground squirrel (*Citellus*). Some rodents such as squirrels (*Sciurus*) and field mice (*Apodemus*) eat mainly seeds, nuts, berries and buds, while others such as the vole *Microtus* feed mainly on grass. Rabbits feed mainly on grass.

Rodents have a single pair of chisel-shaped incisors in each jaw (Fig. 14-5). Lagomorphs have, in addition, a much smaller second pair of upper incisors. There are no canines in either order but a toothless gap, known as the diastema, between the incisors and the premolars. The premolars and molars are bunodont in rats, squirrels, etc., but hypsodont in grazing rodents such as voles, and in rabbits.

The jaw movements of rats (*Rattus*) have been studied by X-ray cinematography. The rats were fed biscuit impregnated with barium sulphate (which is opaque to X-rays) so that the food as well as the bones and teeth could be distinguished in the films. Two main jaw actions were observed, one used for biting and gnawing with the incisors and the other for chewing with the premolars and molars (Fig. 14-5).

The range of jaw movements that is possible in any animal depends on the structure of the jaw articulation. The jaw articulations of the rat are very different from those of primitive mammals such as *Didelphis* (Fig.

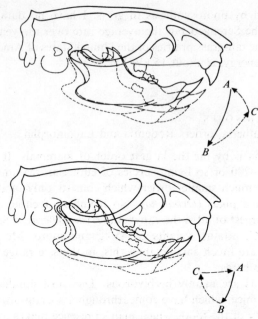

Fig. 14-5. Outlines of the skull drawn from X-ray cinematograph films of rats (*Rattus norvegicus*) feeding. The upper outline shows the sequence of jaw positions involved in biting with the incisors. The lower one shows the sequence of positions involved in chewing. From K. Hiiemae (1971). *Zool. J. Linn. Soc.* **50**, 111–34.

12-10*a*), and even more different from those of typical carnivorous mammals (Fissipedia) which are modified in the opposite direction. In Fissipedia the articular condyles of the lower jaw are transversely elongated and fit into transverse grooves in the squamosals, forming hinge joints which also allow some sliding of the jaw from side to side (Fig. 13-7*a*). In rodents the condyles are narrow and rest in *longitudinal* grooves in the squamosals. In addition to the hinge action of opening and closing the mouth the lower jaw can slide forward and back along the grooves. The front of the jaw can also be swung from side to side: this involves the left condyle sliding anteriorly while the right one slides posteriorly, and vice versa.

The rat slides its lower jaw forward for biting and gnawing. In this position, the upper and lower incisors meet edge-to-edge without bringing the premolars and molars into contact (Fig. 14-5 above). It moves the jaw backward for chewing, so that the lower incisors are posterior to the upper ones and the premolars and molars can make contact (Fig. 14-5, lower). The two jaw positions, forward for biting and backward for chewing, are apparently common to all rocents and lagomorphs, and indeed are shared by man. The reader should find that if he brings his incisors edge-to-edge to bite, his premolars and molars do not make contact. He must move his lower incisors posterior to his upper ones in order to chew.

Fig. 14-5 (upper) shows the cycle of jaw movements made by a rat biting off a piece of food. The incisors meet edge-to-edge, and then the lower ones pass posterior to the upper ones. Each has a thick layer of enamel on its anterior face and none on the posterior face. The cutting edges of the teeth are formed by the layer of hard enamel. The lower teeth scrape the upper ones as they pass from position C to position A in gnawing (Fig. 14-5, upper), and so sharpen them. However, the lower teeth themselves would tend to become blunt in use if there were no mechanism for sharpening them. Sharpening does not seem to have been observed in normal conditions, but it can often be observed when a rat is recovering from anaesthesia. The lower jaw is moved even further anteriorly than in gnawing, so that the lower incisors are anterior to the upper ones. The upper incisors are then used to pare off dentine from the bevels of the lower ones. Both pairs of incisors grow continuously, so wear is made good.

The toothless diastema enables rodents to gnaw without ingesting the shavings, since the lips can be drawn in through it, behind the incisors, to close off the rear part of the mouth. A squirrel can gnaw through the shell of a nut but ingest nothing until it reaches the kernel, and a beaver can fell a tree without filling its mouth with wood shavings. The diastema also enables grazing mammals to eat long lengths of grass. When a rabbit (*Oryctolagus*), for instance, eats long grass it bites it off near the ground, allowing the long ends to protrude from the side of the mouth through the diastema. These ends are drawn into the mouth by the tongue as chewing proceeds.

The rat is unusual among mammals in having upper and lower jaws of equal width, so that the upper and lower molars can be in contact on both sides of the mouth simultaneously. This makes it possible for chewing to occur on both sides of the mouth at the same time. There is evidence that this sometimes happens, but chewing on one side seems to be usual. The possibility has been investigated by studying X-ray films, taken from above, of rats chewing barium-impregnated biscuit. Streams of X-ray opaque material could be seen moving posteriorly from the molars to the gullet. Sometimes there were streams on both sides of the mouth, indicating that chewing was occurring on both sides.

Chewing in the rat is not mainly a side-to-side movement, as it is in *Didelphis* and in many herbivores (including the ruminants). Instead, it is mainly a forward-and-back movement (Fig. 14-5, lower). The jaw moves anteriorly during the active phase of the cycle, while it is actually grinding the food, and posteriorly in the recovery phase. The main chewing muscles are the masseters, which pull the jaw upwards and forwards; they are the largest jaw muscles in rats and in other rodents such as marmots (Table 13-1).

X-ray cinematography has been used to study the jaw movements of rabbits, as well as of rats. The lower jaw is moved forward to bring the incisors into use, and back for chewing, as in rats. However, the chewing action is mainly from side to side.

Rabbits pass their food through the gut more than once. As well as normal, hard faeces they produce soft pellets which they remove from the anus by mouth, and swallow whole. Domestic rabbits which are fed by day produce these soft pellets at night and hard faeces by day. Wild rabbits which feed at night and spend the day in a burrow produce the soft pellets by day and faeces at night. One possible advantage of the second passage through the gut is that it may enable the rabbit to digest some of its gut micro-organisms and so recover some of the food used by the micro-organisms for growth.

Rats have stout necks and keep them more or less horizontal. Rabbits have longer, more slender necks which are generally held at a steep angle, whether the rabbit is sitting or running. This raises the rabbit's head, giving it a better view over long grass and other obstructions and presumably enabling it to spot predators from a greater distance. Though rats and rabbits hold their necks at such different angles, they hold their snouts at much the same angle. There is an appropriate difference in the shape of the skull. The rat skull is essentially as in Fig. 14-4*b* but the rabbit skull bends down at the posterior end as in Fig. 14-4*c*. Thus the relationship between the neck and the posterior end of the skull is much the same in these two cases. There are many other herbivorous mammals with skulls of each of these types. Pigs have more or less horizontal necks and straight skulls as in Fig. 14-4*b*. Squirrels and ruminants generally have sloping necks and bent skulls, as in Fig. 14-4*c*. Dogs and many other carnivorous mammals, and monkeys, also have sloping necks and bent skulls.

A good many species of rodent live in hot deserts. Most of them avoid excessive temperatures in the same way as desert lizards (p. 284) by spending the hottest part of the day in a burrow. Some, like lizards, emerge in the morning and evening. Others spend the whole day in the burrow and emerge only at night, when temperatures are too low for most lizards to be active.

Drinking water is generally not available in the desert. Some rodents depend on succulent plants such as cactus instead. Others can survive on a diet of dry food. For instance, the kangaroo rats (*Dipodomys*) of the Arizona desert maintain the water content of their bodies at a normal level, when kept for many weeks without water, on a diet of dry rolled oats. Two main adaptations keep water losses low enough for this to be possible. Closely spaced turbinals cool the expired air very effectively, reducing losses by evaporation (see p. 338). At low humidities, kangaroo rats actually breathe out cooler air than they breathe in. The kidneys have very long loops of Henle and can produce very concentrated urine (see p. 395). Up to 3.8 mol urea^5 l^{-1} has been found in the urine of *Dipodomys* (and up to 5 mol l^{-1} in *Notomys*, an Australian rodent).

Kangaroo rats kept on a dry diet flourished at moderate humidities, but lost weight at relative humidities below about 10%. This was partly because evaporative losses are greatest in dry air, and partly because the water content of the food depends on the humidity. For instance, pearl barley absorbs 3.7% water at a relative humidity of 10%. Many rodents store food in their burrows and this may help desert rodents to maintain water balance, since the air in the burrow tends to be moister than the air outside. The relative humidity in kangaroo rat burrows has been investigated by tying a miniature humidity recorder to the animal's tail, and letting it go down the burrow. Values averaging about 30% were found.

Though the water absorbed by the food may be critical for the kangaroo rat's water balance, it is not the animal's main source of water. This is the water produced by metabolism (see p. 275), which amounts to about 54% of the weight of dry cereal foods.

LARGE HERBIVORES
Infraclass Eutheria, orders Artiodactyla, Perissodactyla and Proboscidea

It was shown on p. 381 how many carnivorous mammals have increased the effective length of their limbs without a corresponding increase in mass, by adopting the digitigrade stance. They stand with the metacarpals and metatarsals nearly vertical. The hoofed mammals have gone a stage further, adopting what is known as the unguligrade stance. They stand on the tips of their toes with the phalanges, as well as the metacarpals and metatarsals, more or less vertical. The claws have become hooves which support the animals' weight. Carnivores use their claws for seizing and wounding prey, and for gripping food while they tear and cut it with their teeth. Large herbivores do not, in general, have the same need to seize and grip.

There are two orders of hoofed mammals. The Artiodactyla include the pigs, hippopotamuses, camels, deer, antelopes, etc. The Perissodactyla include the tapirs, rhinoceroses and horses. Pigs and tapirs are relatively primitive hoofed mammals. Their limbs have been effectively lengthened by adoption of the unguligrade stance but are not otherwise greatly modified (Fig. 14-6). The relative lengths of the main limb bones are much the same as in dogs and cats, and there is a reasonably full complement of toes. Pigs have four digits on each foot, fore and hind, instead of the five of primitive mammals. Tapirs have four digits on each forefoot and three on each hind foot. The missing digits are the first ones (and the fifth digit in the tapir hind foot). The first digit is also missing from all the feet of some Fissipedia such as *Lycaon*, and from the hind feet of cats (*Felidae*).

Of the digits which remain in pigs, 3 and 4 are about equally strong, and much larger than 2 and 5. In the tapirs, digit 3 is strongest. This is a constant difference between artiodactyls and perissodactyls: whether the number of digits is odd or even digits 3 and 4 are equally strong in artiodactyls, but digit 3 is clearly the strongest in perissodactyls with 2 and 4 smaller and about equal.

The ruminants are more advanced artiodactyls than pigs, and horses are more advanced perissodactyls than tapirs. The advanced hoofed mammals have only one or two effective toes on each foot. Rudiments of digits 2 and 5 remain as dew claws in most deer (Cervidae), and the splint bones of horses are rudiments of metacarpals and metatarsals 2 and 4, but ruminants have only two toes which touch the ground and horses have only one. The metacarpals or metatarsals of the two remaining digits of ruminants are fused together to form a cannon bone.

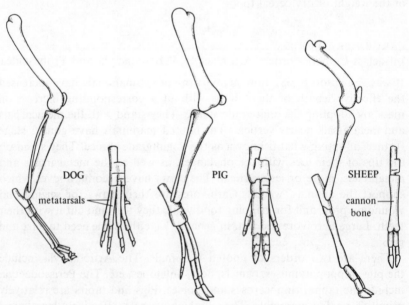

Fig. 14-6. The skeleton of the lower part of the right hind limb of a dog (*Canis*), a pig (*Sus*) and a sheep (*Ovis*).

Reduction in the number of digits can be expected to reduce the mass (and moment of inertia) necessary to give the skeleton the strength it needs. If some digits are lost the remainder must, of course, be strengthened, but a saving of mass can still be expected. The forces that act on the foot have components acting parallel to the bones, tending to compress them, and components at right angles to the bones, tending to bend them. The latter are probably responsible for the major part of the stress that acts in the bones of the foot, so we can expect the thickness of the bones to be determined mainly by the need to withstand bending moments. If a bone of tensile strength T is to withstand a bending moment M its cross-sectional area must (by a standard engineering formula) be at least $k(M/T)^{2/3}$, where k is a constant that depends on the shape of the cross-section. If it is replaced by two bones of the same shape in cross-section each resisting $M/2$, each bone requires a cross-sectional area $k(M/2T)^{2/3} = 0.63k(M/T)^{2/3}$ and the two together must have a cross-sectional area $1.26k(M/T)^{2/3}$. Thus two toes must be heavier than one to resist the same bending moment. Similarly, if ruminants had two separate metacarpals (or metatarsals) in each foot instead of a single cannon bone, their feet would have to be heavier.

There is also a difference in proportions within the limbs, between primitive and advanced hoofed mammals. Consider the ratio of lengths femur: tibia: longest metatarsal. In pigs and tapirs (and also in dogs and cats) this ratio is around $1:1:0.4$. In ruminants and in horses it is more nearly $1:1:0.7$. The metatarsals (and metacarpals) have become relatively longer in the advanced hoofed mammals, as well as being reduced in number.

One might suppose that their highly modified limbs would make one- and two-toed herbivores even faster than the digitigrade carnivores. There is not enough evidence to establish whether they are as a general rule faster, but there are indications that they may be. A lion can catch a zebra if it takes it by surprise, but if the zebra has time to accelerate to full speed it easily escapes. Gazelles (*Gazella*) can keep pace for a while with a car travelling at 70–80 km h^{-1} but the maximum speed of greyhounds and wild dogs is about 65 km h^{-1}. Hunting dogs depend on stamina rather than speed to catch gazelles. A Pronghorn antelope (*Antilocapra americana*) galloping in front of a car reached 98 km h^{-1}, which is a little faster than the 90 km h^{-1} recorded for a cheetah.

The teeth of ruminants and horses have evolved more or less in parallel, as well as their limbs. Pigs and tapirs have relatively simple, bunodont teeth. Pigs eat roots, fruit, insects, etc. and tapirs browse on the leaves and shoots of trees and water plants. Ruminants and horses have more specialized grinding teeth with crowns of the types illustrated in Fig. 14-3 *D* and *F*, respectively. In many ruminants, and in horses, the teeth are strongly hypsodont. Horses and many ruminants feed predominantly on grass, but some deer and antelopes browse more than they graze. The distinctive stomach of ruminants has already been described (p. 440).

The evolutionary history of horses is known in great detail, for the large number of fossils which have been found constitute an exceptionally complete record. Modern horses (*Equus*) seem to have evolved through several other genera from *Hyracotherium*, which lived about fifty million years ago and included several species the size of small to medium dogs. Though much smaller than tapirs, *Hyracotherium* had the same number of toes as tapirs and similar leg proportions (femur length:tibia length:longest metatarsal length). They had bunodont teeth, very like tapir teeth. The genera which intervene between them and *Equus* show successive changes in all these features.

Two families of artiodactyls are particularly successful: they are Bovidae (antelopes, cattle, etc.) and the Cervidae (deer). Bovidae have horns and Cervidae have antlers which are sometimes very large and conspicuous. The horns of a wild American ram (*Ovis*) may be as much as 12% of the body weight. A moose (*Alces*) 2.3 m high at the shoulder may have antlers with a span of 1.8 m. The extinct Irish elk (*Cervus giganteus*) had extraordinary antlers up to 3.7 m across.

Horns are permanent and unbranched. Antlers are shed annually, and are often branched. They are generally shed in the autumn and regrown in spring, and tend to be larger and have more branches each successive season in the life of the deer. A horn is a bony projection of the skull with a sheath of keratin (the material known as horn). An antler is also a bony projection of the skull, but it is covered during growth by soft skin with a rich blood supply. Later in the season the skin is shed and the bare bone exposed.

Horns and antlers are used in fights between members of the same species. Rams charge full tilt at each other from a distance of several yards, colliding horns against horns. Antelopes and deer engage horns or antlers with their opponent and wrestle.

Breakage of antlers is remarkably rare. It has been estimated that less than 5% of shed moose and caribou (*Rangifer*) antlers have any part broken off, and breakage of main branches is very rare indeed. The annual shedding and replacement of antlers seems certainly not to be necessary to repair damage.

Many horns and antlers are effective enough as weapons, but many give the impression of having evolved more for show than for business. The horns of rams which are coiled in such a way that the point is unlikely to strike the opponent seem poorly adapted for inflicting serious wounds, as do the complex, cumbersome antlers of moose. It has been suggested that rams' horns may have evolved to absorb the shock of head to head impact, and so protect the animal from brain damage in fighting, but wart-hogs (*Phacochoerus*) and other hornless animals fight in the same way.

There is a lot of evidence that horns and antlers function mainly for display, rather than as weapons. Males generally do not fight when they meet, but adopt postures which seem adapted to display the size of the horns or antlers. The male with smaller horns or antlers nearly always sub-mits to the other. In a long series of observations of American wild sheep very few exceptions to this rule were observed, and they were cases when the ram with the larger horns was elderly and decrepit. Almost all the breeding was done by rams with large horns. It was clear that horn size was more important than seniority: a young ram might submit to its elders one year but if its horns grew so fast as to overtake theirs in size it would dominate over them the next.

Males often have larger horns than females, and antlers are confined to males in all but one species. The exception is the reindeer or caribou (*Rangifer*), which experiences particularly severe winter conditions. Deep or hard snow may restrict access to the food below. The deer feed on patches of lichen, grass, etc. that have been blown clear of snow by wind, and also clear patches themselves by digging with their forefeet. They frequently try to displace each other from these feeding places, using antler threats. Thus antlers are used in competition for food, as well as for mates. The animal with smaller antlers, or no antlers, submits. The males shed their antlers in autumn, in the usual way, but females and calves retain theirs into the winter. With their antlers, the males lose their dominance: a large bull with no antlers will even submit, in competition for food, to an antlered calf. Barren females shed their antlers in later winter but pregnant ones retain theirs until shortly after calving in late spring or early summer.

It has been suggested that horns and antlers may have another function, in dissipating excess heat. This function requires a good blood supply close to the surface. Horns have plenty of blood vessels immediately under the keratin but antlers only have a good blood supply while they are growing. In any case dissipation of heat seems unlikely to be an important function of antlers, since it is far from obvious why females of most species should have less need than males to get rid of heat. The main evidence that horns have a function in heat regulation comes from experiments by Dr Richard Taylor on goats (*Capra*). He trained goats to run on a moving belt, like lizard shown in Fig. 9-5*b*. He attached thermocouples to their horns, and found that exercise on the moving belt caused a sharp rise in horn tempera-ture, apparently due to dilation of the blood vessels of the horn. The same effect could be obtained without exercise by injecting the drug procaine near the nerves to the horn. The rise in temperature of the horn surface implies faster heat loss, but the loss seems never to be a very large proportion of the total heat loss from the body. The importance of heat loss from the horns is that it cools the brain selectively. Some experiments were carried out on goats with thermocouples inserted in their brains. It was found that in suitable conditions the brain could be warmed or cooled about 0.5 K by covering the horns with heat-insulating material, or by wetting them. The brain is particularly liable to damage by high temperatures: brain temperatures above about 42 °C are fatal to man.

The brain is selectively cooled because much of the venous blood returning from the horns passes through the cavernous sinus (Fig. 14-7). Most of the blood going to the brain is supplied by branches of the carotid artery which pass through the cavernous sinus, breaking up in it into a large number of small vessels (the carotid rete). The blood coming from the horns tends to cool the blood going to the brain.

The cavernous sinus also receives venous blood from the nose, and this blood may be more important than the blood from the horns in keeping the brain cool. In a series of experiments with sheep (*Ovis*), thermocouples

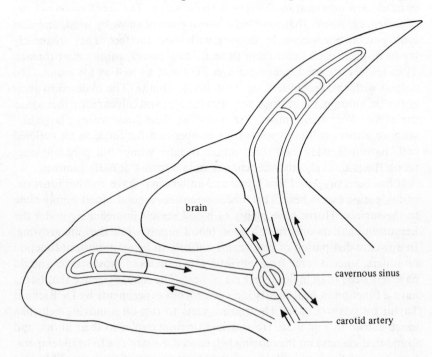

Fig. 14-7. A diagram of the head of a goat showing the paths of blood through the carotid rete and cavernous sinus.

were implanted by surgery in various blood vessels including the cavernous sinus. It was found possible to lower the temperature in the sinus about 1 K by blowing air at room temperature into the nostril. Bovidae in general seem to have cavernous sinuses and carotid rete, but they are not the only mammals to do so. Pigs have them, and cats have a very similar arrangement. In these hornless animals, the nose alone must be responsible for the cooling effect. Most other mammals seem to have no mechanism for selective cooling of the brain.

Blood returning from the nose is much cooler when the animal is panting, than when it is not. In normal breathing, air both enters and leaves by the nostrils. Losses of heat and water are limited because the air which is warmed as it is breathed in is cooled again as it is breathed out (p. 338). In panting, air enters mainly through the nose and leaves mainly through the mouth. This has been demonstrated in experiments with dogs, using hot-wire anemometers to measure air velocities in the nostrils and mouth. The air is warmed as it is breathed in but it is not cooled again as it is breathed out, so more heat and water are lost. Most of the heating of the air, and the evaporation of water, occur in the nose, so the venous blood from the nose tends to be cool.

The effectiveness of the arrangements for selective cooling of the brain is illustrated by an observation on a gazelle (*Gazella*), an animal which has horns and also pants. It was found that after exercise its brain was up to 2.9 K cooler than the central parts of the body.

The lizards and rodents which live in hot deserts spend the hottest part of the day in burrows. Camels (*Camelus* spp.) and oryx (*Oryx* spp.) also live in hot deserts. They are far too large to burrow, but their size is an advantage in another way. A 500 kg camel is 10000 times as heavy as a 50 g kangaroo rat, but its surface area cannot be much more than $10000^{2/3} = 460$ times as much. The rate at which an animal absorbs solar radiation, and long-wave radiation from its surroundings, tends to be proportional to its surface area, but the amount of water it has available for cooling its body by evaporation is proportional to its weight. Hence a big mammal can avoid overheating for longer than a small one could, cooling itself by sweating and panting. Also, a big mammal heats up more slowly than a small one in the same conditions, and if it were big enough it could, in principle, give up sweating and panting altogether, simply allowing its body to heat up during the day and cool down in the cold desert night. Camels do sweat in really hot conditions, but they save a lot of water by allowing their body temperatures to fluctuate. Professor Knut Schmidt-Nielsen measured the rectal temperatures of Arabian camels (*C. dromedarius*) in the Sahara. He found that in winter, and also in summer when plenty of water was supplied, the rectal temperature fluctuated only between about 36 °C and 38 °C. When water was short however, his camels allowed their temperatures to fall as low as 34.5 °C overnight and to rise to 40.5 °C in the course of the day. Such a large increase in body temperature involves the storage of a great deal of heat. The specific heat of intact animals is about 3.4 kJ kg^{-1} K^{-1} (about 0.8 of the specific heat of water), so the energy needed to heat a 500 g camel by 6 K is about 10 MJ. The latent heat of evaporation of water is 2.5 MJ kg^{-1}, so 4 kg water would have to evaporate to get rid of this much heat. The camel can save 4 kg water daily by allowing its body to heat up instead of keeping its temperature constant by sweating.

There is an advantage in allowing the body temperature to rise as high as possible, so as to reduce the net rate of heat uptake. The hotter the surface of the body, the more of the heat it receives will be lost again by radiation and convection. The camel does not let its temperature rise above 40.5°C. It has no horns and does not pant, and is probably unable to keep its brain cooler than the rest of the body. Dr Richard Taylor found that oryx could allow their temperatures to rise much higher. He kept his oryx in hot conditions indoors, and used a thermocouple to measure the rectal temperature. If plenty of drinking water was available rectal temperatures rose only to about 40 °C, even when the temperature of the room was 45 °C. The oryx kept cool mainly by sweating. When water was short they stopped sweating but panted more, and allowed their rectal temperatures to rise to 46.5 °C. This temperature was maintained for up to six hours without apparent ill effects. Presumably the brain was kept considerably cooler.

The largest of the herbivorous mammals (and indeed of all land animals) are the rhinoceroses, hippopotamus and elephants. The rhinoceroses are perissodactyls, the hippopotamuses are artiodactyls and the elephants are given an order to themselves, the Proboscidea. The White rhinoceros (*Diceros simus*) attains weights over 3 tons, the Common hippopotamus (*Hippopotamus*) 2 tons and the Indian and African elephants (*Elephas* and *Loxodonta*) 4$\frac{1}{2}$ to 6 tons, respectively. The problem of supporting a very large body on land has already been considered in the account of brontosaurs (p. 317). The very large mammals, like brontosaurs, have relatively thick limbs, and elephants stand with their legs straighter than most smaller mammals.

MARSUPIAL HERBIVORES
Infraclass Metatheria, order Marsupialia, suborder Diprotodonta

The herbivorous marsupials which are grouped together in the suborder Diprotodonta all live in Australasia. Instead of having many small lower incisors like *Didelphis* and the other Polyprotodonta they have a single pair of large, forward-pointing incisors. Most have three paris of upper incisors, so each long lower incisor bites against three upper teeth.

There are two main groups of diprotodonts. The phalangers (Phalangeridae) are mostly fairly small, arboreal animals which feed on leaves and fruit. The kangaroos and wallabies (Macropodidae) are generally larger, ground-living animals which feed on grass and shrubs. Their shape and manner of locomotion are well known. They have micro-organisms which digest cellulose in the stomach, like the eutherian ruminants.

PRIMITIVE PRIMATES
Infraclass Eutheria, order Primates, suborder Prosimii

The primates have a special interest, as the order of mammals to which man belongs. Most primates are monkeys, and the resemblance of monkeys to man is clear. The primitive primates considered in this section of the chapter are less obviously man-like.

The most primitive of the mammals generally regarded as primates are the tree shrews (*Tupaia* and related genera) which live in Southeast Asia. More species are found in Borneo than anywhere else. They are almost typical primitive mammals and it is not at all clear that they are closely related to the other primates: some zoologists think they should be placed in the order Insectivora. They have the primitive mammalian stance and gait, like the opossum *Didelphis* (Figs. 12-3 to 12-6). They have teeth like those of *Didelphis* (Fig. 12-10) with forward-sloping lower incisors and with molars of the primitive pattern shown in Fig. 12-11. They have long, sensitive snouts like shrews and opossums. Their most striking primate-like character is a bar of bone behind the eye, formed by downward extension of the frontal and upward extension of the jugal. This is not an exclusively primate character (ruminants have it) but it is found in all primates.

Tree shrews are small, between about 30 and 350 g in weight. They have long, bushy tails like squirrels. Some species spend most of their time on the ground and others in trees, running up trunks and along branches and jumping from tree to tree. They take a varied diet which includes insects and fruit. They often sit up to eat, holding food between cupped forepaws, but they cannot grasp food between fingers and thumb in the same way as man and some monkeys. Rats and squirrels hold food between their paws in the same way as tree shrews do.

Several other groups are included among the Prosimii. There are the lemurs (Lemuroidea) which live in Madagascar and the Comoro Islands, in trees or among rocks. They fill on these islands the ecological niches that are occupied by monkeys on the African mainland. There are the lorises and bushbabies (Lorisiformes), small to medium-sized primates which live in trees in Africa and Asia and are active mainly at night. There are also the tarsiers (*Tarsius*) and the Aye-aye (*Daubentonia*).

These prosimians are in some ways intermediate between tree shrews and monkeys. Their eyes face more forward than those of tree shrews, their molars are modified towards the square crushing type found in monkeys and some or most of their claws are replaced by nails. Some of them are highly specialized for peculiar ways of life. The Aye-aye has incisors like those of rodents, and the middle finger of each hand is long and very slender. It gnaws with its incisors to get foods such as bamboo pith, and the larvae of wood-boring insects. It uses the peculiar finger for picking insects out of crevices and burrows. The bushbabies (*Galago* and *Euoticus*) are remarkable jumpers. A 250 g bushbaby could jump from the ground to a perch at a height of 2.3 m. Bushbabies jump from branch to branch in trees and hop like kangaroos on the ground. They have large hind legs with elongated tarsal bones, like frogs (Fig. 8-17c).

MONKEYS

Order Primates, suborder Anthropoidea, superfamilies Ceboidea and Cercopithecoidea

The Ceboidea are the monkeys of Central and South America, and the Cercopithecoidea are those of Asia and Africa. They are closely similar in most respects but there are a few small, consistent differences between the superfamilies that indicate that they have evolved separately over a long period of time. For instance, Ceboidea have three premolar teeth in each side of each jaw but Cercopithecoidea have only two. Fossil evidence indicates that the two superfamilies evolved independently from prosimian ancestors.

Most monkeys spend a lot of their time in trees, and are active mainly by day. They feed largely on leaves, fruit and other plant materials but many also eat insects and small birds and mammals.

Monkeys do not stand and walk with their limbs bent like opossums and tree shrews, but with them relatively straight like dogs. They generally walk in this way, on all fours, along the branches of trees, grasping the branch with hands and feet and using the tail for balance. They also leap from tree to tree.

All monkeys can grasp objects to the extent at least of being able to hold food in one hand, but they differ in dexterity. The least dextrous are the marmosets (Callithricidae) which grasp food simply by bending their digits, so that the food is held between digits and palm (Fig. 14-8a). In this type of grasping there is no functional difference between fingers and thumb: all the digits are on the same side of the object being grasped. Tree shrews and marmosets cannot separate the first digit at all widely from the others but other monkeys can, and this digit functions as a thumb (Fig. 14-8*b*). When an object is grasped the fingers are generally placed on one side of it and the thumb on the other.

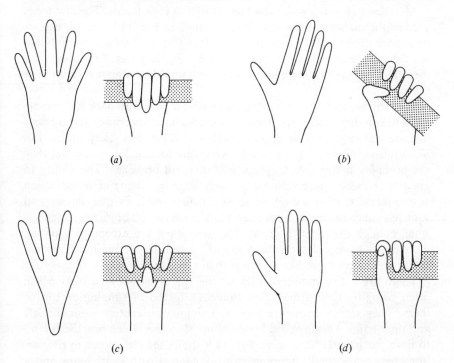

Fig. 14-8. Diagrams of the hands of (*a*) marmosets, (*b*) other American monkeys, (*c*) certain prosimians and (*d*) man, showing their different manners of grasping.

American monkeys (other than marmosets) have hands of the type shown in Fig. 14-8*b*. The thumb can be separated quite widely from the other digits and it can be bent and straightened, but it cannot make other movements. Such a thumb cannot be placed squarely, thumb-print to fingerprint, against a finger unless it lies at 180° to the other fingers in the spread hand as in Fig. 14-8*c*. Some prosimians including the potto (*Perodicticus*) have hands like the one in this diagram. Asiatic and African monkeys, apes and man are capable of an additional type of thumb movement. The joint between the carpals and the metacarpal, at the base of the thumb, allows the thumb to rotate a little about its own long axis. This makes it possible for the thumb to be placed squarely against a finger. Some monkeys including baboons (*Papio*) use the index finger and thumb in this way, for instance to pluck grass (as food) or to remove ectoparasites from each other's bodies. Many others have thumbs which are too short for this to be very convenient. They are apt to pick up small objects between the thumb tip and the *side* of the index finger, instead of using the tip of the index finger as a man or baboon would do.

Monkeys can grasp with their feet, as well as their hands. The first toe is used in the manner of the thumb represented in Fig. 14-8*b*, but it cannot be rotated about its long axis like the human thumb. Tree shrews and man are the only primates which cannot grasp with their feet. Marmosets use their first toes in grasping with their feet in a way they cannot do with their thumbs.

Tree shrews and squirrels have claws which penetrate bark and enable them to run up vertical tree trunks. Monkeys have fingernails and toenails instead, and rely for their grip on friction with the skin. They cannot run like squirrels up smooth tree trunks which are too thick to grasp but they are probably better able to keep hold of small branches. The ability to grasp is probably more valuable to fairly large tree-living mammals (such as monkeys) than it would be to small ones (such as tree shrews and squirrels) since the larger ones must more often run on branches which are small enough for them to grasp. The marmosets are exceptional among monkeys in having claws instead of nails, except on the big toe. They are small, and climb more like squirrels than other monkeys.

There are a few monkeys such as the spider monkey (*Ateles*) which often swing by their arms below branches instead of running on top of them. They travel through a tree by swinging alternately from the left and right hand. This is called brachiation. Monkeys which practise it tend to have particularly long arms. Fig. 14-9 shows the differences in proportions between a small, jumping monkey (*Aotus*) with short limbs and a long trunk, a typical monkey (*Cebus*) with relatively longer limbs and a brachiator (*Ateles*) with even longer limbs. The series shows a greater proportional increase in arm length than in leg length.

As well as having long arms, monkeys which brachiate tend to have small thumbs. They do not grasp the branches as they swing but simply hook their fingers over them, so a long thumb would be apt to get in the way. They also tend to have broad chests, and distinctively shaped scapulae. These peculiarities can be explained as permitting the range of arm movements used in brachiation. Quadrupedal mammals run with their forelimbs below the trunk, and have the scapula appropriately placed against the side of the rib cage with the glenoid cavity (where the humerus articulates) pointing ventrally (Fig. 14-10*a*). Brachiators have to reach out with their arms in all directions. They do not have their scapulae against the side of the rib cage but on the back, as in man. The glenoid points more

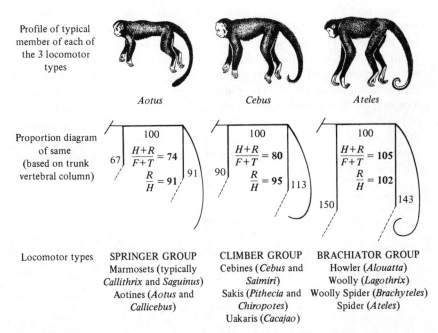

Fig. 14-9. Drawings and diagrams of American monkeys which have different habits of locomotion. From G. E. Erikson (1963). *Symp. zool. Soc. Lond.* **10**, 135–64.

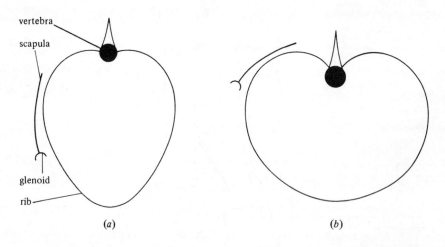

Fig. 14-10. Diagrammatic transverse sections through the thorax of (*a*) a quadrupedal monkey and (*b*) a brachiating monkey or ape.

laterally and the arm can be extended in a wider range of directions. The broad proportions of the rib cage allow this arrangement (Fig. 14-10*b*). The difference in shape between the scapulae of brachiators and quadrupedal monkeys is essentially a bend, as Fig. 14-11 shows. The scapulae which are compared are from a quadrupedal monkey, a baboon (*Papio*) and a chimpanzee (*Pan*). The chimpanzee is of course an ape, not a monkey, but it brachiates and its scapula is very similar in shape to those of brachiating monkeys such as *Ateles*. The bending of the scapula makes the glenoid cavities of brachiators point more anteriorly than those of typical monkeys, which is appropriate since brachiation involves raising the arms above the head.

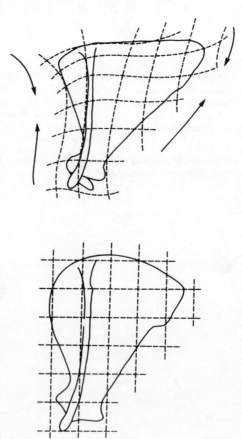

Fig. 14-11. Outlines of the left scapula of (below) a baboon (*Papio*) and (above) a chimpanzee (*Pan*). The grids are superimposed so that corresponding intersections are in homologous positions. From C. E. Oxnard (1969). *Am. Scient.* **57** (1), 75–96.

Most monkeys have long tails but some have none. Some American monkeys such as *Ateles* (Fig. 14-9) have a patch of naked skin under the tip of the tail, and use it to grasp branches. The tail can grip strongly enough to support the monkey's weight.

Monkeys use their hands for manipulating food as well as for grasping branches. They make little use of their teeth for manipulation, and the jaws and teeth become less well suited to this function than in primitive mammals. The jaws have become shorter. Both upper and lower incisors have become square and vertical, like human incisors, and are used for taking bites from fruit and leaves. The molar teeth are of the square, bunodont crushing type as in rats, squirrels and pigs.

It is particularly important for an animal which jumps from branch to branch to be able to judge distance accurately. Monkeys have forward facing eyes with overlapping visual fields and so can make use of the stereoscopic effect. Most monkeys have short snouts, like cats. Baboons (*Papio*, etc.) have longer snouts which do not impede forward vision because they are low: the eyes look forward over the snout. The spinal cords of monkeys enter the skull below so that the relationship between neck and skull is as shown in Fig. 14-4*c*.

Monkeys have larger brains than most other mammals of similar size. Fig. 14-12 shows that their brains are generally nearly twice as large for the same body weight as those of prosimians, which are in turn about four times as large as those of primitive Insectivora such as shrews and hedgehogs. Most mammals have brains intermediate in size between those of primitive insectivores and of prosimians, but seals and toothed whales have even larger brains than one would expect (by extrapolation) in monkeys of the same body weight.

Evolution of monkey brains has involved enlargement of different parts of the brain to different extents. The cerebral hemispheres of monkeys make up a bigger proportion of the brain than in primitive mammals, and within the cerebral hemispheres the part known as the neocortex has grown especially fast. The neocortex seems to be the site of the memory, of pattern recognition, and of conscious control of movement.

APES AND MEN
Order Primates, suborder Anthropoidea, superfamily Hominoidea

The apes are the gibbons (*Hylobates* and *Symphalangus*), the orang-utan (*Pongo*), the chimpanzee (*Pan*) and the gorilla (*Gorilla*). There are six species of *Hylobates* and two of *Pan*. All the apes live in Africa or Asia, and they resemble the African and Asiatic monkeys rather than the American ones in features such as the number of premolars. The earliest known

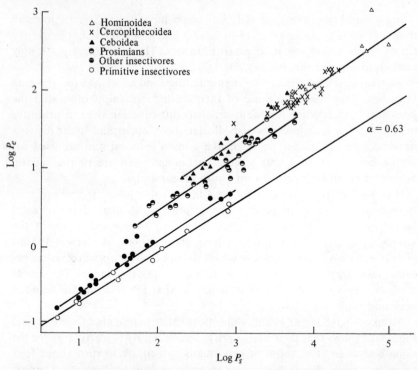

Fig. 14-12. A graph of the logarithm of the weight of the brain in grammes (log P_e) against the logarithm of the weight of the body in grammes (log P_s) for various insectivores and primates. Shrews (Soricidae), hedgehogs (Erinaceidae) and tenrecs (Tenrecidae) are included in the Primitive insectivores group. From R. Bauchot & H. Stephan (1969). *Mammalia*, **33**, 225–75.

human fossils (i.e. fossils placed in the genus *Homo*) are from Africa and Asia, and men also have the characteristic features of African and Asiatic monkeys.

Gibbons are no larger than many monkeys but the other apes, and man, are the largest primates. Apes and men differ from most monkeys in having no external tail. They have relatively short, broad trunks with dorsally placed scapulae, like monkeys which brachiate. Small details distinguish their molar teeth from those of monkeys.

The apes differ considerably from each other in habits. The gibbons, from Southeast Asia, spend nearly all their time in the trees, where they feed mainly on fruit. They are extremely able brachiators, and their delightful brachiating action can often be watched in zoos. When they are travelling fast they can leap considerable distances from one handhold to the next. They sometimes run (bipedally) along the tops of branches. They often hang by one arm when feeding. Typically, they reach out with the other arm for a fruit-bearing twig and hold this twig with the feet while plucking fruit with the free hand. They use their teeth for peeling fruit. Their arms are quite extraordinarily long, considerably longer (relative to the length of the trunk) than those of *Ateles* (Fig. 14-9). Their thumbs are much longer than those of *Ateles* and are folded across the palm during brachiation.

The orang-utan lives in Sumatra and Borneo, and also spends most of its time in the trees. It brachiates sometimes, but does not leap from handhold to handhold, and generally climbs in a rather cautious manner using its feet as well as its hands. Its arms are not as long, relative to the trunk, as those of gibbons, but longer than those of *Ateles*. The African apes, the chimpanzee and gorilla, spend less time in the trees and have relatively shorter arms. Chimpanzees spend half to three-quarters of the daylight hours in trees, and sleep at night in nests built in the trees. They brachiate more than orangs, but not as much or as proficiently as gibbons, and they seldom hang by an arm while feeding. Gorillas spend most of the day on the ground, where they find a wide variety of plant foods. The nests where they spend the night may be on the ground or in trees. They apparently never brachiate in the strict sense of the word, but use their feet as well as their hands. Wild adult gorillas weigh up to 180 kg so they can climb only on strong branches.

When on the ground gorillas and chimpanzees practise knuckle-walking. This is walking on all fours with the fingers bent so that their middle phalanges are placed on the ground.

One of man's most distinctive features is his upright posture. He is not the only primate to stand and walk on his hind legs, but he does it more often than the others and stands much straighter. Captive gibbons usually walk bipedally when on the ground but wild gibbons seldom leave the trees. The other apes sometimes walk bipedally when on the ground but usually walk on all fours. Wild chimpanzees seldom walk or run bipedally except in long grass or when carrying a load. Some monkeys including species of *Macaca* sometimes walk bipedally when they are carrying food. None of these primates walk bipedally with knees straight and trunk erect, like man. All walk with their knees bent and the trunk leaning forward.

Fig. 14-13 has been drawn from X-ray cinematograph films of a chimpanzee walking bipedally. The point of contact of the foot with the ground remains in front of the hip joint throughout the stride, or for nearly all of it, so the ape must lean forward to keep its balance. The femur only approaches a vertical position just before the foot is lifted from the ground. Because of this, and because the trunk leans forward, the range of movement of the femur relative to the pelvic girdle is much more nearly the same as in

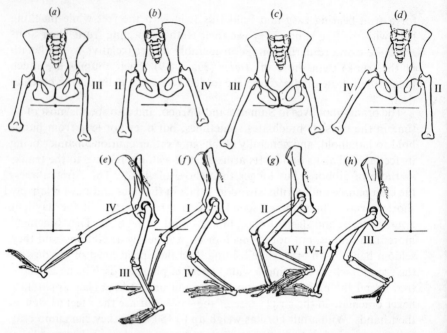

Fig. 14-13. Outlines of bones drawn from X-ray cinematograph film of a chimpanzee walking bipedally. The sequence of movement of each leg is divided into four phases indicated by Roman numerals, to aid integration of front and side views. The dots with straight lines through them represent a fixed point in the background. From F. A. Jenkins (1972). *Science*, **178**, 877–9. Copyright © 1972 by the American Association for the Advancement of Science.

quadrupedal walking of a chimpanzee (or of a dog, Fig. 12-7) than if the ape walked like a man. The chimpanzee has a pelvic girdle of fairly typical mammalian shape (see, for instance, Fig. 12-6). Since the muscles which move the femur have their origins on the girdle, a gross change in the range of movement at the hip requires a corresponding change in the shape of the girdle. The pelvic girdle of man is quite different in shape from those of apes and other mammals; it has a very short, broad ilium.

  Running on two legs might need more energy than running on four, or it might need less. Which is the case? Measurements have been made of the oxygen consumption of a variety of mammals, running at different speeds. Most of them have been made by appropriate modifications of the method shown in Fig. 9-5b. It has been found to be a general rule that for any given animal, a graph of oxygen consumption against speed is approximately a straight line. The slope of the line gives an indication of the cost of running, over and above the cost of standing still. It can be used to calculate the volume of oxygen used in transporting 1 g of animal a distance of 1 km. This quantity is called the minimum cost of running. It tends to be larger for small animals as Fig. 14-14 shows. A line has been drawn on the graph through all the points for quadrupedal species, and the points for man lie well above the line. The cost of running for man is about twice as high as would be expected of a quadrupedal mammal of the same weight.

Is this high value an inevitable consequence of running on only two legs? An attempt has been made to answer this question by experiments with chimpanzees, which use both quadrupedal and bipedal locomotion. They were trained to run on a moving belt, and it was found that the

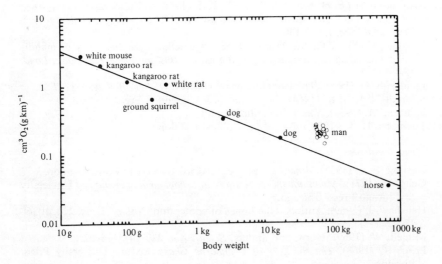

Fig. 14-14. A graph of minimum cost of running (defined as explained in the text) against body weight, for various mammals. From C. R. Taylor, K. Schmidt-Nielsen & J. L. Raab (1970). *Am. J. Physiol.* **219**, 1104–7.

oxygen consumption was the same at any given speed, whether the ape was running on two limbs or on four. Similar results were obtained in experiments with Capuchin monkeys (*Cebus*), which normally run quadrupedally but could be trained to run bipedally. These results seem to show that there is nothing inherently more energetic about bipedal running than quadrupedal running. The cost of running for the chimpanzee falls above the line in Fig. 14-14, but the cost for the monkey falls almost exactly on the line.

The brains of apes are about the size one would expect in monkeys of the same body weight, but the brain of man is about three times as large (Fig. 14-12). The reader will not need to be convinced that his very large brain is capable of very remarkable achievements.

FURTHER READING
General
See the list at the end of Chapter 12.

Feeding on plants
Blaxter, K. L. (1962). *The energy metabolism of ruminants.* Hutchinson, London.
Hungate, R. E. (1966). *The rumen and its microbes.* Academic Press, New York.
Swenson, M. J. (ed.) (1970). *Dukes' Physiology of domestic animals*, 8th edit. Comstock, Ithaca.

Small herbivores
Ardran, G. M., F. H. Kemp & W. D. L. Ride (1958). A radiographic analysis of mastication and swallowing in the domestic rabbit: *Oryctolagus cuniculus* (L.). *Proc. zool. Soc. Lond.* **130**, 257–74.
Hiiemae, K. M. & G. M. Ardran (1968). A cinefluorographic study of mandibular movement during feeding in the rat (*Rattus norvegicus*). *J. Zool., Lond.* **154**, 139–54.
Landry, S. O. (1970). The Rodentia as omnivores. *Q. Rev. Biol.* **45**, 351–72.
Schmidt-Nielsen, K. (1964). *Desert animals.* Clarendon Press, Oxford.
Shorten, M. (1954). *Squirrels.* Collins, London.
Thomson, H. V. (1956). *The rabbit.* Collins, London.

Large herbivores
Darling, F. F. (1937). *A herd of red deer.* Oxford University Press, London.
Geist, V. (1971). *Mountain sheep. A study in behaviour and evolution.* University of Chicago Press, Chicago.
Henshaw, J. (1969). Antlers – the bones of contention. *Nature, Lond.* **224**, 1036–1037.
Modell, W. (1969). Horns and antlers. *Scient. Am.* **220**, (4), 114–22.
Prior, R. (1968). *The roe deer of Cranborne Chase.* Oxford University Press, London.
Schmidt-Nielsen, K. (1964). *Desert animals.* Clarendon Press, Oxford.
Simpson, G. G. (1951). *Horses.* Oxford University Press, New York.
Taylor, C. R. (1966). The vascularity and possible thermoregulatory function of the horns in goats. *Physiol. Zool.* **39**, 127–39.
Taylor, C. R. (1970). Dehydration and heat: effects on temperature regulation of East African ungulates. *Am. J. Physiol.* **219**, 1136–9.

Marsupial herbivores
Ride, W. D. L. (1959). Mastication and taxonomy in the macropodine skull. *Publ. Syst. Assoc.* **3**, 33–59.

Primitive primates, and monkeys
Clark, W. E. Le Gros (1962). *The antecedents of man. An introduction to the evolution of the primates*, 2nd edit. Edinburgh University Press, Edinburgh.
Hill, W. C. Osman (1972). *Evolutionary biology of the primates.* Academic Press, London.

Napier, J. R. & P. H. Napier (1967). *A handbook of living primates.* Academic Press, London.
Oxnard, C. E. (1969). Mathematics, shape and functions: a study in primate anatomy. *Am. Scient.* **57**, 75–96.
Schultz, A. H. (1969). *The life of primates.* Weidenfeld & Nicolson, London.
Tuttle, R. (ed.) (1972). *The functional and evolutionary biology of primates.* Aldine Atherton, Chicago.

Apes and man

Campbell, B. (1966). *Human evolution. An introduction to man's adaptations.* Aldine, Chicago.
Jenkins, F. A. (1972). Chimpanzee bipedalism: cineradiographic analysis and implications for the evolution of gait. *Science,* **178**, 877–9.
Schaller, G. B. (1963). *The mountain gorilla.* Chicago University Press, Chicago.
Taylor, C. R. & V. J. Rowntree (1973). Running on two or on four legs: which consumes more energy? *Science,* **179**, 186–7.
van Lawick-Goodall, J. (1971). *In the shadow of man.* Collins, London.

Index